Veterinary Extension Education

Veterinary Extension Education

Dr. G.R.K. Sharma
M.VSc., Ph.D.
Assistant Professor
Department of Veterinary & Animal Husbandry Extension,
College of Veterinary Science,
Sri Venkateswara Veterinary University,
Tirupati,
Andhra Pradesah.

BS Publications

A unit of **BSP Books Pvt., Ltd.**

4-4-309/316, Giriraj Lane, Sultan Bazar,
Hyderabad - 500 095
Phone : 040 - 23445605, 23445688

Published by :

BS Publications

A unit of **BSP Books Pvt., Ltd.**

4-4-309/316, Giriraj Lane, Sultan Bazar,
Hyderabad - 500 095
Phone : 040 - 23445605, 23445688
e-mail : info@bspbooks.net

ISBN : 978-93-52300-40-2 (HB)

Sri Venkateshwara Veterinary University, Tirupathi

(Established under Act 18 of 2005 of Govt. of Andhra Pradesh)

Dr. Manmohan Singh, I.A.S
Vice-Chancellor

Foreword

Due to lack of comprehensive livestock extension policy and due to variety of other reasons the livestock sector has not blossomed into livestock industry and thus the livestock farmers deprived of economic benefits that would have accrued out of it. The basic concept of veterinary extension is educating the farmers in various aspects of scientific management of livestock farming, natural resources and time, where elaborate knowledge in concept, principles, techniques and activities of extension education turns out to be most important. All these are neglected almost always. To be competitive globally, not only appropriate technology is to be generated but also the same has to be disseminated to actual and users in an effective and efficient manner.

The arrival of book entitled **"Veterinary Extension Education"** is timely and a boon for the students of veterinary, dairy science and fisheries faculties as well as field veterinarians and to all those involved in livestock sector. Considerable thought and efforts have been invested to bring out this book, which will certainly remain forerunner in the field of extension. This book aims to impart the latest concepts, which could reduce the gap and even establish an information network technology generation and technology dissemination to the upcoming graduates.

I sincerely hope that students of veterinary, agriculture, fisheries, dairy technology, business management etc., as well as field veterinarians and practitioners will find this unique book most useful.

I appreciate the efforts made by **Dr. G.R.K.Sharma** in taking up this arduous task to meet the needs of the young graduates, field veterinarians and livestock farmers. I wish him good luck for this future endeavors.

MANMOHAN SINGH

Preface

This book was motivated principally by my interest in attempting to bridge the gap between the basic and applied sciences in order to evolve a common interactive framework. During my veterinary studies and subsequent experience with lecturing in a technology based and farmer oriented university, I observed a limited comprehension of the concept of 'Extension Education ' among students at large. Later, while pursuing doctoral degree, seemingly simple ideas such as 'extension education', 'communication', 'transfer of technology' and 'PRA tools' were transformed into multifaceted concepts with innumerable implications. This intellectual shift from basic science to applied science, combined with my experience led me to undertake the study of fundamental aspects of extension education from a new perspective. My educational background proved to be an asset in this journey, and led me to unexplore areas in the theoretical and practical considerations of extension education.

The present volume extensively examines notions of extension, development, communication, technology transfer and the participatory concept. Most of the earlier works focus on the general aspects of dissemination, treating all types of technology alike; the incompleteness of these analyses in this regard has meant that no rigorous attempt could previously be made to devise a formal approach to the complex process of extension education. It is the purpose of the present book to meet this need.

Extension should take note of the changes that are taking place in and around the country to improve its efficiency and effectiveness. Livestock is shifting from subsistence to market oriented farming, though a large number of farmers are producing for their own consumption. Farmers need a wide range of information for different resource and constraint situations, for production, as well as for pre-production and post-production functions. These require some new approaches in extension. Sustainability issues have become very important, as farmers are to obtain increased return from their enterprises by preserving and enhancing their own and also the country's resources.

There has been a gradual recognition of the fact that farmers do not depend entirely on extension agents for information and advice, but also many other sources, including fellow farmers. Understanding and involving the veterinary knowledge and information system of the farming communities are essential for effective extension work. Information technology is improving and spreading at a rapid rate. The farmers are also diversifying their enterprises. The extension specialists should be prepared to handle the diverse information needs of their clientele at a greater speed and at a higher level of efficiency.

The importance of participatory methods and group approach are gaining ground for sustainable development in livestock. It is imperative that the extension organizations are aware of the application and implications of participatory methods and group approach in various situations. Out of three important branches, extension, which originated as an informal, out-of-school education for farm adults and youth, has now attained the status of a distinct discipline with its own teaching and research activities leading up to higher levels. To be a successful extension worker, one needs to know not only 'How' to extend but also 'what' to extend. It is the later aspect that, this book on 'Veterinary Extension Education' is concerned with. It has been written mainly for undergraduate students of agricultural, veterinary, dairy technology, home science etc. However, it will also useful for extension worker training centers, farmers' training centers and other extension training institutions as well as to all categories of field extension personnel. The inclusion of sociology, teaching methods, survey techniques, information and communication technologies is justified and felt indispensable for effective extension work. All these and much more information have been presented in this book.

I do not consider myself as the most learned indivudual on Extension Education, still I feel confident that my experience in this field can help the readers a lot and I decided to give expression to all my experiences in the shape of a book, which would definitely help the readers. Any queries or suggestions regarding this book are always welcome.

- Author

Acknowledgements

I am indebted to my students and learned colleagues in my university, who provided the necessary stimulus for writing this book. I wish to express my grateful thanks to the several sources from which I have drawn information for writing this book. I am very thankful to Dr. K. Veeranjaneyulu, University Librarian, ANGRAU, Hyderabad for his constant encouragement and meaningful contributions to this book. Where emotions are involved words cease to mean the lexicon could not have the words to express the affection, blessings and encouragement of my parents (Sri G.V.K.Sharma and Smt. Mahalaxmi); my better half Dr. G. Shivani and my daughter Chy. G. Vaishanvi Sharma for their patience and confidence in me to work hard for the completion of this book. I express my gratitude to Dr. Manmohan Singh, Hon'ble Vice-Chancellor, Sri Venkateswara Veterinary University for writing Foreword to this unique kind of book. Above all I thank the almighty 'LORD BALAJEE' who has given me all the moral strength and perseverance to complete this assignment. I have no words to express my appreciation for the painstaking efforts of the publisher M/s B.S. Publications, Hyderabad that have made this desire fulfilled.

- Author

Contents

Foreword .... (v)

Preface .... (vii)

Acknowledgement .... (xi)

List of Tables .... (xxi)

List of Figures .... (xxiii)

CHAPTER 1

EXTENSION EDUCATION : CONCEPT, OBJECTIVES AND PRINCIPLES — 1

Concept of Extension 2

Differences between Formal Education and Extension Education 3

Philosophy of Extension 5

Scope of Extension Education 6

Objectives of Extension Education 6

Principles of Extension Education 7

Extension: Indian scenario 9

CHAPTER 2

GENESIS OF EXTENSION IN INDIA — 11

Stage I : Pre-Independence Era (1866-1947) 11

Stage II : Post-Independence Era (1947-52) 13

Stage III : Community Development and National Extension Service Era (1952-60) 15

Stage IV : Intensive Agricultural Development Era (1960-onwards) 19

Need for Democratic Decentralization 19

Panchayat Raj 20

Gram Panchayat 20

Functions of Gram Panchayat 21

Panchayat Samithi 21

Functions of Panchayat Samithi 22

Mandal System 23

CHAPTER 3

RURAL SOCIOLOGY   25

Scope of rural sociology 27

Concepts of Rural Sociology 28

Classification of Social Groups 35

Social Change 37

Factors of social change 39

Social Stratification 39

Social Class 40

Social Mobility 41

Social Values and Value Systems 43

Types of values 44

Functions of Values 45

CHAPTER 4

TEACHING-LEARNING PROCESS   47

Steps in extension teaching 47

Principles of teaching 49

Requirements of extension teaching 49

Learning, Learning situation and Learning experience 50

Principles of learning in farming situation 53

Principles of adult learning 54

Types of learning 54

Theories of learning 56

CHAPTER 5

EXTENSION TEACHING METHODS   58

Principles of Teaching 49

Requirements of Extension Teaching 49

Learning, Learning situation and Learning experience 50

Principles of Learning in Farming Situation 53

Principles of Adult Learning 54

Types of Learning 54

Theories of Learning 56

CHAPTER 6

AUDIO-VISUAL METHODS & MATERIALS 96

Definitions 96

Purpose of Audio-Visual Aids 96

Selection of Audio-Visual Aids 108

Evaluation of Audio-Visual Materials 108

CHAPTER 7

COMMUNICATION FOR DEVELOPMENT 111

Functions of Communication 111

Communication Theories 112

Mathematical Theory 112

Mathematical Theory and Human Interactions : 113

A Shannon and Weaver Model 113

Social Psychological Theory 113

Linguistic Theory 114

Continuing applications of Communication Theory 114

Models of Communication 115

Elements of Communication System 118

Critical Factors in Communication 130

Concepts Relating to Communication 131

CHAPTER 8

EXTENSION MANAGEMENT 137

Functions of Extension Management Process 141

Organization 148

Organisational Theories 150

Rural Organizations 154

Administration 154

Functional Elements of Administration 155

Principles of Good Administration 156

CHAPTER 9

### ADOPTION AND DIFFUSION PROCESS	157

Adoption Process 157

Factors Affecting Adoption Process 159

Innovation Decision Process 159

Attributes of Innovation 161

Consequences of Innovation 162

Adopter Categories 162

Rejection and Discontinuance 164

CHAPTER 10

### PROGRAMME PLANNING AND EVALUATION	165

Definitions 165

Programme Planning Process 165

Basis for Sound Extension Programme 166

Steps in Programme planning process 167

Importance of Programme Planning in Extension Work 169

Characteristics of a Good Extension programme 170

Programme Evaluation 170

Degrees of Evaluation 171

Important Elements for Evaluating the Extension Work 173

CHAPTER 11

### LEADERS AND LEADERSHIP	174

Types of Leaders 174

Roles of Leaders 176

Qualities or Traits of a Leader 177

Principles of Leadership 178

CHAPTER 12

### LIVESTOCK DEVELOPMENT PROGRAMMES IN INDIA	184

Introduction 184

Dairy Development Programmes 186

Fodder Development Programme 187

Gosadan Scheme 188

Operation Flood 188

Intensive Cattle Development Project 191

Intensive Dairy Development Programme 192

Technology Mission 193

Milk and Milk Products Order (MMPO) 193

Piggery Development Programmes 193

Poultry Development Programmes 194

Sheep Development Programmes 196

Objectives of sheep development programmes 198

Suggestions to overcome the constraints 199

Rabbit Development Programmes 199

Various Other livestock development programmes ... 200

Veterinary Services for Cattle development 201

Importance and status of livestock in India 202

Rural Development Programmes In India 203

Agricultural Technology Management Agency (ATMA) 209

Krishi Vigyan Kendras 211

CHAPTER 13

COOPERATIVES: MEANING & OBJECTIVES   213

Principles of Cooperation 214

Dairy Cooperatives 215

I. Primary Milk producer's cooperative society 216

II. District dairy cooperative union 217

III. State cooperative federation 217

Management of dairy co-operatives 219

The Amul story 222

Structure of the Anand Pattern 224

National Dairy Development Board 226

CHAPTER   14

LIVESTOCK ECONOMICS & ACCOUNTANCY   233

Economic : Meaning, Definitions and Scope 233

Consumption 241

Types of consumption 241

Demand 242

Demand Schedule 243

Elasticity of Demand 246

Degrees of Elasticity 248

Supply 249

Interrelated Supply 251

Interaction of Demand and Supply 252

Cost 253

Price Determination 255

Other Factors Involved In Price Determination 255

Classification of price determination 256

Production 258

Factors of Production / Traditional Classification 258

Are these four factors reducible to two? 258

I. Land 259

II. Labour 259

III. Capital 261

Theory of Production 262

Merchandising and Product Planning 264

Standardisation and grading of products 266

Strengths and Weaknesses of Livestock sectors 267

Accountancy 268

Final Accounts 275

Analysis of Financial Accounts and 280

Different Financial Test Ratios 280

Preparation of Project of Dairy Farm 283

Estimated Annual Income 283

Estimated annual Expenditure 284

CHAPTER 15

LIVESTOCK FARMING:
TYPES, SYSTEMS & MARKETING   285

Large scale farming 285

Advantages of large-scale farming 286

Disadvantages of large scale farming 287

Small scale farming 287

Advantages small scale farming 287

Disadvantages of small scale farming 288

Classification of farming 289

Functions of livestock 293

Marketing methods of urban and rural societies 294

Marketing options and their efficacy 296

Dairy marketing options 297

Meat marketing options 298

Marketing institutions 298

Marketing channels 299

Factors affecting length of marketing channels......................... 299

Marketing channel for live poultry 300

Rural Marketing.. 301

CHAPTER 16

SURVEY TECHNIQUES & SAMPLING METHODS   304

Social Survey 304

Social Sampling and Sampling Methods 308

Reasons for sampling 309

Principles of sampling 309

Simple Random Sampling 310

Multi-Stage Sampling 314

Cluster Sampling 315

Non-Probability sampling... 315

Accidental or Haphazard or Convenience Sampling 316

Purposive or Deliberate or Judgment Sampling......................... 316

Quota Sampling.. 317

Heterogeneity Sampling 317

Snowball Sampling 317

Procedure for selecting a sample 318

Techniques of Data Collection 319

Observation Method 319

Case Study Method 322

Interview Method 322

Tools of Data Collection 324

I. Questionnaire 324

II. Interview Schedule 327

CHAPTER   17

PARTICIPATORY RURAL APPRAISAL   336

Introduction 336

Features of PRA 337

PRA Tools 340

Types of Mapping 341

Transect Walk 343

Why use diagrams? 354

General guidelines in drawing diagrams 354

Concept maps 356

When to use concept diagrams 356

Flow diagrams 357

When to use flow diagrams 358

CHAPTER   18

ENTREPRENEURSHIP   360

Introduction 360

Psychology of the Entrepreneur 361

Sociological Appraoch 362

New Venture Creation 363

A Systematic Approach to the Process of
Technology Transfer 366

Personnel 367

Finance 369

Marketing 370

Operations 370

Outside Assistance 371

Corporate Entrepreneurship 373

CHAPTER 19

PRIVATE EXTENSION: INDIAN EXPERIENCES — 375

What is Private Extension / Privatisation? 375

Stake holders in Private Extension 375

Indian Experiences 376

Existing type of extension and need for change 380

Consequences of privatization of extension service 381

Role of Public Extension 382

Role of Private Extension 383

Strategies for privatizing extension 383

CHAPTER 20

TRANSFER OF TECHNOLOGY — 385

Technology Transfer in Different Disciplines 385

Technology classification 386

Mechanisms of Technology Transfer 389

Modes of Transfer 390

Transactions and Channels for Technology Transfer 392

Stages in the Technology Transfer Process 393

CHAPTER 21

FISHERIES EXTENSION IN INDIA — 396

Emergence of a fisheries extension service 400

Fisheries Administration 402

Research and Development Institutes nformation and
education, training and extension 402

Training Institutes 404

Fisheries cooperatives 405

Fisheries development programmes 406

Extension and Technical Assistance 407

Marketing 407

GLOSSARY 409

REFERENCES 433

List of Tables

Table No.	Title
1.1	Difference between formal and informal education
3.1	Differences between rural, urban and tribal community
3.2	Comparison of primary and secondary groups
3.3	Comparison among class and caste pattern
5.1	Comparison between Result Demonstration, Method Demonstration
8.1	Difference between formal and informal organizations
8.2	Comparision between management and administration
14.1	Individual demand schedule
14.2	Market demand schedule
14.3	Supply schedule
14.4	Schedule for interaction of demand and supply
14.5	Market price versus normal price
14.6	Is land equal to capital?
14.6	Examples of different types of accounts
14.7	Examples of various nature of accounts based on transactions
14.8	Balance sheet
15.1	Differences between types and systems of farming
16.1	Difference between social survey and social research
18.1	Elements of a comprehensive business plan
18.2	Entry wedges

List of Figures

Figure No.	Title
1.1	The three links in the chain of rural development
1.2	The elements of Extension Educational process
3.1	Categorization of culture
3.2	Classification of social change
4.1	Steps in extension teaching
4.2	The elements of a learning situation
6.1	Cone of experience
7.1	Linear Transmission of Messages
7.2	Newcomb's A-B-X or Coorientation Model
7.3	SMCR Model of Communication
8.1	Operational approach of management
8.2	Systems approach of management
8.3	Feedback loop of management control
8.4	Decision making process
8.5	Need-want-satisfaction chain
8.6	Maslow's hierarchy of needs
9.1	Adopter categories on the basis of innovativeness
10.1	Basics for sound extension programme
10.2	Steps in programmes planning process
11.1	Sociogram
13.1	Organization of Anand pattern
15.1	Marketing channel for live animals
15.2	Marketing channel for livestock products
15.3	Marketing channel for live poultry
17.1	Social map

CHAPTER - 1

Extension Education Concept, Objectives and Principles

The term extension was first used in the United States of America to connote the extension of knowledge from the Land Grant Colleges to the farmers through the process of informal education. The term 'Extension' is derived from the Latin roots, 'tensio' means 'stretching' and 'ex' means 'out', thus the term 'extension education' means that type of education which is stretched out into the villages and fields beyond the limits of the schools and colleges. In India, the terms community development and extension education became more popular with the launching of Community Development Projects in 1952 and establishment of the National Extension Service in 1953. Since then, Community development has been regarded as a programme for an all-round development of the rural people and extension education as the means to achieve this objective.

Extension education has now developed as a full-fledged discipline, having its own philosophy, objectives, principles, methods and techniques, which must be understood by every extension worker, and others connected with rural development. It might be mentioned here that extension education, its principles, methods and techniques are applicable not only to agriculture but also to veterinary and animal husbandry, dairying, home science, health, family planning, etc. Based upon its application and use, various names have been given to it, such as agricultural extension, veterinary and animal husbandry extension, dairy extension, home science extension, public health extension, & family planning extension.

Why Extension?

"You can't apply yesterday's method today and be in the business tomorrow." In other words, the rural people should know and adopt useful research findings from time to time and should also transmit their problems to the research workers for solutions. The researchers neither love the time nor are they equipped for the job of persuading the villagers to adopt scientific methods and to ascertain from them the problems. On the other hand, it is

impracticable for the millions of farmers to visit research stations and learn things by themselves. Thus, an agency is required to bridge the gulf between research workers and people at large and to play a dual role of interpreting the results of researcher to the farmers as well as of conveying the farmers problems to the research stations for solutions. This agency is termed as "Extension" and persons manning this agency or organization called "Extension workers".

Rural Development

The three links in the chain of rural development.(Fig. 1.1).

- Research institutions
- Extension organizations
- For efficient communication between researchers and farmers, the extension workers must undergo adequate training through formal Teaching institutions.

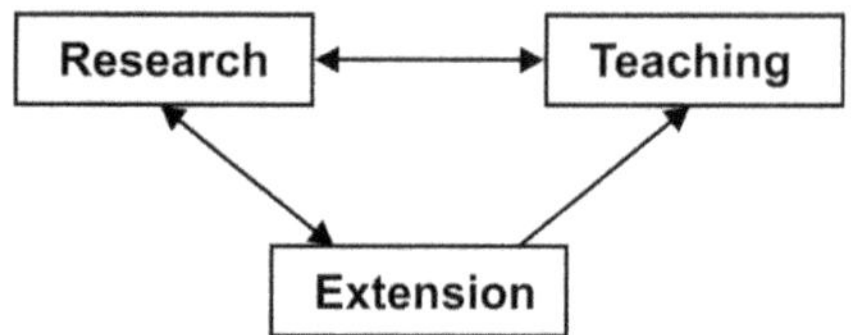

Fig. 1.1 Chain of rural development.

Concept of Extension

The basic concept of Extension is 'Education', which is defined as the production of desirable changes in human behaviour i.e. bringing about of desired changes in knowledge (things known), skills (things done) and attitudes (things felt) either in all or one or more of them.

Education : It is the process of bringing desirable changes in the behaviour of human beings and with the development of society, education has taken many shapes, such as :

- Child education
- Adult education
- Technical education
- Health education
- Physical education
- Education in Humanities and Social Sciences

The broad classification of education is formal, non-formal and extension education.

Formal Education : It is basically an institutional activity, uniform, rigid, full-time, sequential, structured, subject-oriented and leading to certificates like degrees and diplomas.

Non-formal Education : It is not uniform, flexible, part-time, unstructured, learner-oriented and enriches human and environmental potentials.

Extension Education : It is a science which deals with the creation, transmission and application of knowledge designed to bring about planned changes in the behaviour of the people with a view to help them to live better by learning the ways that improve their farm, home and community situation.

Veterinary and Animal Husbandry Extension : It is defined as an applied techno-social discipline developed for the improvement of production and health aspects of livestock through educational means. The agricultural extension and veterinary extension may be same with respect to philosophy, principles, approaches and contents but they differ with regard to objectives, strategies, methods, subject matter specialists, clients, applications, situations and services.

Differences between Formal Education and Extension Education

It may, however, be mentioned here that when extension education is put into action for educating the rural people, it does not remain formal education. In that sense, there are several differences between the two. (Table 1.1).

Table 1.1 Difference between formal and informal education.

Formal Education	Extension education
1. The teacher starts with theory & works up to practicals.	1. The teacher (extension worker) starts with practicals and may take up theory later on.
2. Students study subjects.	2. Farmers study problems.
3. Students must adapt themselves to the fixed curriculum offered.	3. It has no fixed curriculum or course of study and the farmers help to formulate the curriculum.
4. Authority rests with the teacher.	4. Authority rests with the farmers.
5. Class attendance is compulsory.	5. Participation is voluntary.
6. Teacher instructs the students.	6. Teacher teaches & also learns from the farmers.

Table 1.1 *Contd....*

Formal Education	Extension education
7. Teaching is only through . instructors	7. Teaching is also through local leaders.
8. Teaching is mainly vertical.	8. Teaching is mainly horizontal.
9. The teacher has more or less homogeneous audience.	9. The teacher has a large & heterogeneous audience.
10. It is rigid.	10. It is flexible.
11. It has all pre-planned and pre-decided programmes. based on the needs and expressed	11. It has freedom to develop programmes locally and they are desires of the people.
12. It is more theoretical.	12. It is more practical and intended for immediate application in the solution of problems.

The concept of Extension Education process developed by John Paul Leagans identifies five phases, which shows the sequence of steps in a cycle that is expected to result in progress from a given situation to a new desirable situation (Fig. 1.2)

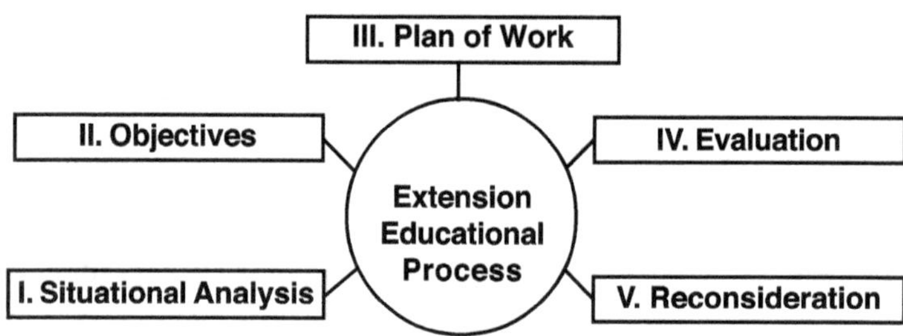

Fig. 1.2 The elements of Extension Educational process.

(i) *Analysis of the Situation* : In this phase facts are needed about the people, their interests, education, social customs, habits, folkways and about types of farming, marketing, size of holding and communication and transport facilities and others such as problems of the organizations as well as the resources available with the organization. A thorough analysis and comparison of 'what is' with 'what should be'.

(ii) *Deciding upon objectives* : This phase must enable the people to select a limited number of problems and to state their objectives clearly. The solutions to be offered must give satisfaction and the objectives should express the behavioural changes in people as well as social or economic outcomes which are desired.

(iii) *Teaching plan of work* : This phase involves content to be taught, methods and techniques of communication in order to create learning situation. The ability to choose and use those methods best adopted to particular objectives is the measure of an extension worker's effectiveness.

(iv) *Evaluation* : This phase determines to what extent objectives have been reached. This is also a test of how accurately and clearly the objectives have been stated. The process of evaluation may be simple or complex and informal or formal.

(v) *Reconsideration* : This phase consists of review of previous efforts and results, which reveal a new situation. If this new situation shows the need for further work, then the whole process may begin again, with new or modified objectives.

Philosophy of Extension

The philosophy of extension education has been described and interpreted in different ways by various authors, thus a clear picture cannot be drawn.

According to Ensminger, the philosophy of extension can be expressed in the following terms :

1. Extension is an education process.
2. Extension is a continuous process.
3. Extension is a 'two-way-channel'
4. Extension is 'helping people to help themselves'
5. Extension is development of individuals and their society.
6. Extension is teaching people what to want and how to satisfy those wants.
7. Extension is 'learning by doing' and 'seeing is believing'
8. Extension is a living relationship, respect and trust for each other.
9. Extension is working together to expand the welfare of the people and
10. Extension is working in harmony with the culture of the people.

According to Mildred Horton, the philosophy of extension can be expressed in the following terms:

1. The individual is supreme in a democracy.
2. The home is the fundamental unit in a civilization.
3. The family is the first training group of the human race.
4. The foundation of any civilization rests on the partnership of man and land.

Scope of Extension Education

Kelsey and Hearne (1967) given emphasis to nine areas, which indicate the scope of extension, these are:

    (i)   Efficiency in agricultural and allied productions

    (ii)   Efficiency in marketing, distribution and utilisation

    (iii)   Conservation, development and use of natural resources

    (iv)   Management on the farm and in home

    (v)   Family living

    (vi)   Youth development

    (vii)   Leadership development

    (viii)   Community development and rural area development and

    (ix)   Public affairs.

Extension is an integral part of livestock and rural development programmes in India. Extension is an integral part of teaching in agriculture, animal husbandry, rural development, social work etc. There is a scope to introduce extension in other disciplines, particularly which involve educating and motivating people and assisting them in their profession. The principles and methods in extension which originated with the need of agriculture and rural development are gradually being applied in other disciplines with success. The methods adopted in popularizing family planning, health and nutrition education etc. are such examples. Application of the extension principles and methods need not be confined to the rural people and rural areas only. It may very well be applied with the urban people and in and in urban situations as well. It appears that there is considerable scope for systematically applying the principles and methods of extension, wherever there is need for creating awareness among the people and developing human resource.

Objectives of Extension Education

The objectives of extension education are the expressions of the ends towards which our efforts are directed. In other words, an objective means a direction of movement. Before starting any programme, its objectives must be clearly stated, so that one knows where to go and what is to be achieved. The fundamental objective of extension education is the development of the people.

The Extension in our country is primarily concerned with the following objectives:

    (i) To assist the people to discover and analyze their   problems, felt and unfelt needs.

    (ii) To develop leadership among people and help them to organize groups to solve their problems.

(iii) To motivate the people to accept the technology and put it into practice.

(iv) To keep the researchers informed about the people's problems and to ensure solutions.

Principles of Extension Education

The extension work is based upon some working principles and the knowledge of these principles is necessary for an extension worker. Some of these principles, as related to veterinary and animal husbandry extension, are mentioned below.

1. *Principle of Interest and Need :* Extension work must be based on the needs and interests of the people. These needs and interests differ from individual to individual, from village to village, from block to block and state to state and therefore, there cannot be one programme for all people.

2. *Principle of Cultural Difference* : Extension work is based on the cultural background of the people with whom the work is done. Improvement can only begin from the level of the people where they are. This means that the extension worker has to know the level of the knowledge, and the skills of the people, methods & tools used by them, their customs, traditions, beliefs, values, etc. before starting the extension programme.

3. *Principle of Cultural Change* : It is obvious that the change agent works personally with the villagers must know what they think. With this attitude, the change agent must seek to understand and discover the limitations, taboos and cultural values in order to select an accepted approach. (Fig. 1.3 and 1.4).

Fig. 1.3 In one culture, the problem has been solved in the same old way for several years.

Fig. 1.4 But, it is doubtful, whether it can be used in same way in other culture.

4. ***Principle of participation*** : Extension helps people to help themselves. Good extension work is directed towards assisting rural families to work out their own problems rather than giving them ready-made solutions. Actual participation and experience of people in these programmes creates self-confidence in them and also they learn more by doing.

5. ***Principle of adaptability*** : People differ from each other, one group differs from another group and conditions also differ from place to place. An extension programme should be flexible, so that necessary changes can be made whenever needed, to meet the varying conditions.

6. **P***rinciple of grass roots* : A group of rural people in local community should sponsor extension work. The programme should fit in with the local conditions. The aim of organising the local group is to demonstrate the value of the new practices or programmes so that more and more people would participate.

7. ***Principle of leadership*** : Extension work is based on the full utilisation of local leadership. The selection and training of local leaders to enable them to help to carry out extension work is essential to the success of the programme. People have more faith in local leaders and they should be used to put across a new idea so that it is accepted with the least resistance.

8. ***Principle of democratic approach*** : The extension worker translates the scientific findings of the research stations in such a way that the farm families can adopt voluntarily to satisfy their own needs. However, extension work is democratic both in philosophy and procedure. It aims to operate through suggestions and discussion. All possible alternative solutions are placed before the participants and ultimately people are left free to decide their line of action and the methods to be adoped.

9. ***Principle of whole-family :*** Extension work will have a better chance of sucess if the extension workers have a whole-family approach instead of piecemeal approach or seperate and unintegrated approach. Extension work is, therefore, for the whole family, i.e. for male, female and the youth.

10. ***Principle of co-operation :*** Extension is a co-operative venture. It is a joint democratic enterprise in which rural people co-operate with their village, block and state officials to pursue a common cause.

11. ***Principle of satisfaction***

The end-product of the effort of extension teaching is the satisfaction that comes to the farmer, his wife or youngsters as the result of solving a problem, meeting a need, acquiring a new skill or some other changes in behaviour. Satisfaction is the key to success in extension work. "A satisfied customer is the best advertisement."

12. ***Principle of evaluation***

Extension is based upon the methods of science, and it needs constant evaluation. The effectiveness of the work is measured in terms of the changes brought about in the knowledge, skill, attitude and adoption behaviour of the people but not merely in terms of achievement of physical targets.

Extension: Indian scenario

India has the largest extension system in the world with 1,17,603 paid agricultural extension personnel catering to the farming and allied needs of over 90 million farm families. Among these an overwhelming majority were small and marginal farmers with an average land holding of 1.63 ha, scattered and fragmented over different agro-climatic zones.

There are five well-specified extension systems prevalent in our country. They are First-Line Extension Education System (ICAR/SAUs), National Agricultural Extension Service / Training and Visit System, Special Extension Programme on specific crops, Rural Development Programmes and Extension Programme of Non-Governmental Organizations, each having its own mandate on poverty eradication. The First-Line Extension System of Ministry of Agricultural Research (ICAR) and State Agricultural Universities (SAUs) have only limited jurisdiction in coverage, while compared to Extension System of Ministry of Agriculture having a broader network. The non-government organizations engaged in extension activities include mostly voluntary agencies (partly or fully funded by government for extension activities), business houses, agricultural processing firms, producer's co-operatives, input agencies and private consultants. In India and many of the developing countries extension is not merely transfer of technology and information. It takes care of the broad perspective of human resource development of its clients. Speaking of the Indian context, two crucial issues need to be considered: First the ability of the farmers to pay for the extension services and secondly, how to demarcate the benefits of extension as 'private' and 'public'.

Agriculture and livestock development in India for the time begin attained a stage where more reliance has to be emphasized on rain fed / dry land areas for national food security, correction of nutritional and regional imbalances and generation of widespread rural employment opportunities in hinterlands. These situations therefore necessitate the need for more public support to extension services in those areas and to put it briefly, the time is not yet ripe in India to think over the complete withdrawal of public support to extension.

Extension in India should continue to focus on good activities such as technology transfer, education (about agriculture, livestock production, animal health, water conservation, natural resource management, and disease forecasting) and human resource development. Extension in India should concentrate on policy issues, quality control, and on channeling assistance and establishing mechanisms to benefit resource poor farmers and farm women. Policy issues related to land holding, supply of credit, input and marketing, prices are decisive and have not received the needed importance till. In India, systems and institutions should co-exist while addressing the needs of farming community so as to derive synergistic advantages of both. The policies should not at all be influenced by vested political interests, as there will be a clear danger of moving away from reality, if, in this case it is the farmers. It's applicability and appropriateness to a particular situation / context may be fully recognized so that extension services will be fully geared to meet twin global goals of food security and sustainable food production. Extension services have been traditionally organized and delivered by the public sector all over the world, which led to a situation wherein, whenever one refers to extension, it denoted public extension service.

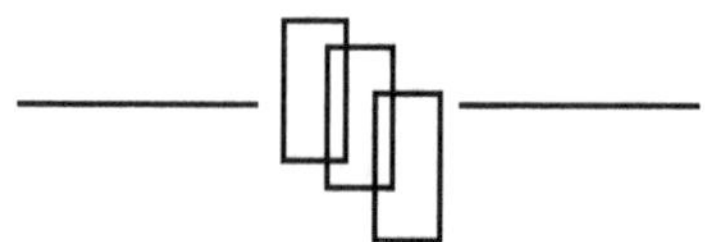

CHAPTER - 2

Genesis of Extension in India

The community development programme in India aiming at the all-round development of the rural people and the Extension Service as a nation-wide organization to achieve these aims are of relatively recent origin in India. This new programme and administrative set-up of Extension Service is the outcome of several years efforts and reforms made over the years. In India, the extension work started with a few outstanding individuals of a philosophic and philanthropic bent of mind, as a nation wide organization, the new set-up are described in four stages:

- Stage I- Pre-Independence Era (1866-1947)
- Stage II- Post-Independence Era (1947-1952)
- Stage III- Community Development and National Extension Service Era (1952-1960)
- Stage IV- Intensive Agricultural Development Era (1960-onwards)

Stage I : Pre-Independence Era (1866-1947)

During the pre-Independence era, individuals or some organizations made various scattered and short-lived efforts towards rural development in various parts of the country. It was during this period the Department of Agriculture came into existence in June 1871 under Government of India, and by 1882 agricultural departments in most of the provinces started functioning in skeletal form. Recognizing the need for new and improved methods of cultivation based on agricultural research, the then government of India also set up an Institute of Agricultural Research at Pusa in Bihar in 1905. Though valuable work was done, villages continued to remain neglected and disorganized. In fact, a number of reforms took place in India during this period. In order to avoid the recurrence of famines, Famine Commissions were appointed from time to time and they made very valuable recommendations. But it was the Royal Commission on Agriculture's recommendation (1928), which established a firm foundation for co-coordinated research and effective agricultural administration. As a result, the Imperial Council of Agricultural Research (now Indian Council of Agricultural Research) was established in 1929. Besides the agricultural departments, there were other development departments, such as animal husbandry, health, and education, but all these departments were working in isolation and reaching the people directly, without any co-ordination among themselves.

During 1903, Seaman A. Knapp has started teaching the farmers through demonstrations: Thus he is often referred as 'Father of Demonstrations'. In 1914 O.H. Benson had initiated the 4-H club:

1ˢᵗ H refers to Head	:	Pledged for clear thinking
2ⁿᵈ H refers to Heart	:	Pledged for greater loyalty
3ʳᵈ H refers to Hands	:	Pledged for longer services
4ᵗʰ H refers to Health	:	Pledged for efficient working

In the same year, the U.S. government has passed a law i.e. Smith Lever Act, to coordinate the extension work nationally. Based on this a separate extension department was established with eight divisions and they are: Division of business administration, Extension information, Field cooperation, Subject matter area, Field trips and training, Agricultural economics, Agricultural research and Agricultural extension.

Marthandam Project : As early as 1903, David Hamilton has started his experiment in model villages along with cooperative lines at Sunderban in Bengal. This experiment continued with cooperative marketing and later offered training to the villagers in cottage industries.

Shantiniketan Institute : One of the pioneers, Rabindranath Tagore, desired that each villager should work to his/her capacity and should help the fellow villagers and for which every villager has to be educated.. He believed in 'self help' and 'mutual help' and initiated the 'Shantiniketan Institute' in 1914 with the following objectives:

- To identify problems and find out suitable solutions
- To develop the resources and
- To improve village sanitation

Following are the approaches to achieve the objectives :

- Conducting demonstrations
- Organizing mobile librarary, Bharati balikas & shows
- Creating the spirit of self help
- Developing the village leadership
- Running night schools

In the year 1914 the 'Christian mission' was initiated with the objectives of spread of literacy and character building and for which several demonstration centers had been established.

In 1921, Gandhiji has started a micro laboratory at Sewagram in Vadodara District of Madhya Pradesh for conducting experiments on social, economic and spiritual aspects of the villagers. He has also introduced a basic system of school education for children and adults, called 'Naitalim' and in the same year, Dr. Spencer Hatch had initiated an all-round project called Marthandam Project in Kerala and Acharya Vinobha Bhave had initiated Bhoodan movement, which laid emphasis on moral and spiritual background.

Gurgaon Project : F.L. Brayne has initiated a fairly extensive experiment in rural reconstruction in Gurgaon district of Punjab. Succeeded in arousing considerable enthusiasm among people. He introduced the idea called 'Village guide', who are non-technical in nature and served as channels for percolation of information from outside. However, Dr. Brayne could not succeed in making his experiment self sustaining and after he left, the programme received a set back stating 'Entirely dependent on a single person', this helped to pave way for comprehensive efforts and in 1933 Dr. Baryne was appointed as commissioner of rural reconstruction in Punjab.

Sri V.T. Krishnamachari, Dewan of Baroda state, conducted a rural reconstruction programme and simultaneously 'Firka development scheme' was started under the dynamic leadership of Sri T. Prakasam.

Stage II : Post-Independence Era (1947-52)

Etawah Project : The idea of intensive all-round development work in a compact area was put into practice as a Pilot Project in Rural Planning and Development in the Etawah District in Uttar Pradesh, which can be regarded as a forerunner of the Community Development Project in India. LT.Col. Albert Mayer, an American engineer, played the key role in the initiation and implementation of the project. The programme was based on the principle of self-help, democracy, integrated approach, felt needs of the people, rigorous planning and realistic targets, institutional approach, co-operation between governmental and non-governmental organisations, close co-ordination between the extension service and the supply agencies and collaboration by technical and social scientists. After an initial period of trial and error, a new administrative pattern was evolved. It percolated to the village level; the activities of different nation-building departments were channeled through one common agency and a multipurpose concept of village level worker was introduced. Each village level worker looked after 4-5 villages. A district-planning officer assisted by four specialist officers and other supporting staff supervised the project. This project was conceived in 1947 and the pilot project was inaugurated in September 1948 under the guidance of LT. Col. Albert Mayer. Initially the project was started in 64 villages with the objectives: 1. To observe the initiation and augmentation, 2. To assess the social development and 3. To see what is the degree of production. 'Education and Persuasion' were the approaches made in achieving the objectives.

Grow-More-Food Campaign *:* The urgent need for stepping up food production was realised even in the pre-Independence era and a Grow-More-Food Campaign was started. Under the campaign, targets for increased agricultural production were laid down for the first time on an all-India basis. But the campaign failed to achieve its targets. Soon after Independence (1947), the Central Government re-defined the objectives of the Grow-More-Food Campaign as the attainment of self-sufficiency in food grains by 1952, and simultaneously increased the targets of production of other crops to meet the shortfall as a result of the partition of the country. At the same time, arrangements were made for integration and co-ordination of the entire campaign for increasing agricultural production. Some state governments associated the public with working of the campaign by setting up non-official committees at the village, taluka, district and state levels. The plans were revised from time to time to make the campaign more effective.

An Enquiry Committee was constituted to study the performance of this programme. In its recommendations, the Committee proposed the formation of development blocks, each consisting of 100 to 120 villages, and the appointment of revenue officers as development officers or extension officers, assisted by technical officers for agriculture, animal husbandry, co-ordination and engineering. For actual work in villages, the Committee suggested the appointment of one village level worker for every five or ten villages.

The Committee also described broadly the functions of the extension service, the manner in which the extension organization would operate, the arrangements required for training the required staff, and the way in which non-official leadership should be associated with the work of village development at the village, taluka, district and state levels. The need for setting up an independent organization of the suggestions made, the manner in which the assistance should be rendered to the state governments as well as to villagers for development work, the role of the central and stage governments in this effort was also emphasized by the committee. Based on these recommendations, the Planning Commission, which was set up earlier by the government of India to prepare a plan for development consistent with the available resources, gave the highest priority to the development of agriculture and irrigation in the First Five-Year Plan. The Commission fixed substantially high targets of internal production and decided, as recommended by the Enquiry Committee, that the drive for food production should form part of plans for overall agricultural development, and that agricultural improvement in its turn should form an integral part of the much wider efforts for raising the level of rural life. The Commission prescribed "Community Development" as the method for initiating the process of transformation of the social and economic life of villages and "Rural Extension" as its agency.

Stage III : Community Development and National Extension Service Era (1952-60)

Community Development Project : As a result of the Grow-More-Food Enquiry Committee Report and the successful experience of the Etawah Project, 15 Pilot Projects were started in 1952 in selected states with financial assistance received from the Ford Foundation. Besides helping to increase agriculture production and bettering the overall economic condition of the farmers, these projects were meant to serve as a training ground for the extension personnel. It was soon realized that for the creation of an urge among the rural population to live a better life and to achieve permanent plentitude and economic freedom in the villages, a much bolder and dynamic effort was called for. It was recognized that the success of this new effort depended upon the whole hearted co-operation of the beneficiaries, government officials and non-officials at every stage, the education of rural masses in the technique of rural development and the timely provision of adequate supplies of the needed inputs and other requirements. Initially, 55 Community Development Projects were started in different parts of the country on 2 October 1952 for three years.

The Projects covered nearly 25,260 villages and a population of 6.4 millions. Each project, in turn, consisted of about 300 villages covering 400-500 square miles and having a population of about two lakhs. The project area was divided into three development blocks, each comprising 100 villages and a population of 60,000 to 70,000. The development blocks, in turn, were divided into groups of 5-10 villages, each group being in the charge of a multipurpose village-level worker. The main aims of these projects were: to increase agricultural production by all possible means, to tackle the problems of unemployment, to improve village communications, to foster primary education, public health and recreation, to improve housing, to promote indigenous handicrafts and small-scale industries and to improve the villagers, lot through their own primary effort. In short, the programme aimed at achieving all-round socio-economic transformation of the rural people. In view of a very enthusiastic response received from the rural population in all projects and the keenness shown and the demand made by the people for Community Development Programmes, the Government of India decided to extend it rapidly to other parts of the country.

Starting with 55 Community Development Projects in 1952, the entire country was covered with the Community Development Programme by 1963. In all, there are 5,263 blocks, besides 101 tribal development blocks. A block is the unit of planning and development. This is a new administrative unit meant for tackling the problems of the rural people in entirety and in a concerted and co-coordinated manner. In this new set-up of Community Development Blocks all the Nation-building government departments were brought together and in order to ensure co-ordination at the block level, a new post of a Block Development Officer was created. This officer is the co-coordinator of the programme and team leader and is supported by 8 extension officers drawn from the development departments. They were one each from the fields of agriculture, animal husbandry, co-operative, panchayat, rural industry, rural development, social education and welfare of women and children. Each normal block was provided with 10 village level workers and two-gram sevikas. Under this new set-up, the block is treated as an administrative unit for all the development departments, and the village-level worker is the contact person between these departments and the people.

The Community development projects had to be completed in 3 years period and was divided into 5 stages.

1. *Conception stage (3 months)* : Survey and Planning were given importance at this stage.
2. *Initiation stage (6 months)* : Emphasis was given in providing Infrastructure and communication facilities.
3. *Operation stage (18 months)* : Approved activities were taken up.
4. *Consolidation stage (6 months)* : Completion of the entire operation was taken up in this stage.
5. *Finalization stage (3 months)* : Final touches were given.

Organizational Set-up for Community Development Project : The organizational set-up for Community Development Programme runs from the national level through state, district and block levels to the village level and there are three main constituents of this new set-up.

(a) The direct-line staff such as State Development Commissioner, Block Development Officer (BDO) and Village Level Worker (VLW).

(b) The auxiliary or specialist staff, such as different heads of technical departments at the state and district levels and extension officers at the block level.

(c) Panchayat Raj System - The Zilla Parishads, Block Samithis and Village Panchayats.

A. *National Level*

At the National level programme, the policies are formulated by the National Development Council presided over by the Prime Minister of

India. Membership of the Council consists of the Central ministers of the concerned ministries, chief ministers of all states, and members of the Planning Commission. The Planning Commission provides guidance for Plan formulation and gives it approval to annual and Five-Year Plans of the states as well as of the Centre. The Ministry of Agriculture and Irrigation is responsible for giving national guidance, policy formulation and technical assistance in regard to Agriculture Extension and Community Development (now Rural Development Programmes. In the Agriculture department, the Agricultural Commissioner, Government of India, assisted by a number of assistant commissioners and directors, with the supporting staff, is in charge of all agricultural development programmes at the national level. Within this Department, special mention may be made of the Directorate of Extension Training responsible for the training of Extension officers, VLWs, instructors of Village-Level Workers Training Centres and others and the Directorate of Farm Information which is concerned with the dissemination of new agricultural technology and innovations through various media.

B. *State Level*

At the state level too, there is usually a State Development Committee presided over by the Chief Minister of the state with the other concerned ministers as its members. This Committee is responsible for the state's plan and programmes and for fixing the targets for regions and districts. Besides this committee, there are usually a number of other advisory or technical committees. As regards the actual administrative functioning, the State Development Commissioner is the top-level executive responsible for directing, co-coordinating and providing overall guidance for development programmes and maintaining a two-way channel of communication between the state governments and the Central government. He co-ordinates the activities of different development departments, such as agriculture, animal husbandry, co-operation, panchayat Raj, health, education, irrigation, power and electricity. The heads of these technical departments are responsible for planning and implementing the technical programmes and for providing the necessary technical guidance, manpower and support.

C. *District Level*

At the district level too, there is usually a District Development or District Planning Committee presided over by the District Collector or Deputy Commissioner. The other members of this committee are the heads of the departments in the district, chairman and vice-chairman of the district boards, representatives of voluntary organizations, local bodies and members of parliament and state legislatures. In the states, where the Panchayati Raj is operating, the Zila Parishads are responsible for planning, co-coordinating and consolidating the development programme in the district. The District collector is the key official who co-ordinates the activities of all development departments at the district level. The district-level technical heads of agriculture, animal husbandry, co-operation, Panchayats, public health, irrigation, education and rural industries are responsible for planning and implementing the development programmes relating to their departments. Administratively, they are responsible to the district collector on one hand and to their state heads of development departments on the other.

D. *Block Level*

A district is subdivided into a number of community development programmes. The Block development officer is the head of the block team, and co-ordinates all the activities of the development departments at the block level. He is assisted by eight extension officers from different fields namely, agriculture, animal husbandry, health, co-operation, panchayats, engineering, social education and rural industry. At the non-official level in the states, where the Panchayati Raj has been implemented, the Panchayat samiti (also called the Block Samiti) has the statutory powers for formulating and executing development programmes. The B.D.O and the extension officers assist the Samiti. Wherever the Panchayat Raj is not working, there are block development advisory committees.

E. *Village Level*

At the village level, the multi-purpose village-level worker is the main extension staff. He is the last extension functionary in the administrative hierarchy and is the main contact person. He is responsible for all developmental work at the village level, and forms a connecting link between the various technical departments and the rural people. Usually, in a normal community development block, there are 10 village-level workers. On the non-official side, usually there is a Panchayat in every village or for a cluster of villages, and is responsible for planning and implementing the community development programmes and ensuring people's participation in them.

National Extension Service

Along with Community Development Projects, it was also decided to launch a programme, which was somewhat less comprehensive, and to integrate both under one comprehensive service! The National Extension Service, inaugurated on 2nd October 1953. The scheme of National Extension Service was designed to provide the essential basic staff and a small fund for the people to start the development work essentially on the basis of self-help. The operational unit of this service was a N.E.S. block comprising about 100 villages and 60,000 to 70,000 people. The N.E.S. blocks were later converted into Community Development Blocks, which had higher budget provisions in order to take up more intensive development programmes.

Stage IV : Intensive Agricultural Development Era (1960-onwards)

Intensive agricultural District Programme (Package Programme). Under the Community Development Programme, the production efforts and the available resources were diffused over the entire country. The educational and extension efforts among the millions of farmers to be tackled remained thin and restricted. As a result of the report of the Ford Foundation team, known as 'India's Food Crisis and Steps to Meet It', a significant departure took place during the period and a new programme known as the 'Intensive Agricultural District Programme', based on the principles of concentration and better management of resources and efforts in potential and responsive areas, with assured water supply, was started in 1960. Since then, at least one district in each state has been covered under this programme. The scheme not only involves the adoption of a package of new practices, but also ensures the availability of credit and production inputs, adequate research information, training and education of farmers and extension personnel, storage and marketing arrangements and price assurance which would encourage the farmers to adopt scientific methods of farming. District collector is the Project Officer who is assisted by an Assistant Project Officer, four or five Subject Matter Specialists and other supporting staff. At the block level, besides a block development officer, there are four Agricultural Extension Officers, two co-operative extension officers and a few officials from other fields. The number of Village Level Workers is 20 in each block. A little later, a similar programme, but less in intensity and thinner in staffing pattern, was started in 1964. This programme is known as the 'Intensive Agricultural Area Programme'. Now about 10 per cent of the total cultivated area in the country is under the Intensive Agricultural District Programme and Intensive Agricultural Area Programme.

Need for Democratic Decentralization

Community Development (CD) programme was aimed at socio-economic development of rural sector with the cooperation of the people. It was soon evident that peoples' participation remained dissatisfactary and the administration was described as "Democracy at the top and dictatorship at the bottom". In view of this, a study team headed by Balwanth Rai Mehta recommended democratic decentralization which was approved by "National Development Council" in 1958. Rajasthan was the first state to adopt democratic decentralization. In 2nd Oct'1959 and then Andhra Pradesh on 1st Nov'1959. Democracy is derived from Greek words "demos: meaning 'people' and 'cratos' meaning 'rule of'. As a whole 'Democracy' means the governance of the people, by the people and for the people. The emphasis is on people as distinct from officers. Although democratic development was suggested, was not easily understood by rural people. Hence Panchayat Raj evolved.

Panchayat Raj

Panchayat Raj means, it is a system of government; Horizontally it's a network of village panchayat, vertically it's a organized growth of panchayats rising up to national level".

Philosophy of Panchayat Raj : The philosophy of panchayat raj system is based on two aspects such as :

1. Concept of self-help and
2. Functions of local self-government

Objectives of Panchayat Raj : The objectives of panchayati raj are:

- To serve as an institutional device for creating self help to meet common ends on an organized basis.
- To make people of each area achieve intensive and continuous development in the interest of the entire population.

This system envisages a three-tier system at the district, block and village levels, as indicated below :

- District level - Zilla Parishad
- Block Level - Panchayat Samiti
- Village Level - Village Panchayat

Gram Panchayat

It is the first formal democratic institution at village level and is the principal unit of local self-government, also designated as the cabinet of village elders. The members of gram panchayats are directly elected from the wards by taking income, population and area into consideration. While sarpanch and upa-sarpanch are elected by the members. Each gram panchayat is required to convene meeting of members of panchayat once in a month and Gramsabha at least once in six months.

Functions of Gram Panchayat

It has three basic functions, such as

(a) *Representative Functions*

The gram panchayat's main role is voice and represents the community's opinion.

(b) *Administrative Functions*

1.Collecting taxes

2. Improving the measures of safety and sanitation

3. Registration of birth and death

(c) *Developmental Functions*

4. Provision for elementary school

5. Public health

6. Street lighting

7. Agriculture, animal husbandry and irrigation

8. Communication *etc.,*

The following agendas are normally laid down and discussed in the Gram Sabha:

(a) Accounts of income and expenditure of the Gram Panchayat.

(b) Annual Administrative Report for the last financial year.

(e) Details of the receipt of funds and developmental activities taken up and completed during the last financial year.

(f) Likely receipt of funds and proposed developmental activities to be taken up during the current financial year.

Panchayat Samithi

It is the second tier of the Panchayat Raj system at the block level and It includes the Sarpanches of all panchayats; local MLAs and MLCs (with a right to vote but not to hold office); one person nominated by a collector and one person from every panchayat, where no sarpanch has been elected. During the selection of members of panchayat samithi, the reservation pattern to be observed:

(i) One from Schedule Caste

(ii) One from Scheduled Tribe

(iii) Two women members

(iv) Two persons with experience in administrations and public life

President and vice-president of panchayat samithi are elected from among the members of the samithi and Block Development Officer (BDO) appointed by government is the chief executive officer of the samithi.

Functions of Panchayat Samithi

    (i)   Agricultural development.

    (ii)   Development of animal husbandry sectors and fisheries.

    (iii)   Rural health and sanitation.

    (iv)   Primary education

    (v)   Social education.

    (vi)   Development of communication.

    (vii)   Development of co-operatives and cottage industries.

    (viii)   Programme for women welfare and social welfare.

    (ix)   Rural housing.

    (x)   Publicity.

    (xi)   Supervision and guidance of panchayats.

Zilla Parishad a it is the third tier of panchayati raj at district level. It includes the presidents of all panchayat samithis; District collector and MLAs, MLCs and MPs of the district. During the selection of members of Zilla Parishad, the reservation pattern to be observed is :

    (i)   One from Scheduled Caste.

    (ii)   One from Scheduled Tribe.

    (iii)   Two women members.

    (iv)   Two persons with experience in rural development and public life.

Functions of Zilla Parishad

- Scrutinizing and approval of budget to panchayats and samithis.
- Distribution of grants to panchayats and samithi.
- Co-ordination and consolidation of plans of panchayats and samithis.
- Advice to the government on implementation of developmental schemes.
- Promoting secondary school education.

Standing Committees : Each Panchayat Samiti and Zilla Parishad consist of 7 Standing Committees.

Standing Committee 1 : It deals with finance, taxation, audit and planning committee and coordination with other standing committees (to be known as Finance Committee).

Standing Committee 2 : It deals with education, environment, cultural, health and sports affairs and other related activities (to be known as Education and Health Committee).

Standing Committee 3 : It deals with road, water supply, communication, irrigation, Rural Electrification and non-conventional energy etc. (to be known as Works Committee)

Standing Committee 4 : The Rural Development Committee shall perform functions relating to poverty alleviation, area development programme, employment, housing corporation, thrift and small-scale industries, including cottage (to be known as development Committee).

Standing Committee 5 : It includes the promotion of educational, economic, social, cultural and other interests of the SC/ST and BCs; protecting them from social injustice and all other forms of exploitation; amelioration of the SC and ST women and other weaker sections of the society; securing social justice to the SC and ST women and other weaker sections.

Standing Committee 6 : It deals with agricultural, food, irrigation, cooperation, fisheries and animal husbandry activities (to be known as Agriculture Committee).

Standing Committee 7 : It deals with poverty alleviation, social and farm forestry, rural housing and drinking water facilities (to be known as Poverty alleviation Committee).

Each standing committee consists of such number of members as may be arrived at by dividing the total population by the number of pancahayt samithis. The panchayat samithi president presides over the finance committee meeting; whereas members are over the elected to preside other committee meetings.

Mandal System

It was hoped that through the Panchayat Raj system agriculture, animal husbandry, co-operatives, communication, cottage industries *etc.,* would be improve. In the beginning, the panchayati raj was successful because peoples' participation was more in developmental activities. But, later on it was diminished to positions were occupied by rich and higher caste people. The factions and non-cooperation between different departments has destroyed the aim of the panchayati raj system and withheld all developmental programmes. Later on the developmental schemes of panchayt raj system were handed over to SFDA, MFAL, DRDA, and ITDA, so that weaker sections may be benefited.

In 1978, the central government had appointed a committee led by Ashok Mehta to review and rectify the weaknesses of the Panchayat Raj system, whch has recommended mandal system and suggested making small administration units instead of larger units at taluka level. Karnataka is the first to adopt mandal system and Andhra Pradesh was the second state to adopt mandal system (25th May 1985), where each mandal head-quarters are provided with the facilities of bank, bus-stand, Primary Health Center (PHC), Veterinary Hospital (VH), police station, telephone exchange, high school, marketing, communication and agricultural godowns. The Mandal System has got Four Stages :

(i)   *Gram Panchayat :* The members of gram panchayat are directly elected from the wards by taking income, population and area into consideration. While sarpanch and upa-sarpanch are elected by the members.

(ii) *Mandal Praja Parishad :* The members of mandal praja parishad were constituted by including all the presidents of panchayats, Mandal praja territorial constituency members, MLAs, MPs and members from linguistic / religious minorities. During the selection of members the reservation pattern to be observed is 15% for Scheduled Castes, 6% for Scheduled Tribes, 20% for Backward Class and 9% for women.

(iii) *Zilla Praja Parishad :* The members of zilla praja parishad were constituted by including all the presidents of mandal praja parishad, Mandal praja territorial constituency members, MLAs, MPs, member of rajya sabha and Members from linguistics / religious minorities. During the selection of members the reservation pattern to be observed is 15% for Scheduled Castes, 6% for Scheduled Tribes, 20% for Backward Class and 9% for women.

(iv) *District Development Council :* It is otherwise known as Zilla Abhivrudhi Samiksha Mandali. It supervises the developmental activities and co-ordinates different developmental programmes. The members of district development council consists of Zilla parishad chairman, MLAs, MPs, Collector is the secretary, five expert members nominated by government. Chief Minister can nominate any minister as chairman of the district development council.

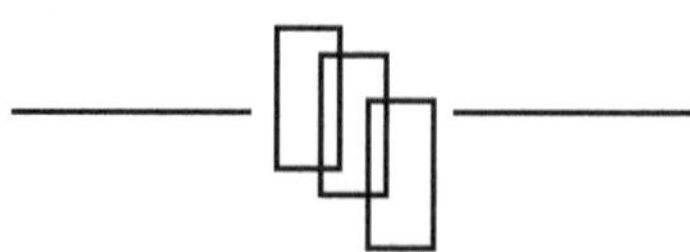

CHAPTER -3

Rural Sociology

Through history man has made efforts to gain increasing control over his physical environment and to gain freedom from its limitations. He has sought to gain knowledge about his environment and the myriad phenomena that surround him in natural and other inanimate objects and in animal life. From this knowledge he has developed fields of study and disciplines. But man himself also been object of study; man has tried to understand his behaviour in relation to others. Emerging from this attempt are the social sciences, which study man and his works from various standpoints. It is the scientific study of man's behavior in relationship to other individuals and groups with whom he/she interacts. Sociologists have defined sociology in many ways. It was 'Auguste Comte' (1789 – 1857), often referred to as the father of sociology, who named the field of study from two words; Latin word 'sociotus' means society & Greek word 'logos' means study of science.

Rural Socialogy is a specialized field of sociology, which deals with the scientific study of social, cultural and economic problems of rural aggregates.

Because of the several misconceptions about sociology, it is useful to describe what it is not, thereby dispelling misconceptions and clarifying the major concerns of the field of sociology.

(i)   *Sociology is science, not philosophy :* A social philosophy is a system of ideas and values that tell people how to behave and lays down procedures, norms and rules of behaviour according to which people are expected to act. Sociology studies what is, not what should be.

(ii)   *Sociology is science, not socialism :* Though the two words are similar, one is a science interested in systematic determination of facts concerning human behaviour; the other is a political and economic ideology.

(iii) *Sociology is science, not social reform or social welfare :* Sociology seeks knowledge about people and their behaviour. It is therefore knowledge, not direct treatment. While, social reforms or social welfare is largely the responsibility of reforms and community development by fully utilizing such knowledge and principles to make their work more effective.

(iv) *Sociology is not based on general impressions and points of view derived from overall experience in working with people :* It is a science based on empirical evudence from organized and verified facts. A person does not become a sociologist by virtue of his contact with people no matter how frequest and how varied his experience, any more thn a women becomes a gynaecologist by virtue of having given birth to a large number of children.

Despite variation in definition the basic focus and theme of sociology remains the same and the broad contents or elements of the field of sociology are:

(i)   It is a study of man
(ii) It is the scientific study of man
(iii) It is the study of man's behaviour
(iv) It is the study of man in relation to groups with whom he interacts.

What is Rural Sociology?

It is a specialized field of sociology. What has been stated above about sociology refers to man regardless of whether his residence is urban or rural, a specialization of the field of sociology has therefore emerged. It has been stated that sociology is the scientific study of man's behavior in relationship to other individuals and groups with whom he/she interacts. Simply enough, when this study focuses on man living in rural areas, then it is the field of rural sociology. It seeks to understand the caste, social groups, religious beliefs, social and economic background etc.,

Importance of Rural Sociology

1. Helps the people to understand himself and his own social nature.
2. Seeks to understand human affections and actions without praising and blaming.
3. Studies man and his institutions.
4. Interprets rural behavior.
5. Eliminates the tragedies and emotions of individuals.
6. Diagnoses the rural evils and social evils.
7. Emphasizes the growth and change in the society.

8. Apprises the happenings in the society.

9. Interprets the role and relevance of rural sociology in community development.

Scope of Rural Sociology

- It seeks to understand about social class and caste.
- It seeks to understand about religious believes.
- It seeks to understand about social and economic background of rural people.
- It seeks to understand about involvement of formal and informal groups.
- It seeks to understand about the differentiation between rural and urban communities.

Characteristics of Rural Society

1. Village is the unit of rural society.
2. Caste is the dominant institution in rural society.
3. Traditional way of living and caste occupation prevails mostly.
4. Cooperative functions of several castes exist in socio-religious life.
5. The village is said to form a self-sufficient economic unit.
6. Characteristics are generally governed by local/regional traditions.
7. Leisurely attitude towards life and low standard of living are the chief characteristics of rural society.
8. The character and structure of rural society largely depends on its religion, caste and linguistic compositions.

Significance of Rural Sociology in Extension

1. It deals with human beings in rural setup.
2. Social change is brought about not in vacuum but in a structure of human relations.
3. Extension worker has to understand the culture in which he has to bring about change.
4. It helps to overcome the cross cultural barriers.
5. It helps to understand the value systems of individuals, families and groups.
6. It helps to lay emphasis on increasing the quality and quantity of production.
7. It helps to understand the existing pattern of behaviour and its importance.
8. It helps in reconstruction of social, cultural and economic fields of rural society.

9. It provides scientific knowledge about society and laws governing its development.
10. It helps to describe the rural trend, which is the product of time and society.
11. It aids to study the social situations, processes and problems of rural population.
12. It helps to introduce the innovative technologies so as to bring about overall socio-economic development.

Concepts of Rural Sociology

1. *Society* : is a group of people who have lived together, sharing common values and general interests, long enough to be considered by others and by themselves as a unit.

2. *Community* : is group of people living within a contiguous geographical area, sharing common values and a feeling of belonging to the group, who come together in the common concerns of daily life. A society may include many communities and neighbourhoods. 'Permanency' is the basic element of community, while the other elements are locality, community sentiment and naturality. A larger community like a nation provides peace and protection and a smaller community like neighbourhood, provides friends and friendship. (Table 3.1).

Table 3.1 Differences between rural, urban and tribal community.

Criteria	Tribal Communty	Rural Community	Urban Community
Size	Small	Large	Small
Density of Population	Low	High	Low
Occupation Agriculture	Mainly	Different profession	Take up any
Environment	Open	Closed i.e. man made environment	Close to nature
Type of family	Joint	Nuclear	Joint
Size of family	Large	Small	Large
Economy	Subsistance	Cash	Subsistence
Material possession	Less	More	Less
Communications Transportations	Less	More	Meagre
Homogeneity	More	More	More
Heterogeneity	Homogenous	Heterogeneous	homogenous
Hierarchy	Less in number	More in number	Less in number

Table 3.1 *Contd....*

Social Contact	Less Number of contacts Primary contacts are predominant	More number of contacts Secondary contacts are predominant	Less number contacts Primary contacts are
Social Diffe-rentiation	Less	More	Less
Social Control	More informally	More Legally	More informally
Social Change/ Conservatism	Slower rate	Faster rate	More conservative
Social institutions	Less	More	Less
Social mobility	Less intensive	More intensive	Less intensive
Familism/ Individualism	More attached to family	More individualized	More attached to family
Sacred/Secular	More sacred	More secular	Mainly sacred
Literacy	Less	More	Less
Standard of	Lower living	High	Primitive way of life

3. **Group** : Two or more people who are in reciprocal interactions with one another. It is not incorrect to say that the entire sociological field revolves around group relations.

4. **Organisation** : are groups with special concerns and interests that have developed a structure involving specific roles for various members, and that have a more or less formal set of rules and regulations for operation.

5. **Culture** : It is the expression of our nature, in our own mode of living and thinking. It stands for the moral, spiritual and intellectual attainments of man. In sociology the word culture denotes, acquired behaviour, which are stored and transmitted sub points of among the members of the society between generations and shared.

Characteristics of Culture

- It is acquired but not innate.
- It is a social product.
- It is a total social heritage (i.e. it passes from one generation to another through traditions and customs)
- It fulfills both ethical and social needs of those groups, which are ends in themselves.

- It is idealistic and normative (i.e. it embodies all the ideas and norms)
- It is an integrated system
- Language is the chief vehicle of culture (i.e. man lives not only in present but also in past and future).

Organizational Structure

Culture is not simply an accumulation of behaviour patterns; it has a structure, which is made up of various units.

(a) ***Cultural Trait*** : It is the smallest unit of culture and the irreducible unit of learned behaviour pattern or material products, which is categorized into (i) *material cultural traits* such as wearing dhotis and sarees; (ii) *Non-material cultural trait* such as bowing slightly in respectful greetings to elders and puling hem of saree over the head to give respect to elders.

(b) ***Cultural Complex*** : it is a group of related cultural traits. For example, thread ceremony in Bhrahmins is a cluster of various number of related cultural traits to form a particular religious cultural complex.

(c) ***Culture Pattern*** : It is a group of cultural complexes. For example, existing pattern of rural Hindu society. It embodies *specialists*, who are the elements of culture shared by some groups but not all; *Ethos*, are the norms and values that characterizes a culture and serves to distinguish one culture from another; *Ethocentricism*, is the tendency of man to consider his own culture of higher value and superior to others; *Culture area*, the geographical area within which a common culture exists and *Cultural lag*, the time difference between some technological change and the resulting change in culture.

Categorization of Culture : (Fig. 3.3)

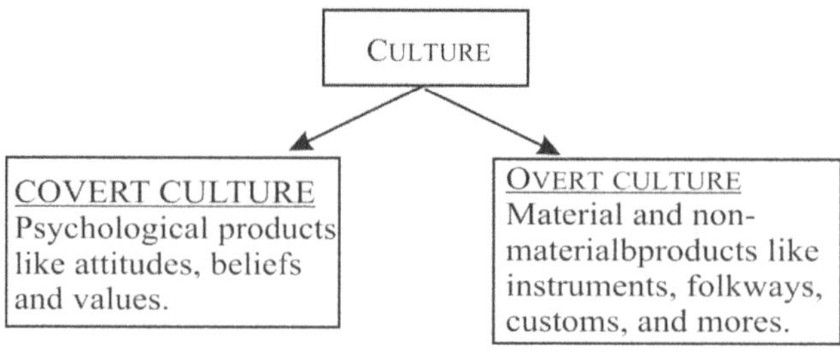

Fig. 3.3. Categorization of Culture.

6. *Institution* : is an organized system of relationships, which embodies certain common values and procedures and meets certain basic needs of the society. Institutions are culturally approved patterns of behaviour including prescribed roles and procedures, and are grouped some basic social needs. These have persisted long enough to be considered as permanent.

Classification of Social Change (Fig. 1.4)

A. Social Institutions

Five basic social institutions are generally recognized these are as follows:

(a) *Family* : is the most basic social institution providing for procreation, upbringing and socialisation of the young generation. It provides economic security, love and affection to the family members. Members of the family have different roles in decision making as well as participation in physical tasks relating to earning a living and home making.

(b) *Religion*

is belief in supernatural, It constitutes a set of beliefs regarding the ultimate power in the universe, the ideal and proper pattern of behaviour, and ceremonial ways of expressing these beliefs. Religion also provides a foundation for the mores of the society and taboos in various cultures have religious sanction. Religion provides a means by which individuals can face crises and ups and downs in life with strength and fortitude.

(c) *Government*

as a political institution, administers the regulatory functions of law and order, and maintains security in society. Development work is now a major responsibility of the government, and for effective implementation of prgrammes, government may decentralise its functioning by creating local self-Government institutions like panchayats at village level.

(d) *Economy*

provides basic physical sustenance of the society by meeting the needs for food, shelter, clothing and other necessary supply and services. Economic institutions include agriculture, animal husbandry, marketing, credit and banking system, cooperatives etc.

(e) *Education*

is the process of socialisation, which begins informally at home and then formally in educational institutions. Education as an institution helps develop knowledge, skill, attitude and understanding of the people and strives to make them competent members of the society. Education has a tremendous influence on the behaviour of individuals. Education widens the mental horizon of the people and makes them receptive to new ideas.

B. Rural institutions

Five basic rural institutions are generally recognized and these are as follows:

(a) *School* : There is a natural bondage between school and rural community. It is the home, where the child is first introduced to social life, till the child is introduced to school education, where the unconscious process of education is closely associated with the gradual growth of the child.

(b) *Gram Panchayats* : Panchayats are the local governments, which undertake various activities such as: to increase production in both agriculture and livestock, bring waste land under cultivation, provide irrigation facilities, provide drinking water facilities; arrange cooperative management and implement the state government's rural developmental activities.

(c) *Cooperatives* : Cooperation starts with human life itself. When man began to associate with each other cooperation came into existence. Cooperation has became the one aspect of the vast movement which promotes voluntary association of individuals having common needs and ends.

(d) *Mahila Mandals* : Livestock farming is a family enterprise. All members of the family are involved in the process of increased production. In addition to this, women have traditional responsibilities i.e. taking care of family, certain agricultural tasks and almost all activities of livestock farming. Besides this, women need to be helped in spending income towards family health and living standards. For these education is highly essential and the same could be provided through mahila mandals. These are independent voluntary organizations of local women.

(e) *Youth Clubs* : Rural development programmes would be effective, when it has sound, competent and enlightened leadership and for whom training is obligatory. Youth clubs prepare farmers for future responsibilities. These trained young farmers will grow up as scientific farmers, with responsibilities of helping their brethren to change their attitudes and to take up new practices.

7. *Association* : It is a group of social beings attached to an organization with a view to secure a specific end or ends.

8. *Social Norms* : Techniques, customs, folkways, mores, taboos and laws are rules based on social values, that control and direct interpersonal relationships in society. Techniques are ways of doing

things in which technical efficiency is the criterion of operation. *Customs* are socially prescribed forms of behaviour and transmitted by tradition, these are the accepted ways in which the people do the things together. *Folkways* are the approved forms behaviour for specific situation. *Mores* are those things, which one ought to or ought not to do. *Taboos* are the unwritten laws of the society and the purpose of writing such laws may be explained in three forms such as production, protection and prohibitive. *Laws* are formalized norms with legal and/or political enforcement, such as acts and statute of a nation.

9. *Social Roles* : is the expected behaviour of one member of society in relation to others. A single person in society may play number of roles such as father, teacher, citizen, sociologist etc; some roles are temporary; others are permanent and some may conflict with others.

10. *Religion* : It is the human response to powers, which are beyond human control, supernatural or supersensory. It has two main components (i) *Rituals* are the prescribed manner of certain actions designed to establish link between the performing individual and supernatural power (ii) *Belief* realize or ensure that the rituals will be observed and there are some supernatural powers and are different from place to place and people and people.

11. *Social Process* : are the interactions of groups and individuals with one another, and these may take four basic forms: *(i) Competition,* where the object is to outdo another in achievement of a goal; *(ii) Conflict,* where the object and goal is to eliminate the other; *(iii) Cooperation,* where persons or groups unite, in an effort to achieve a common goal; *(iv) Accommodation,* is a temporary working arrangement between groups to achieve a goal; *(v) Assimilation,* where a mutual cultural diffusion occurs, which makes the persons or groups culturally alike.

12. *Social Control* : It is the process by which the group makes use of its various resources, to bring about conformity to its norms and values. Social action requires social control, informally through a group's action, folkways, mores etc. and formally through government.

13. *Socialisation* : is the process by which an individual is inducted into his social and cultural world. It commences the day he/she is born. Society and its various components, starting with the mother and the home, teach and instill the rules and regulations of the society to which he/she belongs. Socialisation involves the development of personality, attitudes, habits and expected social roles.

14. **Crowd :** A temporary collection of people reacting to stimulii.
15. **Category :** It is a collection of individuals have at least one common characteristic
16. **Aggregation :** It is the collection of individuals in physical proximity of one another.
17. *Social Groups* : It is a unit of two or more people, who are in interaction and in communication with each other.

Characteristics of Group

1. Groups always consist of at least two persons and can extend to many more.
2. Groups do not form by mere plurality of numbers, communication and interaction is essential.
3. While communication and interaction are essential, it must be reciprocal i.e. two way.
4. Groups may be long-lived or of brief duration. Group exists only as long as there is reciprocal psychological interaction, it ceases to be group when active relation between the minds of the two or more involved ceases.
5. Groups and group life are also to be found among non-human beings. Apes and livestock exhibit evidence of organized group activity in various ways. However, the continuity of culture from one generation to another makes man's group experience unique.
6. Common interest, shared values and norms turn out to be important ingredients in social groups.

The formation of groups may be based on various kinds of situations that cause people to unite. The following has been suggested as classifications based on kinds of situations in which people unite:

1. Physiological kinship and community of blood
2. Marriage
3. Similarity in religious and magical beliefs
4. Similarity in native language and mores
5. Common possession and utilisation of lands
6. Similarity in territorial proximity
7. Common responsibilities, acquisitions and privileges
8. Similarity in occupational interests
9. Similarity in economic interests
10. Subjection to same social institution or agency of social service and social control.
11. Common enemy or danger and
12. Mutual aid

Classification of Social Groups

Although the social group categories are not inclusive and they do overlap, it is important to know the nomenclatures used as well as the categories.

A. Groups based on Type of Relationship

Primary and Secondary Groups : Primary groups are characterised by intimate face-to-face associations and informal, personal relationships. There will be a strong feeling of belonging to the group on the part of members. Members are interested in one another as individuals and they share experiences, hopes, plans, problems and fulfill their need for companionship. Primary groups are relationship-directed and the decisions are more traditional and non-rational. While, few ties of sentiments, formal and impersonal and utilitarian relationships, or characterize secondary group. In secondary groups, members are not concerned with other member as individuals but as their roles or address. Secondary groups are goal-oriented and the group decisions are more rational and the emphasis is on efficiency. For example, family, friendship and play groups are primary groups and cooperative societies and labour unions are the secondary groups. (Table 3.2).

Table 3.2 Comparison of primary and secondary groups.

Criteria	Primary Group	Secondary group
Size	Small	Large
Relationship	Personal, informal and face-toface	Impersonal, formal and specialised
Purpose	Not organised for specific purpose	Organized for specific purpose
Permanency	Permanent	Temporary
Social distance	Less	More
Group decision	Non rational	Rational
Organization requirement	Do not require to achieve a purpose	Do require to achieve a purpose

B. Groups Based on Type of Organization

Formal and Informal Groups : The degree of formality or informality that exists in the group may be regarded as a continuum with formal groups at one end and informal groups at the other. It should be recognized that no group is 100% formal or informal in its relationships although definite marked differences exist in the operation end of these two types of groups. The formal groups, like village council or labour unions involves definite roles, rules of operation and rigidly enforced behaviour of its members. Informal groups, like play groups, friendship groups have no such organization, rigidity or formality.

C. Groups based on Social Class

Horizontal and Vertical Groups : Horizontal group is used to describe members who are alike in the status or position in the class system of society. Thus, all the farmers, blacksmiths, goldsmiths would be the members of a horizontal group in the villages, in addition those with like incomes, but different occupations may belong to horizontal group. Vertical groups are those, composed of members from different social strata and whose membership cuts vertically across the horizontal grouping in society.

D. Groups Based on Structure

Voluntary and Involuntary Groups : Voluntary groups are those, where one can become member according to one's choice. The same could be organised like youth clubs or unorganised like play groups. While, involuntary groups are those, where membership is by birth and the choice does not rest with the individual.

E. Groups Based on Personal Feeling or Belongingness

In and out Groups : An in-group is a group to which a person feels he belongs and with which he identifies strongly, such as 'my family', my 'neighborhood', 'my temple'. Other groups to which he feels he does not belong are his out-groups. The relationship in these groups, as in other cases, may be represented on a continuum with in-group at one end and out-group at the other, with intermediary groups in between which in-group and out-group feelings vary in degree.

F. Groups Based on Compulsion

The membership in certain groups may be as a result of compulsion of participation i.e. social pressure or pressure from other members like a child in school and a prisoner in jail.

G. Reference Groups

It is a group with which an individual feels identified, the norms and objectives of which he accepts. A reference group may therefore be any group: primary, formal, horizontal or otherwise and strongly influences the individual's behaviour. An individual may not be a member of the groups. Influenced by it. To understand the behaviour of human beings, one must know their reference groups. For example, a family or neighborhood reference group may influence a farmer to accept or reject an improved farming practice.

Social Change

The difference between social and cultural changes is largely theoretical, the two are so closely interwoven that distinction is very difficult. Social change refers to the process by which alteration occurs in the structure and functioning of social system. Cultural change refers to the changes in the culture of the society. A distinction also exists between social change and progress; the later term progress is the change in a direction that is considered desirable. Different people may define desirability differently and what is desirable for one may be undesirable for another.

Characteristics of Social Change

- Social change invariably results from the interaction of a number of factors and not simply from a single factor. This happens because of the mutual interdependence of social phenomena.

- Social change in one aspect of life gives rise to a chain reaction of changes affecting other aspects. For example an increasing number of women are employed in various types of positions and have entered into professional and non-professional areas of society, occupying themselves increasingly in roles other than those in the home.

- Social change takes the form of replacement i.e. specific changes may occur in various ways and may be grouped broadly as modifications and replacements of material goods in society and social relationships. One illustration is the modification of diet and food habits; availability of transportation facilities and disorganisation of rural families i.e. from joint family to nuclear family.

- Social change is environmental. It must take place within a geographic or physical and cultural context. Both these contexts have impact on human behaviour and in turn man changes them.

- Social change is temporal, but at the same time mere passage of time does not cause change as in the biological process of ageing. Society and culture do not get worn out through time. During the passage of time, interaction of various physical, cultural and social factors and conditions bring about social change.

- Social change essentially involves human aspects. The composition of society is not constant, but changing. People constantly move and over a period of time considerable variation in size and composition of society is normally experienced. *The environmental, temporal and human aspects of social change are necessary conditions under which change takes place.*

- Social change involves both tempo and direction of change. The notion that the direction of change is inevitable and universal laws govern the tempo of change has been abandoned by most sociologists.

Conditions of Social Change

1. New need in society creates a situation, making it conducive to change. These new needs may be created or actual or imaginary.

2. Dissatisfaction with the status provides a condition favourable for social change. Dissatisfaction with status quo has led to innovation, invention, new systems of education, social reform movements and various other changes in society.

3. The accumulation of knowledge and techniques with society is an important condition of change. The extent to which these knowledges, experiences and are techniques are available, determines the extent of change in the society.

4. The dominant values, general orientation and attitudes towards change in a culture condition the circumstances of change, making them less or more conducive to change.

5. The degree of complexity of culture and structure within society. Complex society with differentiation of statuses, roles, class and specialisation of functions, together with efficient and effective systems of communication and transportation, encourage, facilitate and expedite change.

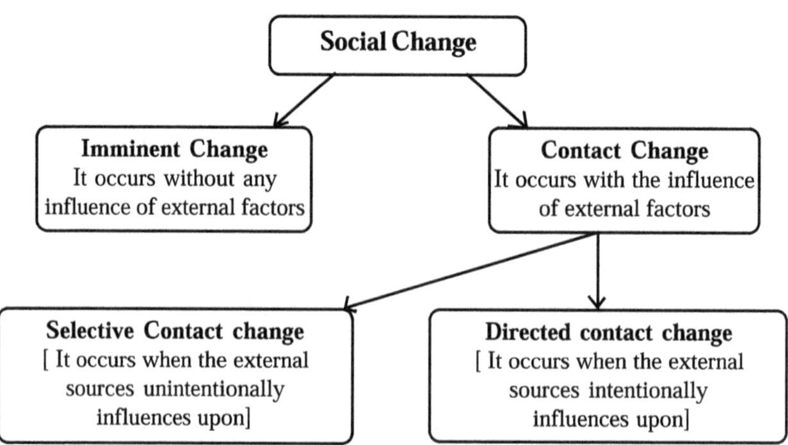

Fig. 3.4 Classification of social change (Based on source).

Factors of Social Change

Social change results from the interplay of several factors, classified under three major factors i.e. geographic, biological and cultural.

A. *Geographic Factor* : Climate and all other factors of earth's surface influence man's livelihood. The abundance or scarcity of topographical factors including the soil, its resources and other minerals influence the possible development of man and make up the physical context within which he lives. Man does not passively accept geographical factors as they are. For instance, he uses scientific knowledge and experience to make the desert bloom into 'high producing agricultural lands'. Man modifies and adds to the natural landscape by constructions, such as dams, canals and highways creating a cultural landscape which changes with far greater rapidity.

B. *Biological Factor* : Plants and animals form a part of man's non-human environment and man in turn utilises them for his various purposes such as food, shelter and clothing. For example, he has taken a wild bird of the jungle and converted it into 'egg producing machine' and also it is evident in some areas of the world, rapid population growth interacting with other factors stimulate rapid social change.

C. *Cultural Factor* : Contact with different cultures, diffusion, invention and discovery are sources from which some changes in society emanate and emerge. Technological inventions of many kinds are constantly disturbing the older ways of living. Interplay of factors is significant in processes of social change. Changes in material aspects of culture stimulate changes in the non-material aspects of culture. For example, a new invention in educational technology will stimulate the educational process itself.

Social Stratification

The division of population into two or more layers, each of which is relatively homogenous and between which there are differences in privileges, restrictions, rewards and obligations'. Social stratification, to Sorokin, means The differentiation of a given population into hierarchically superposed classes."

Such stratification, he held, is a permanent characteristic of any organized social group. Stratification may be based on economic criteria—for example, when one focuses attention upon the differentials between the wealthy and the poor. But societies or groups are also politically stratified when their social ranks are hierarchically structured with respect to authority and power. If, however, the members of a society are differentiated into various occupational groups and some of these occupations are deemed more honourable than others, or if occupations are internally divided between those who give orders and those who receive orders, then we deal with occupational stratification. Though there may be other concrete forms of stratification, of central sociological importance are economic, political, and occupational stratification.

Social Class

It is defined as the abstract categories of persons arranged in levels according to the social status they possess (social status is the basic criterion for social class).

To be recognized as social class a group requires a feeling of :

- Equality in relation to members of its own
- Inferiority in relation to those who stand above in social class and
- Superiority in relation to those who stand below social class.

- The movement of individuals from one stratum to other is possible.
- Class is not hereditary in nature
- Class can be acquired or changed

Classification

Class may be categorised based on power or prestige or wealth etc.

- Social Classes are   :   Upper, Middle and Lower
- Cultural Classes are   :   Hindu, Mohammedian and Christian
- Economic Classes are   :   Business, Service and Farmers
- Political Classes are   :   Congress, B.J.P., Janata Dal etc.
- Self-identific Classes are   :   Rotary clubs, Lion clubs
- Participatory Classes are   :   Cricket, Hockey, Football etc.

Caste

The word caste is derived from a Portuguese word 'casta' which means complex of hereditary qualities and is defined as a social category or system of stratification, whose members are assigned a permanent status within a given social hierarchy and whose contacts are restricted accordingly.

Characteristics

- Caste is determined by birth.
- Caste is most rigid and often referred as the extreme form of closed class system.

- Each individual caste has its own laws, which governs the food habits of its members.
- It is the dominant institution of village.
- Villagers are governed to a great extent by traditional caste occupations like carpentry, goldsmith, blacksmith, etc.
- Endogamous in nature i.e. majority of the individuals marry their own caste members only.

Identification Methods of Class/ Caste

- *Subjective approach :* In this approach the class or caste of an individual can be identified through self identification i.e. by asking the members of the society to identify or rank themselves
- *Objective approach :* In this approach the identification of class or caste of an individual can be done with the help of measures/indications viz., income, education, occupation, type of house.
- *Reputation approach :* This particular approach is restricted to smaller communities. (Table 3.3).

Table 3.3 Comparison among class and caste pattern

Criteria	Class pattern	Caste pattern
Openness	It is more open and elastic in nature, thus social mobility between classes is easy	It is more closed and strongly rigid
Secularism	Its is secular in origin, but not spiritual and are not founded on religious dogmas	It is divinely ordered and the hold of religious belief is essential for the continuance of the system
Endogamy/ Exogany	Members may marry outside the class i.e. one class member	Members do not marry outside caste
Class conscious- ness	Feeling of class consciousness is necessary to constitute a class	Feeling of any subjective consciousness is not necessary to become a member of caste
Prestige	There is no rigidity fixed order of prestige	Relative prestige of different caste is well established

Social Mobility

Social mobility is understood as the transition of people from one social position The first concerns movements from one social position to another situated on the same level. The second refers to transitions of people from one social stratum to one higher or lower in the social scale, as in ascendant movements from rags to riches or in the downward mobility of inept children of able parents. Both ascending and descending movements occur in two principal forms: the penetration of individuals of a lower stratum into an existing higher one, and the descent of individuals from a higher social position to one lower on the scale; or the collective ascent or descent of whole groups relative to other groups in the social pyramid. But what distinguished Sorokin's orientation from that of many contemporary students of stratification and mobility was that his main focus was upon collective, not on individual phenomena. As he puts it, "The case of individual infiltration into an existing higher stratum or of individuals dropping from a higher social layer into a lower one are relatively common and comprehensible. They need no explanation. The second form of social ascending and descending, the rise and fall of groups, must be considered more carefully".

Groups and societies may be distinguished according to their differences in the intensiveness and generality of social mobility. There may be stratified societies in which vertical mobility is virtually nil and others in which it is very frequent. We must therefore be careful to distinguish between the height and profile of stratification, and the prevalence or absence of social mobility. In some highly stratified societies where the membranes between strata are thin, social mobility is very high. In contrast other societies with various profiles and heights of stratification have hardly any stairs and elevators to allow members to pass from one floor to another, so that the strata are largely closed, rigidly separated, immobile, and virtually impenetrable. In regard to degrees of openness and closure, Sorokin holds to his usual position. No perpetual trend toward either increase or decrease of vertical mobility can be discerned in the course of human history; all that can be noticed are variations through geographical space and fluctuations in historical time.

Attempting to identify the channels of vertical mobility and the mechanisms of social selection and distribution of individuals within different social strata, Sorokin identifies the army, the church, the school, as well as political, professional, and economic organisations, as principal conduits of vertical social circulation. They are the "sieves" that sift individuals who claim access to different social strata and positions. All these institutions are involved in social selection and distribution of the members of a society. They decide which people will climb and fall; they allocate individuals to various strata; they either open gates for the flow of individuals or create impediments to their movements.

In considering the impact of actual rates of social mobility, as well as the ideology of social mobility, modern societies, offers a fresh approach in the light of current experience. Far from indulging in unalloyed enthusiasm about high degrees of social mobility, was at pains to highlight its dysfunctional and its functional aspects. Sorokin stressed, among other things, the heavy price in mental strain, mental disease, cynicism, social isolation, and loneliness of individuals cut adrift from their social moorings. He also stressed the increase in tolerance and the facilitation of intellectual life that were likely to occur with more frequency in highly mobile societies.

Social Values and Value Systems

Values: are the attitudes held by individuals or group or society as a whole, as to whether material or non-material objects are good or bad or desirable or undesirable.

Norms : The rules that govern action directed towards achieving the values are called norms. Value system of society: Social value to constitute for society's preferences or estimates of worth in respect of material or non-material objects in society, taken together as a set of attitudes go to form a system, called the value system of the society.

Norms are the accepted and approved forms of behaviour that are based on and consistent with dominant social values in society;thus values and norms go together. The fulfillment of three conditions is necessary before something can develop into a value for an individual: (a) He must be aware of its existence. (b) His awareness must become a matter of concern to him so that he develops an emotionalized attitude towards it, regarding it as good or bad in some degree and is not indifferent to it. (c) His awareness and attitude must not be merely transitory but must endure in time.

Characteristics of values

1. Values are socially created rather than determined biologically or inherited.
2. Values are socially shared, while individuals in society may have individual values, the set of values that constitute the value system of society are shared and transmitted among members and accepted by them. While social values are shared, it does not follow that they are held by individuals with equal intensity. Intensity with which values are held may vary considerably. Thus, some individuals may easily violat social values and norms while others adhere to values with greater intensity.
3. Values are learned; they are acquired and not inherited. The process of learning and acquisition of social values commences from childhood in the family and through the process of socialisation. Values can be transmitted from one group to another within a society through various social processes.
4. Values are abstract attitudes and assumptions on which there is social consensus about the relative worth of objects in society.

5. Values are gratifying to people and have an important part in meeting social needs. Three elements involved in consideration of social values have been identified. The first is the object itself; second is the capacity of the object to satisfy social needs and third is the appreciation of the people for the object and for its capacity to provide the desired satisfaction.

6. Values tend to be linked together harmoniously to form patterns; these patterns form the value system in society. When harmonious integration of values in society does not exist, social problems arise.

7. Value systems vary from culture to culture in accordance with the relative worth attributed by each culture to its patterns of activity and its goals.

8. Values frequently represent alternatives and value systems consist of ranked alternatives. For example, a college student may not have sufficient income to feed himself and pay his tuition fee and other dues. On the basis of his values, he may deicide to forego all except one meal a day and meet his other necessary expenses.

9. Values may differ in their effects upon the individual and society as a whole. The values of a subgroup within society may be in conflict with those of society as a whole and work against its interests and welfare.

10. Values involve emotions and people often sacrifice and even enter into conflict to uphold them. For example, the fighting for a cause involves values charged with emotions.

11 Values exert strange influence on the development of individuals and society in at least two important ways: First, by making it easy or difficult for rural people to accept new practices, to form new types of organisations and operate in new ways: these are the illustrations of the strong influence of values. Second, by influencing the scientific findings of rural sociologists and other social scientists. The sociologist not only has personal values of his own but is also subject to conformity with the values of the society within which he lives.

12. Values are virtually automatic. For example, respect for women and aged persons are automatic responses.

Types of values

A. *Ultimate Value* : Often called as dominant values, they constitute the core a of society's value system. Every society has a unique set of ultimate values, which forms the general framework within which the behaviour of individuals and groups is controlled or influenced. Ultimate values are abstract and often not attainable. For example, the democratic procedures expressed in the system of government.

B. *Intermediate Value* : These are derived from ultimate values and are actually ultimate values that have rephrased into more reasonably attainable categories. I.e. within the framework of social institutions such as religion, government and education are intermediate values such as freedom of speech, religious freedom, free education, non-discrimination etc.

C. *Specific Values* : The sub-divisions of intermediate values are called specific values and are almost unlimited in number. Specific values must be in conformity with the total value system of which they form the smallest unit. They constitute the personal and group preferences expressed in daily life.

Taken together, specific, intermediate and ultimate values form the value system in society, which serve as a basic determinant of human behaviour. For example, to a farmer, the construction of a house is an ultimate value, while intermediate value may be represented by brick construction with a slab roof and a specific value may be denoted by the provision for livestock shed.

Functions of Values

Values are not ends in themselves and do not serve as objectives or goals towards which social action is directed. The following have been identified as the functions of social values:

1. Values provide a ready means for judging the social worth of persons. They help the individual himself to 'know where he stands' in the eyes of his fellow men.
2. Values focus the attention of people upon material cultural items that are considered desirable, no useful and essential.
3. The ideal ways of thinking and behaving in any society are indicated by values. They form a kind of blueprint of socially accepted behaviour so that people can almost always discern the 'best' way of acting and thinking.
4. Values are guideposts for people in their choice and fulfilment of social roles. They create interest and provide encouragement so that people realise that the demand and expectations of the various roles are functioning towards worthwhile objectives.
5. Values act as a means of social control and social pressure. They influence the people to conform to the mores, encourage them to do the right things and give them a feeling of merited esteem. On the other hand, they act as restraints against disapproved behaviour.

6. Values function as a means of solidarity. It is an axiom among social scientists that groups cluster around and are united by common shared values of a high order. People are attracted to others who cherish the same values; and it may be said that common values are among the most important of the factors that create and maintain social solidarity.

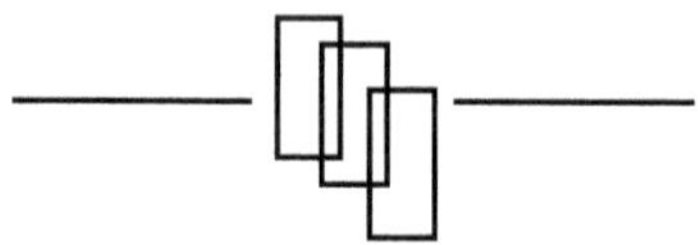

Teaching and Learning Process

Extension is an educational process for bringing about desirable changes among the people, which involves both learning and teaching. It is, therefore, necessary here to understand what is meant by learning and teaching. 'Teaching' is the process of arranging a situation in which learning can take place i.e., attention of the individuals drawn, their interest developed, desire aroused and action promoted.

Can Teaching be Learned? Yes. The ability of individuals engaged in teaching varies. There are no born teachers as there are no born lawyers, doctors, engineers, fishermen, or carpenters. Everyone is gifted by nature. Anyone with good intelligence and the will to study, practice, plan, and revise ways of doing the job can gain the skill for effective teaching. Hence, there is no mystery about learning how to be a good extension teacher. It is simply a matter of hard work, practice, concentration and the will to achieve proficiency. This is the price one must pay for acquiring real skill in any profession.

Steps in Extension Teaching

It is a planned and deliberate act on the part of the extension worker, where the role of an extension worker is more of facilitator and motivator rather than teacher. (Fig. 4.1).

In order to bring about the desired changes in the behaviour of people, the extension worker should organize activities so that there would be repetition of the desired behaviour. This conscious attention to organization of teaching activities in a sequence greatly increases the efficiency of learning. This is the advantage of an educational programme over incidental or occasional exposure to learning. The extension worker arranges the activities whereby things to be learned undergo one or more steps of the following:

Fig. 4.1 Steps in extension teaching.

Attention : Farmers are not always aware of the improvements, in such instances, the first task is to draw the attention of the farmers. Until the individual's attention has been focused upon the change, that is, it is considered desirable, there is no reorganization of a problem to be solved or a want to be satisfied. Attention is the starting point to the arousing of interest, in addition to supplying information to those desiring to learn, the extension worker creates desire for information on the part of those who are indifferent to improvements in concerned sectors.

Interest : Once attention has been captured, it becomes possible for the extension worker to appeal to the basic needs of the individuals and arouse their interest in further consideration of the idea and in easy stages the extension worker reveals to the learners how the new practices will contribute to the learners' welfare. The presenting of one idea at a time, which is definite and specific, is another important factor in building interest.

Desire : The extension worker must continue the stimulation of the learner's interest in the new idea or better practice until that interest becomes desire, or unfreezing the existing behaviour, or motivating force to compel action. The extension worker explains to the learner that the information applies directly to the learner's situation; that the innovation will satisfy a significant want or need of the learner.

Conviction : The extension worker sees to create a fixed belief that makes the learner to know what action is necessary and just how to take that action. He also makes the learner visualies the action in terms of his own peculiar situation and acquire confidence in his own ability to do the thing.

Action : Unless the conviction is converted into action, the teaching effort is fruitless. At this stage, the job of an extension worker is to make it easy for the learner to act. Teaching farmers how to control ectoparasites in sheep by using a new chemical in a particular way will not be practiced, until its price is reasonable and safe to use. If action does not follow soon after the conviction has been created, the new desire fades away and people continue to act as before.

Satisfaction : The end product of extension teaching effort is satisfaction that comes to the farmer as a result of meeting a need, solving a problem or acquiring a skill. Follow up on the part of extension worker helps the learner to evaluate the progress made, strengthens the satisfaction and minimises the annoyance. The satisfaction resulting from successful completion of a job will lead the farmers logically to the accomplishment of the difficult job. "It is believed that a satisfied customer is the best advertisement".

It must be understood that the above six steps in extension teaching often blend in with each other and lose their distinct identity.

Principles of Teaching

1. The farmer should subscribe to and understand the purposes of the course. The teacher at the first meeting should introduce what is to be discussed.

2. The farmers should want to learn.

3. The teacher or extension worker should keep a friendly and informal atmosphere so that the learners can ask things they do not understand.

4. The physical conditions should be favourable and appropriate

5. The teacher should involve the learners so that they participate and accept some responsibility for the learning process.

6. The teacher should make use of the learners' experience

7. The teacher should prepare well for the class, should keep his teaching aids handy and should be enthusiastic about teaching.

8. The method of instruction should be varied and appropriate.

9. The teacher should always be willing to be a learner and change his teaching material with availability of knowledge on the subject and he should be keen to learn more and prepare material for his teaching.

Requirements of Extension Teaching

Good extension teaching requires carefully planned programs, procedure, and technique. Designing good teaching plans is a highly professional job, and one that pays well as an achievement. A number of conditions must be met and actions must be taken to make extension teaching procedures and methods effective. Extension teaching requires specific and clearly defined objectives. All purposeful teaching has specific objectives derived from the broader program objectives. Before extension teaching can attain maximum effectiveness, changes desired in the behavior of people must be identified. Teaching objectives properly stated, contain four different aspects: people to be taught, behavioural changes to be developed in people, content or subject matter to which the behaviour is related, and the life situation in which the action is to take place. It is true that one can accomplish some favourable results without a clear definition of what he is trying to do, but if improvements are to result from the total program, extension workers must have well-defined teaching objectives. Achievements, therefore, can be adequate only in terms of some standard, and that standard can only be derived from one's concept of the objectives he wishes to attain through teaching effort.

Extension teaching must accomplish certain kinds of educational changes in relation to the subject matter taught. Among these are:

- Changes in knowledge, or things known, amount of knowledge, and kinds of knowledge. Examples: stocking rate (by farm size), kinds and amounts of fertiliser to use, kinds and amounts of pesticide to use.
- Changes in skills, or ability to do things. How easily and effectively one can do a specified task, and the number of things one can do, are all reflections of skill.
- Changes in attitude or feelings, for or against things and issues, points of view, beliefs, reactions and the like. Attitudes are important in determining what a person does and how he does it. They must become strongly positive (favourable) before desired changes in behaviour will give him satisfaction if the interest is met.
- Changes in interest: Interest is a specific form of attitude, but educational interest may be defined as a desire to learn, or to gain information, or to understand, or to gain skill, pertaining to some object in one's environment that he believes will give him satisfaction if the interest is met.
- Changes in understanding: Understanding has to do with gaining insight into the relationship of facts and issues, usually involving cause and effect. It has to do with the development of a deeper and broader vision of how various elements, important facts and principles operate in a situation. To gain understanding requires knowledge and thinking skill.

Extension teaching usually requires a combination of teaching methods. Because not all extension methods will reach the people or influence them, a combination of various teaching methods must be considered. By and large, the changes people make on their farms, in their homes, and in their communities are in proportion to the number of times they are exposed to information through personal visits, meetings, demonstrations, and the like. Obviously, if wide response is desired, rural people must be exposed to changes. Extension teaching requires careful evaluation of results. Evaluation is useful in guiding teaching effort and educational programmes. Extension teaching is complex because observation alone cannot be a basis for evaluation. Pre-testing is more precise than casual observation in determining the outcome of an educational activity.

Learning, Learning Situation and Learning Experience

It is the process by which a relatively permanent change or modification in behaviour occurs as a result of practice or experience. It is an active process on the part of the learner. From the above definition of learning, the following three major aspects were highlighted:

(i) ***Changes or Modifications in Behaviour***

Learning is a process by which a change occurs in behaviour. The changes in behaviour characterize the learning to be good or bad; desirable or undesirable. All changes in behaviour do not necessarily lead to efficiency and improvement; at least in terms of values. For example, a child may learn to keep the room clean or dirty.

(ii) ***Changes or Modifications are Relatively Permanent***

Learning generally results in relatively permanent change in behaviour. For example, after learning to ride the bicycle, one will never forget it. Similarly, certain changes in behaviour due to some temporary conditions cannot be referred to as learning. For example, changes in behaviour due to illness, physical injury, or drug habituation cannot be considered as normal behaviour, therefore, are not the results of learning.

(iii) ***Changes or Modifications are due to Practice or Experience***

This aspect of learning limits the kind of change in behaviour that can be considered to represent learning. It is true that all changes in human behaviour are not due to learning. Some of them may be attributed to process of maturation or biological inheritance, which are said to be unlearnt.

The essential role of an extension worker is to create effective 'learning situations'. An effective learning situation requires the following essential elements: (Fig. 4.2).

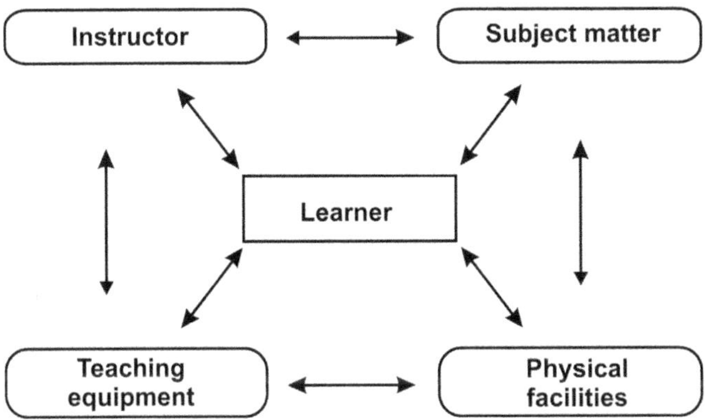

Fig. 4.2 The elements of a learning situation.

Instructor (Extension Worker or a Village-Level Worker)

- Should have in mind that teaching objectives are clearly significant to the learners, and are attainable through the educational process within the mental and physical limitations of the learners.
- Should have a thorough knowledge of significant subject matter related to the learner's needs.
- Be personable, enthusiastic, and interested in the subject matter.
- Use democratic instructional procedures and approaches.
- Be prepared, prompt, and courteous in every teaching-learning session.
- Arrange and manage the learning situation to prevent or minimise distractions within and outside the learning situation.
- Be skillful in the use of teaching material and equipment such as blackboard, visual equipment and other reading materials.
- Always prepare and use a carefully developed teaching plan.

Learner (farmers, women workers and youth)

- Should have need for information
- Be interested
- Be capable of learning and
- Must use the information gained

Subject-matter (recommended improved practices, such as artificial insemination, feeding balanced diet, etc.)

- Should be pertinent to learners' need
- Applicable to real life situations
- Well organized and logically presented
- Presented clearly
- Challenging, satisfying and significant to the learners
- Fit into overall objective

Teaching equipment (flannel-board, blackboard, charts, models, samples, slides, film strips, etc.)

- Should meet the needs effectively
- Readily available
- Each item must be used skillfully

Physical facilities (such as sitting accommodation, good visibility, etc.)

- Should be free from outside distractions
- Well lighted and ventilated
- Adequate space for the group
- Well-arranged comfortable furniture

The extension worker should skillfully manipulate the elements of the learning situation and provide satisfactory learning experiences for the people. The farmer, the women or the youth are the focal points in the learning situation. The main aim of an extension worker is to bring about a change in the behaviour of the people with the help of a judicious combination and use of different elements. All teaching should be carried out according to the needs and resources of the local community or group.

Learning Experience : Is the core of the educational process. It is the mental and/or physical reaction of a learner to seeing, hearing, or doing the things to be learned. There is a constant reaction by the learners with each of the other four major elements in the learning situation. For example, a learner may at one time be reacting to the dress of the instructor, to his mannerisms, or to his voice; at another time to his teaching equipment, or to the manner in which he handles the subject matter; later to some aspect of the physical facilities such as the hard chair, or poor light, or excessive heat. In addition to the mental focii on these elements, and many others not mentioned here, learners react to such items as outside obstructions, individual interpretation, members of the group, and personal problems. The great task of the extension worker is to minimise the almost infinite number of possible distractions to the mental process. The effectiveness of a learning experience is therefore, related directly to the manner and extent of mental concentration on the subject matter. Learning experience should contribute towards the achievement of the objectives.

Principles of Learning in Farming Situation

The principles of learning in farming situation are enumerated as below :

1. What is relevant and meaningful is decided by the learners and must be discovered by the learners.
2. Learning is a consequence of experience. People become responsible when they assume responsibility and experience success.
3. Cooperative approaches are enabling farmers in farming situation. As people invest in collaborative group approaches, they develop a better sense of their own worth.
4. Learning is an evolutionary process and is characterised by free and open communication, engagement, acceptance, respect and the right to make mistakes.
5. Each person's experience of reality is unique. As people become more aware of how they learn and solve problems, they can refine and modify their own styles of learning and action.

Principles of Adult Learning

1. *Learning is personal :* It will be an active process, provided the learner react to the message.
2. *Involvement :* Individuals are more likely to remember a solution, if we have worked it out by ourselves rather than one which has been thought for us.
3. *Readiness :* Learning takes place more quickly if we want to learn and are ready to learn.
4. *Association :* Learning that is related to our experiences is more likely to be remembered.
5. *Conditioned principle :* Apart from stimulating and encouraging, one should aim at physical and psychological climates that will be conditioned to learning.
6. *Comfort assists learning :* Individuals learn more, when they are comfortable and at ease without much distraction.
7. *Adopt teaching :* Teaching should be adopted as per the needs of the audience.
8. *Distribution :* Learning that is distributed over several short lessons is more effective than a single long session.
9. *Capacity :* Most of us learn effectively up to a certain stage. After that the rate of learning decreases. This is called the capacity of the individual.
10. *Arousing interest :* Arousing interest is necessary before starting the learning process.
11. *Enough practice :* Allowing enough practice time for becoming familiar with recently acquired knowledge and skills.
12. *Encouragement :* Giving continuous encouragement to the learners will improve the learning.

Types of Learning

Conditioned Response Learning : Ivan Pavlov was the pioneer of the study on conditioning. Pavlov noticed that dogs salivated in anticipation of receiving food. He performed an experiment around this nature of dogs and discovered conditioned reflex. Salivation by dogs expressed as unconditioned response to an unconditioned stimulus i.e., arrival of food. Pavlov found that a neutral stimulus like sound of buzzer when accompanied repeatedly with the known stimulus, that is arrival of food, created the stable response; that is salivation. The accompanying neutral (conditioned) stimulus becomes identical to that of original (unconditioned) stimulus by way of limiting an identical (conditioned) response. The implication of Pavlov's finding is that an organism reacts to a new experience on the basis of identical past experience. Future learning is based on and referred to past-learnt experience.

Verbal Learning : Ability to manipulate symbols, as in language, makes it possible for us to learn things.

Motor Learning : Often known as skill learning. It involves primarily the use of muscles of the body. In this, the individual learns muscular coordination as a mode response to some situation.

Perceptual Learning : As a result of past experience people perceive the situation differently. One-way of changing a person's habit of responding is to change the way in which the individual perceives the environment.

Attitude Learning : An attitude is an emotionalized system of ideas, which predisposes an individual to act in a certain way under certain conditions. Much of our earning involves change in our attitudes, our disposition to give favourable or unfavourable response to objects, persons, situations or abstract ideas.

Laws of Learning : Thondike's four laws of learning are :

1. ***The law of readiness*** : The learners will become ready when they feel that learning the new behaviour will satisfy their motives. Therefore, the teacher has first to study the felt needs of the learners and help them to focus their attention on the problems which block the satisfaction of those needs. In this way the learner will become ready for learning.

2. ***The law of exercise*** : Continued practice is considered necessary for retention of what is learnt. The teacher must help the learners to practice and review the desirable behaviour as many times as is found necessary. When a response is repeated several times it tends to become habitual i.e. learning is a self-activity.

3. ***The law of effect*** : People learn more rapidly and permanently when the learning experience is pleasant or enjoyable. As a teacher, one has to see that the effect of learning experience is desirable to the students.

4. ***The law of belongingness*** : The law indicates that the teacher should help the learner to perceive the relationships. The relationship between the elements may be: cause and effect relationship; known and unknown relationship; old and new relationship; specific and general relationship etc. Whenever a new behaviour is to be taught to the learners it must be related to the situation of the learners or with the background of the learners.

Theories of Learning

Learning Theories are classified into three major orientations, which are :

A. *Behavioural Theories :* Which include the following

(i) Classical Conditioning Theory

Classical learning is a form of association learning where a connection between a stimulus and a response is established. For example, the behaviour of a child who avoids burning fire wood after being burnt by it once. Classical conditioning also involves substitution and association of one stimulus for another, which was first observed by Russian psychologist Pavlov. It is of two types, (a) classical reward theory: Here, the conditioned stimulus is paired with reward, e.g. food is paired with bell and (b) classical aversive theory: Here, the conditioned stimulus is paired with an aversive stimulus, e.g. electrical shock is paired with bell.

(ii) Trial and Error Theory

This was developed by E.L. Thorndike, putting a cat which was hungry in a cage, and a fish was kept outside the cage. The cage was so constructed that, if a lever was pressed the cage would open. Initially, the cat indulged in random movements, exploratory in nature; and in the process, it touched the lever and cage and on subsequent trials, the cat improved slowly and pressed the lever immediately. During the process of learning wrong movements are dropped out and right movements are strengthened.

(iii) Operant or Instrumental Conditioning Theory

It is an active process in which learner responds to stimuli according to the way in which responses affect the stimuli. The term 'operant' was proposed by B.F. Skinner. The respondents behaviour is an unlearnt reaction to a specified stimulus such as the papillary constriction to light. It is of two types, such as: (a) Instrumental reward conditioning: This is a case of positive reinforcement, e.g. food is a positive reinforcer; (b) Instrumental aversive conditioning: This is a case of negative reinforcement, e.g. shock is a negative reinforcer.

B. Cognitive Theory

Here, the emphasis is on information storage and processing without explicit building up of stimulus-response association or manipulation of reinforcers. Such learning is referred to as cognitive learning. It also includes:

- Cognitive Development theory
- Assimilation theory
- Discovery learning theory
- Hierarchical moral of learning theory
- Gestalt theory

C. Humanistic Theory

It includes hierarchy of needs, transactional analysis and androgogy and pedagogical aspects of human life.

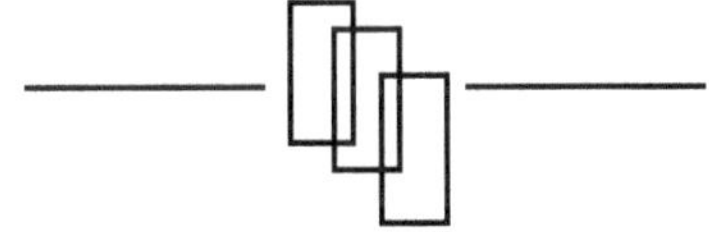

CHAPTER - 5

Extension Teaching Methods

The extension-teaching methods are the tools and techniques used to create situations in which communication can take place between the rural people and the extension workers. They are the methods of extending new knowledge and skills to the rural people by drawing their attention towards them, arousing their interest and helping them to have a successful experience of the new practice.

Functions of Teaching Methods

1. To provide communication so that the learners may see, hear and do the things to be learnt.
2. To provide stimulation that causes the desired mental and/or physical action on the part of the learner.
3. To take the learner through one or more steps of the teaching-learning process.

Classification of Extension Teaching Methods

I. According to use

(a) *Individual contact method*

- Farm and Home visit
- Office calls
- Telephone calls
- Personal letters
- Result Demonstrations

(b) *Group contact method*

- Method Demonstration
- General meetings
- Group Discussion
- Tours
- Field trips

  (c) *Mass contact method*
- Television
- Radio
- Exhibition
- Posters
- Leaflets
- Bulletins
- Circular letters

II. According to Form

  (a) *Written*
- Personal letters
- Circular letters
- Leaflets
- Bulletins

  (b) *Visual*
- Result Demonstration
- Posters
- Exhibits
- Slides
- Charts

  (c) *Spoken*
- Farm and Home visit
- Office calls
- Telephone calls
- General meetings
- Radio

  (d) *Spoken and Visual*
- Method Demonstration
- Television

Individual Contact Methods

Farm and Home Visits

It is a face-to-face type of individual contact by the extension worker with the farmer and / or the members of his family on the latter's farm or at his home for one or more specific purpose connected with extension.

Objectives or Purposes?

1. Obtain and/ or give first hand information on matters relating to farm and home conditions.

2. Give advice or otherwise assist to solve a specific problem; or to teach skills etc.,

3. Arouse the interest of those not reached by other methods.

4. Select local leaders, demonstrators or co-operators.

5. Promote good public relations

6. Other wise contribute to strengthening the extension organisation or facilitate extension programme.

Principles or Procedure to be Followed

1. *Decide upon the place of the farm and home visit in the teaching plan outlined to advance a particular phase of the extension programme.*

    (a) Consider alternative methods which might be employed.

    (b) Decide whether the visits are primarily for direct teaching or are needed to increase the effectiveness of group methods and mass media.

2. *Clarify the purpose of the visit – Which of the purposes mentioned above are expected to be achieved by the visit?*

3. *Plan the visit :*

    (a) Review previous contacts with members of family.

    (b) Check subject matter information likely to be needed – leaflets or bulletins etc

    (c) Work out schedule of visits in the community to save time.

    (d) Remote and unfrequented farms and homes should always be kept in view.

    (e) Consider best approach in view of individual family situation.

4. *Make the visit:*

    (a) Punctuality and consideration for the time of the farmer should always be borne in mind.

    (b) Be friendly, sympathetic and complimentary.

    (c) Gain and deserve interviewee's confidence.

    (d) Let the farmer do most of the talking.

    (e) Speak only when he is willing to hear.

    (f) Talk in terms of his interest.

    (g) Use natural and easy language, speak slowly and cheerfully.

    (h) Be accurate in your statements.

    (i) Don't prolong arguments.

    (j) Compliment the farmer for good ideas.

    (k) Be sincere in learning as well as teaching.

(l) Arouse interest and create a desire to take action.

(m) Render the farmer a real Service.

(n) Leave clear impression as to object of visit.

(o) If possible, hand over a folder or bulletin etc., pertaining to the topic discussed, or a packet of seeds if necessary. This will help in developing friendship.

(p) Leave the farm or home as a friend.

5. *Record the visit:*

(a) Date, purpose of visit, what was accomplished, and follow up commitments made.

(b) Make sure through appropriate office device that follow up at appropriate time is not overlooked.

6. *Follow up the visit:*

(a) Send applicable literature or other things by post.

(b) Extend invitation to attend a meeting; if any; on the concerned topic.

(c) Make subsequent visits as and when required.

Advantages

1. Provides extension worker with first–hand knowledge of farm and home conditions, and the view points of farm people.
2. If made on request, the farmer or home-maker is likely to be ready to learn.
3. The ratio of 'takes' (acceptance) to 'exposures' (efforts) is high.
4. Builds confidence between the extension worker and the farmer.
5. May increase greatly the effectiveness of group methods and mass media.
6. Contributes to selection of better local leaders, demonstrators and co-operators.
7. Develops good public relations.
8. Useful in contacting those who do not participate in extension activities and who are not reached by mass media.

Limitations

1. Requires relatively large amount of extension worker's time.
2. Number of contacts possible is limited.
3. Comparatively costly.
4. Time of visit may not be always opportune from the stand point of farmer.
5. Danger of concentrating visits on the progressive farmers, and neglecting those who are most in need of such personal contacts.

Office calls

It is a call made by a farmer or a group, on the extension worker, at his office for obtaining information or other help needed or for making acquaintance with him.

Objectives

1. To facilitate quick solution to farmer's problems, by saving the time of extension worker.
2. To enable the farmers to bring specimens of diseased plants or insect pests etc., so that the extension worker can identify them and give necessary advice to the farmers.
3. To arrange for or ensure timely supplies and services.
4. To promote close contact between farmers and extension organization.

Principles to be Followed (i.e. General Arrangements)

1. Office should be located conveniently so as to facilitate large volume of calls.
2. Space and furniture should be arranged to permit orderly routing of callers.
3. It should be possible for caller to confer privately with the extension worker.
4. Office room should be kept attractive with bulletin boards, leaflets etc.,
5. Office should be open during usual working hours.
6. Extension worker should regularly attend office, while at headquarters.
7. Arrangements should be made to provide information to the callers in the absence of the extension worker.
8. Cordial, sincere interest shown in visitor's problem.
9. Applicable reference material, including record of previous contacts readily accessible.
10. Unhurried consideration of entire problem without undue waste of time.
11. Caller made to feel welcome to call again.
12. As a follow up act, see that unfinished work connected with the call is completed as promised.

Advantages

1. Visitor likely to be highly receptive to learning.
2. Economical use of extension worker's time.
3. Good indication of farmer's confidence in extension.

Limitations

1. Extension worker cannot be always at headquarters .
2. Callers in his absence may not be satisfied with the information or guidance obtained.

3. Office contacts removed from actuality of farm or home situation may not reflect the real problem or accurately reveal pertinent conditions.

4. Visitors likely to be limited to those participating in other extension activities.

Personal Letters

It is a personal and individual letter written by the extension worker to a farmer in connection with extension work. Under the existing conditions of high percentage of illiteracy this extension method is relatively unimportant in India. Nevertheless, instances are not wanting when a few farmers write to the extension worker for advice. Moreover with the obvious increase in the number of literates in rural India, and the involvement of educated youth in extension activities, this method (Personal letters) may assume more importance in future than at present.

Objectives

1. To answer enquiries from the agriculturists regarding specific farm problems, or supplies and services etc.,

2. To seek the farmer's co-operation in extension activities.

Principles to be Followed

1. Promptness

A letter asking for information should be answered promptly, because the person writing the letter has more than passing interest in the matter and will be likely to use information which provides a satisfactory solution to his problem. Remember that information delayed is information denied.

2. Put Yourself in the other Fellow's Shoes

Have a genuine concern for the other fellow's interest, view point, limitations and desires.

3. The Letter Should be:

(a) Complete – give all necessary information to accomplish its purpose.

(b) Concise – Say what you have to say in the fewest words consistent with clearness, completeness and courtesy.

(c) Clear – so that it not only can be understood but cannot be misunderstood.

(d) Correct – containing no mis-statement of facts, or grammatical mistakes

(e) Courteous – tone appropriate for the desired response. How something is said as important as what is said.

(f) Neat – free from over – writings, strikings etc.,

(g) Readable – short sentences, short words, and human interest make for easy reading.

Result Demonstration

A result demonstration is a method of teaching designed to show by example the practical application of an established fact, or group of related facts. In other words, it is a way of showing people the value or worth of an improved practice whose success has already been established in the research station. In this method the new practice is compared with the old one on farmer's holdings so that the villagers may see and judge the results for themselves. Such demonstration requires a substantial period of time and records need to be maintained. It is in no sense an experiment or a trial except perhaps in the mind of the co-operator (demonstrator).

The result demonstration may (i) deal with a Single practice, such as the use of improved strains of paddy seed; or (ii) it may be concerned with a *series of related practices* as in the case of Japanese method of paddy cultivation; or (iii) in some instances it may include the entire farm, as in the case of balanced farming. (i.e., Whole Farm Demonstration).

The result demonstration may be (i) varietal (ii) manorial (iii) cultural (iv) combination of two or more of the aforesaid three types, or (v) composite demonstration in which all the essential improved practices in respect of any crop are included as a package of improved practices.

There are two common sense principles underlying this method.

(a) Whatever a farmer himself does or sees, he is likely to believe.

(b) What is good for one person will have general application to others (under similar conditions).

Objectives or Purposes

1. To show the utility and feasibility of a recommended practice under village conditions.

2. Chiefly to establish confidence on the part of the farmer as well as the extension teacher.

Procedure or Technique

1. *Analyze Situation and Determine Need* : (Determine the place of the result demonstration in your teaching plan).

   (a) Is it necessary to establish further confidence in local application of research findings and results of observation plots?

   (b) What has been the experience of the extension worker in guiding the practice under similar conditions?

(c) Is it possible to locate good illustration of the practice locally, eliminating the necessity of expensive result demonstrations?

(d) Is the need for result demonstration felt by the farmers?

2. *Decide upon Specific Purpose*

(a) Which particular audience should have the learning experience?

(b) What specifically do you want them to learn?

(c) Is it to give confidence to the extension worker and provide him with teaching material?

(d) Is it to establish confidence of farmers in the new practice?

(e) Is it to develop confidence in extension on the part of a community or of a minority group with whom the extension worker is not known well and favourably?

3. *Plan the Result Demonstration*

(a) Consult subject matter specialist.

(b) Make its simple and clear-cut as possible. (The more complex the demonstration, the greater the difficulty in evaluating the results attributable to each of the practices involved).

(c) Decide upon evidence needed and how local proof will be established.

(d) Determine number of demonstrations needed to accomplish purpose.

(e) Locate sources of material.

(f) Reduce plans to writing (calendar of operations etc.)

4. *Select Demonstrators*

(a) Consult with local leaders and select a demonstrator who commands the confidence and respect of his neighbours, and who is interested in improving his practices.

(b) Visit the prospective demonstrator to make sure that all conditions for success of demonstration are favourable.

(c) The demonstrator should be conscious of his responsibility for the successful completion of the demonstration and its effect upon the community.

(d) The demonstrator should be willing for the demonstration to be used for teaching purposes such as publicity; pictures, meetings, tours and personal enquiries.

(e) The demonstrator should have to secure the necessary physical equipment, supplies and materials to carry the demonstration to a successful conclusion.

(f) Explain and agree upon procedure with demonstrator and leave written instructions preferably.

5. Select the Plot

(a) The plot should be located preferably in a roadside field for easy accessibility and publicity.

(b) The field should be representative or typical of the soils in the village (neither too rich nor too poor).

6. Start the Demonstration

(a) Give wide publicity before starting the demonstration.

(b) Get all the materials ready

(c) Start the demonstration in the presence of the villagers.

(d) Assist in getting the demonstration under way to make certain that the omission of some key points will not make later work fruitless.

(e) Arrange for a method of demonstration meeting where a skill may be involved in the beginning stage of demonstration, or later.

(f) Mark the demonstration plots with large signs, so that all can see.

7. Supervise the Demonstration

(a) Visit the demonstration plot with sufficient frequency to maintain demonstrators' interest, check on progress, and see that succeeding steps are performed as outlined.

(b) Maintain records and assist the demonstrator also in keeping proper records.

(c) Give publicity to the demonstration and the farmer at suitable stages.

(d) Conduct tours to successful demonstrations at proper times.

(e) Let the demonstrator himself explain to visitors, as far as possible.

(f) Mention in news stories, circular letters, radio talks etc., at critical stages.

8. Complete the Demonstration

(a) See that final steps to complete the demonstration are taken.

(b) Take photographs.

(c) Hold meetings at demonstration where visual evidence will contribute to confidence.

(d) Summarise records. Analyze and interpret data.

9. Follow-up

(a) Give wide publicity to results of demonstration.

(b) Encourage demonstrator to report at meetings.

(c) Prepare visual aids based on the results of demonstration.

(d) Get other farmers to agree to demonstrate during the next season.

Advantages
1. Gives the extension worker extra assurance that recommendation is practical and furnishes local proof of its advantages.
2. Increases confidence of farmers in the extension worker and his recommendations.
3. Useful in introducing a new practice.
4. Contributes to discovery of local leaders.
5. Provides teaching material for further use by the extension worker.

Limitations
1. Requires lot of time and preparation on the part of extension worker.
2. A costly teaching method.
3. Difficult to find good demonstrators who will keep records.
4. Teaching value frequently destroyed by unfavourable weather and other factors.
5. Few people see the demonstration at the stage when it is most convincing.
6. Unsuccessful demonstrations may undermine the prestige of extension, and entail loss of confidence.

Group Contact Methods

Method Demonstration : It is a relatively short-time demonstration given before a group to show how to carry out an entirely new practice or an old practice in a better way. It is not concerned with proving the worth of a practice but how to do something. It is definitely not an experiment or trial but a teaching effort. In contrast to the result demonstration conducted by the farmer (demonstrator) under the supervision of the extension worker to prove that the recommended practice will work locally, the method demonstration is given by the extension worker himself or a trained leader for the purpose of teaching a skill to a group.

In the role of a skilled technician the extension worker or leader shows the step-by-step procedure in the operation, explaining each succeeding step as he proceeds. The learners watch the process, listen to the oral explanation, and ask questions during or at the close of the demonstration to clear up points about which there is uncertainty. Where practicable, as many members of the group as possible, repeat the demonstration in the presence of the others. This helps to fix the process in the minds of the audience and increases confidence in their ability to master the technique. The method demonstration is the oldest form of teaching. Long before language was developed, men taught their children how to hunt, how to cultivate etc., through demonstration method.

Objectives or Purposes

1. To enable the people to acquire new skills.
2. To enable people to improve upon their old skills.
3. To make the learners do things more efficiently, by getting rid of defective practices.
4. To save time, labour and annoyances and to increase satisfaction of learners.
5. To give confidence to the people that a particular recommended practice is a practicable proposition in their own situation.

Procedure or Steps to be Followed

1. Analyze the Situation and Determine the Need

   (a) Determine that the subject matter practice involves skills, which need to be demonstrated to many people.
   (b) Is the demonstration for new skills developed through research, or for old skills not being performed successfully?
   (c) Is it suitable for visual presentation to a group?
   (d) Can the demonstration be repeated satisfactorily by local leaders?
   (e) Is the practice really important from the farmer view point?
   (f) Can people afford to follow the practice?
   (g) Are supplies and equipment available in sufficient quantities to permit wide spread use of the practice?

2. Plan the Demonstration

   Gather all the information about the practice. Familiarise yourself with the subject matter. Check on research findings.

   (a) Talk over the problem with a few village leaders. Let the villagers help you plan the demonstration. Let them provide land and other requisites.
   (b) Have a timetable, depending on how much skill is required and how soon it is to be acquired.
   (c) Have a job break-down or a demonstration outline giving the operations in logical steps.
   (d) Identify the key points to be emphasized under each step.
   (e) List out and select demonstration materials and equipment most likely to be available or readily obtainable.
   (f) Arrange for diagrams, directions, and other teaching materials to be distributed.

(g) Prepare kits of special material needed by local leaders if they are to repeat the demonstration.

(h) Make sure that the work place is properly arranged: (lighting; no colours, no distracting noises).

3. Rehearse the Demonstration

(a) Practice demonstration until you are thorough with all the steps and know exactly what you should say or do at each step, so that the operation can be performed in a manner to inspire confidence.

(b) Make sure steps and points will be clear from audience's point of view.

(c) Check time required, to make sure there is opportunity for audience's questions and other expected participation.

4. Give the Demonstration

(a) Prior publicity should be given about the place and time.

(b) Be at the spot early to check equipment and material.

(c) Make physical arrangements so that all participants can have a good look at the demonstration and take part in the discussion.

(d) Explain purpose, and how it is applicable to a local problem.

(e) Find out what they already know about the practice.

(f) Show each operation slowly step by step, repeat where necessary.

(g) Use simple words to explain each step of the operation.

(h) Make sure the audience can see and hear clearly.

(i) Emphasize key points and mention why they are important.

(j) Solicit questions at each step before going on to next step.

(k) Give opportunity to learners to practice the skill

(l) Distribute supplemental teaching materials (bulletins, leaflets etc.) pertaining to the demonstration.

(m) Summarize steps covered in demonstration.

(n) Get the names of participants who propose to adopt the practice. This helps followup.

(o) If demonstration is given before local leaders who will repeat it; emphasise teaching points to be made. Explain contents of demonstration kit.

5. Follow up

(a) Give publicity on the demonstration through press, radio, meetings etc.

(b) Arrange for reports on number of participants and attendance at demonstration given by local leaders

(c) Make a sample check to assess the extent of use of the skill and satisfaction derived by those attending the demonstration.

Advantages

1. Peculiarly suited in teaching skills to many people.
2. Seeing, hearing, discussing and participating in a group stimulates interest and action.
3. The costly 'trial and error' procedure is eliminated.
4. Acquirement of skills is speeded.
5. Builds confidence of extension worker in himself, and also confidence of the people in the extension teacher, if the demonstration is performed skillfully.
6. Simple demonstrations readily lend themselves to repeated use by local leaders.
7. Introduces changes of practice at a low cost.
8. Provides publicity material.

Limitations

1. Suitable only for practices involving skills.
2. Needs good deal of preparation, equipment and skill on the part of the extension worker.
3. May require considerable equipment to be transported to the work place.
4. Requires a certain amount of showmanship not possessed by some extension workers.

Comparision between Result Demonstration and Method Demonstration :
(Table 5.1).

Table 5.1 Comparison between Result Demonstration, and
Method Demonstration

Particulars	Result Demonstration	Method Demonstration
Purpose	To show locally the work or value of a recommended practice	To teach how to do a job involving skill, (to teach doing skills)
Conducted by	Farmer (demonstrator) under the guidance of extension worker	Extension worker himself or local leader specially trained for the purpose
For the benefit of	The demonstrator as well as other farmers	Persons present at the demonstration
Comparison	Essential (Not necessary to have replications in the same field	Not essential
Maintenance of records	Necessary	Not necessary
Time required	Substantial period	Relatively very little
Cost	Costly	Relatively cheap
Inter relationship	Usually follows observation plots; may involve one or more method demonstrations.	Often paves the way for result demonstration

Basis for Demonstration

1. Most people retain 10-15% of what they READ; if the subject is explained in clear and simple language or in particular technical terms.
2. The majority remember about 20-25% of what they HEAR, if their concentration is not limited through listening "with one ear" to a speaker who perhaps fatigues them with a tedious lecture.
3. About 30-35% of what they have SEEN is kept in mind by the majority; even more if what is offered is well arranged and selected.
4. The majority remember 50% and more of what they have SEEN and HEARD at the same time, provided both presentations complement one another.
5. Upto 90% of what is taught is kept in mind by the majority of people, if they participate actively, and if ALL THE SENSES are involved.

"Only the Demonstration can make Teaching Perfect".

General Meetings

The term "General meetings' includes all kinds of meetings held by extension workers. Especially large in size held during special occasions, melas or festivals attended by thousands. Geographically, the meetings may be held in a neighbourhood, a community or village, a block, a district or State. The meetings may be held in a hall, home, field, shandy and so on. They may be held periodically or sporadically. The method of presentation may be the lecture or formal talk, informal or formal discussion, or the showing of slides, or a motion picture film. Special kinds of meetings often take the name of the meeting objective; e.g. Programme planning meeting. Evaluation meeting, Annual meeting, Vanamahotsava meeting, Farmers' Day meeting, meeting at result or method demonstration etc.

Essential Elements

It is obvious that elements which make for a successful meeting will vary greatly with the kind of meeting being held. Nevertheless, there are certain elements, which are essential practically in all meetings. They are detailed below:

1. *Determine the place of the meeting in the teaching plan* : **based on the following**
   (a) Is it felt desirable to reach many people quickly?
   (b) Is group action required? Will the group–approach contribute to learning?
   (c) Will it serve to focus attention on the problem, and provide material for news articles; radio talks, circular letters etc., as additional means of teaching.

2. ***Define the specific purpose of the meeting*** **: based on the following**
   (a) To disseminate subject matter information.
   (b) To develop interest in a new subject.
   (c) To change attitudes towards a problem.
   (d) To deepen understanding of public problems.
   (e) To determine a programme or plan of action.
   (f) To develop leadership and local responsibilities.
   (g) To provide an opportunity for social contacts.
   (h) To evaluate the progress made under a project or scheme.

3. ***Plan in advance for meeting***
   (a) Decide number of meetings, places, and tentative dates.
   (b) If the time and place are to be selected; it is important to select the season time of the year, day of week, and time of day in terms of the work cycles of those persons expected to attend and select the place in terms of its accessibility to the majority of the persons who are to attend.
   (c) After selecting the tentative date, check that there are no important competing events that will affect attendance.
   (d) Select meeting place, which will provide suitable lighting, seating arrangement, ventilation and other necessary facilities.
   (e) Encourage participation of local leaders in arranging and conducting the programme. Agree upon the part each will play and approximate time each will take.
   (f) Outline a tentative programme or agenda.
   (g)   As far as practicable, hold day time meetings; to reduce number of night meetings.
   (h) Secure speakers or resource persons as needed.
   (i) Inform speaker regarding local conditions and suggest subject matter to be adapted to the needs of local audience.
   (j) Select the audio-visual aids best suited to the occasion.
   (k) Provide for social and recreational features.
   (l) Utilise the methods of publicising the meeting that are necessary to ensure satisfactory attendance of those people intended to reach the meeting.

4. ***Conduct the Meeting***
   (a) Start the meeting on time. Chairman (usually a local leader) should open meeting promptly.

(b) State the purpose, and programme of the meeting (Programme is developed in an orderly manner, the procedure, of course, depending on the kind of meeting).

(c) Make the introduction brief.

(d) Focus attention on the central theme.

(e) Keep meeting moving on schedule.

(f) Use appropriate audio-visual materials.

(g) Watch reaction of audience. Encourage audience participation when desirable.

(h) At appropriate time, take action on matters calling for decision.

(i) Take advantage of group psychology and employ appeals that arouse interest; create desire and stimulate action.

(j) Close meeting on time with brief summary by chairman.

(k) Give recognition to individuals and groups that have actively participated.

(l) Hand out relevant folders or pamphlets at the time of break off.

(m) Take names of those interested in further information or follow up.

5. *Follow up the Meeting*

(a) Evaluate the meeting, to see if you can make any improvements in meetings to be arranged in future.

(b) Utilise what happened at meeting, in news articles, radio broadcasts etc.

(c) Make farm or home visits, or send additional information to persons requesting for it.

(d) Make sample check to determine satisfaction with meeting, and the extent to which the information is being used.

Advantages

1. Reaches a large number of people.
2. Adopted to practically all lines of subject matter.
3. Recognises basic urge of individuals for social contacts.
4. Group psychology stimulates conviction to act.
5. Promotes personal acquaintance between extension worker and village people.
6. Supplements many other extension methods.
7. Has great news possibilities and publicity value.
8. Influences change in practice at low cost.

Limitations

1. Suitable meeting place and facilities may not always be available.

2. Wide diversity in character and interests of audience may create a difficult teaching situation.

3. May require undue amount of night work on the part of extension worker.

4. Circumstances beyond the control of the workers, such as conflicting attractions, unfavourable weather etc., may result in poor attendance.

5. Meetings which are poorly arranged or conducted may have a reaching unfavourable effects.

6. The holding of meeting may become the 'real' objective, rather than the purpose the meeting was intended to advance.

Although the above elements relate to all kinds of meetings in general, it will be useful to understand the special features of some of the important kinds given below.

Lecture Method

The lecture method is extensively used to present authoritative or technical information to develop background and appreciation and to integrate ideas. The range of subjects that can be covered by this method is unlimited. But the speaker at a given meeting presents a specific subject to a particular audience. The lecture is an excellent method for presenting information to a large number of persons in a short period of time. Its weakness is that people are not likely to master as much of the information as the speaker is likely to assume; because for the most part it is a one-way communication. Members of audience listen in terms of their interests and remember in terms of their motivation and memory. To compensate somewhat for this weakness a discussion or question-and-answer session may be held following the lecture, this is generally called a forum.

Characteristics of the Lecture Method

1. Usually it is an organised presentation
2. It can be used to cover thoroughly the subject matter
3. It is adaptable to large groups
4. It appeals to the "ear marked"
5. It conserves time
6. Results are easy to check
7. Listeners sometimes absorb information without thinking
8. Material gained through lecture is not really learned and
9. The lecturer may "lose" his group or go over the heads of his group.

Lecture Method can be used Advantageously

1. with large groups where the individuals have some common background of information and experience

2. when it is necessary to cover a large quantity of material in a given time
3. when it is necessary to arouse enthusiasm in initiating a new programme or in further development of a programme
4. when giving factual information
5. when providing a common background of information as a basis for further study
6. where there is need to supplement other methods.

Lecture Method is not Effective
1. when skills are to be developed
2. when no testing is done
3. when group participation is desired
4. when problems are to be solved
5. when "doing" ability is to be acquired.

Debate : The common pattern is to have two teams, one representing the affirmative, and the other the negative side of the question. Usually there are two speakers for each side. Each speaker is allowed a definite amount of time to make his main speech and rebuttal after the main speeches have been completed. In this case, there is two-way communication between the debaters, but one-way communication for the audience. The range of subjects for debates is limited to controversial topics. The big advantage in a debate is that more than one side of a question is presented. There is however, one danger. If it is a decision debate there is the temptation for the debate to become highly antagonistic. In such a case, the motive to win the debate by any means may lead to distortion of information, ignoring the primary need to inform the audience. This objection to the debate is overcome by holding non-decision debates or by having a forum after the debate.

Symposium : This is a short series of lectures; usually by 2 to 5 speakers. Each one speaks for a definite amount of time, and presents a different facets or subdivisions of a general topic. The topic should be large enough or general enough to permit two or more subdivisions that are sufficiently significant to justify separate discussion by speakers. The subject may or may not be controversial. It is important that the speakers are of approximately equal ability, to avoid one speaker dominating the meeting or giving the audience a distorted view of the subject. A forum to facilitate mastery of information may follow the speeches. The advantage of symposium over a lecture is that two or more experts present different facets of the topic. It also has an advantage over the debate, as it is possible to escape the antagonism that may accompanies the latter.

Panel : It is an informal conversation put on for the benefit of the audience, by a small group of speakers, usually from 2 to 8 in number. They are selected on the basis of the information and experience they have. Members are seated so that they can see one another and also face the audience. The panel is generally rehearsed before it is presented to the public. The leader introduces the members of the panel to the audience and announces the topic. He has the responsibility to see that the conversation keeps going, by asking questions or making brief comments, and encouraging the less talkative members. There are usually 3 types of panel: (a) Question-answer panel in which the presentation is actually a series of questions by the leader (or Chairman) and answers by the members; (b) Set speech panel, each one making prepared speech; (c) Conversational panel in which members hold a conversation among themselves on the topic, with questions and comments going from one member to another. This third type is more nearly in line with the definition of a panel than the other two, and is the pattern to be achieved. The panel may be used to present almost any topic that may be used for a lecture, debate or symposium. The special advantage of a panel is that a spontaneous conversation about some subject may have more interest for the audience than a lecture. For better mastery of information a forum should follow the panel.

Forum : It is a discussion period that may follow any one of the above methods of presentation. It consists of a question period in which members of the audience may ask questions or make brief statements. The forum provides an opportunity for the audience to clear up obscure points and to raise questions for additional information. It also gives individuals an opportunity to state briefly their understanding of a point and see whether they have interpreted correctly the material presented. It is primarily a means of understanding information.

Buzz Sessions : Also known as Phillips 66 format. With large groups when there is limited time for discussion, the audience may be divided into smaller units for a short period. Groups of 6 to 8 persons get together after receiving instructions to discuss about a specific issue assigned. The secretary of each small group will report the findings or questions to the entire audience when they are reassembled. This is actually a device to get more people to participate in a forum than would be the case otherwise.

Brain Storming : Is a type of small group interaction designed to encourager the free introduction of ideas on an unrestricted basis and without any limitations to feasibility. It is a form of thinking in which judicious reasoning gives way to creative initiative. Participants are encouraged to list for a period of time all the ideas that come to their minds regarding some problem and are asked not to judge the out come. At a later period all the contributions will be sorted out evaluated and perhaps later adopted.

Workshop : It is essentially a long meeting for one day to several weeks, involving all the delegates in which the problems being discussed are considered by delegates in small private groups. There must be a planning session where all are involved in the beginning. There must be considerable time for work sessions. There must be a summarising and evaluation sessions at the close. The workshop as the name implies must produce something in the end a report, a publication, a visual or any other material object.

Seminar : It is one of the most important forms of group discussion. The discussion leader introduces the topic to be discussed. Members of the audience discuss the subject to which ready answers are not available. A seminar may have two or more plenary sessions. This method has the advantage of pooling together the opinions of a large number of persons.

Conference : Pooling of experiences and opinions among a group of people who have special qualifications in an area.

Group Discussion

It is that form of discourse which occurs when two or more persons, recognising a common problems to exchange and evaluate information and ideas, in an effort to solve that problem. Their effort may be directed towards a better understanding of the problem, or towards the development of a programme of action relative to the problem. Discussion usually occurs in a face-to-face or co-acting situation, with the exchange being spoken. And when more than two people are involved, it usually occurs under the direction of a leader.

Purposes
1. To solve a problem (decision making)
2. To exchange information (improve understanding)
3. To motivate.
4. To plan a programme of action.
5. To elect or select a person for a position etc.
6. To entertain.
7. To hear and discuss a report.
8. To form attitudes.
9. To release tensions.
10. To train individuals.

Procedure

1. ***Understand and adopt the proper Technique*** : The technique of a problem solving group discussion consists of the following six steps based on the "reflective thinking" pattern.
   (i)   Recognition of the problem as such by the group.
   (ii) Definition of the problem, its situation and diagnosis.

    (iii)   Listing of as many solutions as possible.

    (iv)   Critical thinking and testing of these hypotheses to find the most appropriate and feasible solution or solutions.

    (v)   Acceptance or rejection of the solution or solutions by the group.

    (vi)   Lastly. Considering how to put the accepted solution into practice.

2. ***See that one of the group members takes up the role of the discussion leader (or Chairman).*** Extension worker should avoid this role as far as possible, because in such a case, a situation is likely to develop where the group listens and the chairman does all the talking.

3. *Size of the group should never exceed 30 persons :*

4. ***Role of the Chairman***

(a) Make physical arrangement for the meeting, so that all members feel comfortable. Seating arrangement should be such that every one can see the faces of all other members. Circular seating is preferable. (Square, rectangle, U or V shape is also used sometimes).

(b) Introduce members, if they are new to one another.

(c) Announce the topic and purpose of discussion.

(d) Follow a plan.

(e) Hear all the contributions made, and from time to time give short summaries of the discussion upto that particular moment, especially when the group moves from one step to another (of the reflective thinking pattern.)

(f) Build a permissive climate.

(g) Keep the group moving at the rate at which their thinking progresses.

(h) Give or get clarification of vague statement.

(i) Promote evaluation of all generalisations.

(j) Protect minority opinion.

(k) Try to get balanced participation.

(l) Promote group cohesion.

(m) Remain personally neutral

(n) Give a final summary of discussion.

Some Don'ts for Chairman

(a) Never ask questions that suggest answers or can be answered with a yes or No. (Put only thought – provoking questions).

(b) Don't favour one view against another when there is a conflict or difference of opinion among members.

(c) Never become emotional about the discussion.

(d) Don't become impatient with the group.

(e) Don't dominate the discussion or answer all the questions raised by the members.

5. *Roles of Members*

(a) Talk one at a time. No private conversation with neighbours. No speech making.

(b) Supply as much pertinent information as possible.

(c) Contribute one point at a time.

(d) Answer questions directly, specifically and briefly.

(e) Test all thinking by critical analysis.

(f) Listen attentively

(g) Stay on the subject.

(h) Exhibit willingness to change opinion when change is justified (i.e., open minded). A person may hold opinions, but opinions should not hold a person.

(i) Support the needed leadership.

(j) Promote group harmony even while criticizing or disagreeing.

6. **Role of the Expert (Extension Worker)**

There may be occasions when a group confronted with a problem does not have sufficient information to enable them to discuss intelligently. In such cases, the role of the expert is not to dominate the meeting, nor to suggest his own solution. He should only supply information the group does not have; furnish technical information, present ways other groups have met in similar situations, and present the immediate problem in its larger setting, with implications for integrating the solution of the problem with other group policies and action programmes.

Advantages

1. It is a democratic method, giving equal opportunity for every participant to have his say.

2. It appeals to the practical type of individuals.

3. It creates a high degree of interest.

4. The strength of group discussion lies in the fact that the discussants approach the problem with an open mind and suspend judgment in a spirit of enquiry.

5. It is a co-operative effort and not combative or persuasive in nature.

6. Combined and co-operative thinking (pouring of wisdom) of several persons is likely to be superior to that of isolated individuals.

7. A small group can think together on a problem in an informal fashion and works out solutions better and faster by using this method than by following rigid parliamentary procedure. (Even parliament and legislatures recognise this when they appoint ad hoc committees).

8. Develops group morale. When a group discusses a question and then comes to a decision, that is "our" decision for the group and they will see that "our" decision is carried out. (Group action is encouraged).

9. It is a scientific method (employing the reflective thinking pattern).

10. Participants need not be good speakers or debaters.

11. Continued experience with such group discussion improves one's capacity for critical and analytical thinking.

Limitations

1. Factions in villages may hinder the successful use of this method.

2. The ideal discussants with self-discipline (open mind and suspended judgment) are difficult to find. So also, it is difficult to find an ideal chairman or leader for a group discussion.

3. It is not suitable for dealing with topics to which discussants are new.

4. In large groups especially, and even in small groups to some extent, it is difficult to achieve group homogeneity or cohesion.

5. The size of the group has to be limited, because the success of the method is perhaps inversely proportional to the size of group; other factors being constant.

6. It is not a good method for problems of fact.

7. It is not suitable for taking decisions in times of crisis or emergency as it is a slow process.

8. Due to its informal conversational style, the scope for orderly or coherent arrangement of ideas is limited.

Field Trips

It is a method in which a group of interested farmers accompanied and guided by an extension worker, goes on tour to see and gain first hand knowledge of improved practices in their natural setting (whether on research farms, demonstration farms, institutions or farmer's fields). It is a series of field and demonstration meetings arranged in a sequence.

Purposes

1. To stimulate interest, conviction and action in respect of a specific practice; e.g. preparing rural compost. The cumulative influence of several ideal compost pits is more likely to provide such stimulation than a single illustration.

2. To impress the group about the feasibility and utility of a series of related practices; e.g. proper preservation of farmyard manure, rural composting, urban composting, green manuring – which are all included under the item "development of local manorial resources".
3. To induce a spirit of healthy competition by showing the accomplishments in other villages.
4. In short, to help people to recognise problems, to develop interest, generate discussion and to promote action.

Procedure

1. Provide for field trips at opportune time in the over all teaching plans.
2. Prepare an outline of specific aims of the trip.
3. Plan the Trip
   (a) Decide upon the places to be visited and the things to be seen and learnt. Do not crowd the programme.
   (b) Then arrange for necessary permission from the concerned authorities to make the trip.
   (c) Fix up date, time, and means of transport, number of participants to be taken, number of stops, arrangements for rest and refreshments in consultation with the village leaders.
   (d) Accompanying staff should pay an advance visit to the actual sites before conducting the party.
   (e) Give definite instructions to participants where and when to meet, insist on punctuality in arrival and departure timings at each stop.
4. Conduct the Trip
   (a) Give guide sheets in simple language (if the majorities are literate).
   (b) Focus attention on the purpose of the trip.
   (c) Let every one see, hear, discuss and if possible participate in the activities at the places of visit.
   (d) Allow time for questions and answers.
   (e) Help them to make notes of interesting information.
   (f) Follow the general instructions regarding conversation applicable to all direct contact methods.
   (g) Avoid accidents.
   (h) Adhere to schedule all through.
5. Record the Trip
   Accompanying staff should note the details of the trip, the names of participants etc. to facilitate follow up.
6. Follow up
   (a) Contact the participants individually and in groups

(b) Arrange for necessary supplies and services.

(c) See that the desired action results.

(d) Build up publicity material

Advantages

1. Participants gain first hand knowledge of improved practices, and are stimulated to action.
2. Eminently suited to the "show me" type of people.
3. Percentage of "takes" to exposures is high.
4. Widens the vision of farmers.
5. Caters to group psychology and leadership.
6. Have incidental values of entertainment and sight seeing.

Limitations

1. It is costly.
2. Difficult to fix a season and time suitable for all.
3. Bottlenecks of transport and accommodation at halting places.
4. Possibility of subordinating educational aspect to the sight seeing aspect.
5. Risk of accidents.

Mass Contact Methods

In addition to the personal contact methods and the face-to-face group teaching methods, mass media enable extension workers to greatly increase their teaching efficiency. Publications, newspaper articles, circular letters, radio, television, exhibits, posters etc, provide helpful repletion for those contacted personally or through groups. They also facilitate dissemination of information to a much larger and different clientele. Even though the intensity of the teaching contact through mass media is less, the large number of people reached and the low cost per unit of coverage more than offset the lack of intensity. The extension teaching plan which neglects communication possible through mass media fails to fully capitalise on what has already been invested in the more intensive contact methods.

Publications

(Extension journals, Bulletins, News letters, Pamphlets, Folders, Leaflets)

General Purpose : Your purpose in writing is to communicate information. Therefore your first consideration is your reader audience. If you were writing for a scientific paper, you would use a vocabulary and style different from what you would use when writing for the general public.

Principles to be followed :

How clearly you communicate information to average readers depends on how well you select, sift, and sort your facts.

1. ***Select the Facts***

    (a) *Suitable subject matter :* Does it meet a need? Is it timely? Is it of current interest? Does it apply to your area? Is the information practical?

    (b) *Readers :* Who are the people you want to reach? What are their problems, interest, and educational levels? Do they have the environment and capacity to make use of the information?

    (c) *Purpose of Publication :* What do you want to teach and accomplish? Do you want to stimulate interest in a programme or do you want to influence the people to do something?

2. ***Sift the facts***

    (a) Sift essential facts necessary to give information clearly.

    (b) Screen out difficult concepts, which are beyond reader's experience or understanding; e.g. pH. Value; calorific value.

    (c) Give the layman an appreciation of subject rather than a detailed explanation.

    (d) Express highlights.

    (e) Don't try to impress the lay reader with all you know.

    (f) Don't document everything.

3. ***Sort the facts***

    (a) Arrange facts in logical order.

    (b) Set out important points in 1-2-3 order (step by step)

    (c) Guide reader with attractive subjects and suitable illustrations and pictures.

4. **Remember the ABC's of Journalism :**

    Accuracy, Brevity, and Clarity, which are the fundamentals of all good writing.

5. **Adopt the following tips for readability**

    (a) Short sentences, clear in meaning, simple in construction, with few prepositional phrases and dependent clauses give an idea in each sentence.

    (b) Use simple, familiar and concrete words

    (c) Personal, or human-interest words.

Advantages
1. Can reach a large number of people quickly and simultaneously.
2. Can be read at leisure, and kept for future reference.
3. Generally people have confidence in the printed page.
4. Necessary supplement to other teaching methods.
5. Information usually definite, well organized, and readily understood
6. Influences adoption of practices at relatively low cost.
7. Provides scope for recognizing achievements of individuals and groups
8. May promote literacy.

Limitations
1. Not suited for illiterate audience.
2. Frequent revision may be necessary to keep abreast of current research.
3. Information prepared to general distribution may not fit local conditions.
4. Impersonal; lacks social value of personal contacts and meetings.

Circular Letters

It is a letter reproduced and sent to many people by the extension worker, to publicise an extension activity (like meeting, exhibit etc.) or to give timely information on farm and home problems.

Purposes
1. To attend a meeting.
2. To stimulate interest in a subject.
3. To adopt a new practice.
4. To perform a service to community or block.
5. To answer a questionnaire.
6. To maintain interest and cooperation of youth club members, local leaders, co-operators etc.
7. To prevent spread of pests and diseases.

Procedure and Principles
1. Determine the place of the circular letter in the teaching plan.
2. Determine specific purpose of the circular letter and the segment of extension clientele to be reached.
3. Plan the use of the circular letter
   (a) Plan letter to serve definite purpose.
   (b) Should be important, timely, and related to specific needs and interests.
   (c) Indicate for each subject matter the number of letters, nature of contents, and approximate date of distribution.
   (d) Organise letters on a series basis, when desirable.

 (e) Have an up-to-date classified mailing list, according to problems and interests of people.

 (f) Check duplicating and mailing equipment in advance.

4. Write circular letters

 (a) Appeal immediately to personal interest with snappy statement in the first paragraph, pointing out the importance of the problem to the person addressed.

 (b) Give a cartoon or illustration containing the central idea. Have a single purpose.

 (c) State the facts concerning the nature or seriousness of the problem (how it affects the locality and the individual).

 (d) Suggest what the person can do to alleviate or solve the problem.

 (e) Letter must be neat and appealing to the eye, and free from errors.

 (f) Above all, personalize your letter by using.

- Expressions you use in every day contact.
- Direct statements.
- Simple sentences.
- Action words with few affixes.
- Personal references
- Appropriate anecdotes.
- Courteous conclusion.

Advantages

In addition to the advantages given in the case of "Publications", the following are the special advantages.

1. Convey timely information effectively to special interest groups.
2. Eminently suited to make announcements to get attendance.
3. Unlike news articles, circular letters have the advantage of making more direct appeal; (not surrounded by other reading matter and head lines to distract attention)
4. Helpful in maintaining interest and co-operation of local leaders, demonstrators or co-operators etc.
5. The author's enthusiasm and personality can put life into the information carried in such letters.

Limitations

1. Special equipment and clerical help necessary.
2. Too frequent use may minimise effectiveness.
3. Not suited to illiterate clientele.
4. Does not have the advantage of personal letters in catering to the needs of a particular individual.

News Articles

Also known as News Stories, News is any timely information that interests a number of persons, and the best news is that which has the greatest interest for the greatest number. It is an accurate, unbiased account of the main facts of a current event that is of interest to the readers of a newspaper.

Purposes

(a) To develop interest

(b) To inform general public

(c) To disseminate information on subject matter.

(d) To create favourable attitude

(e) To reinforce other extension methods like meetings and demonstrations.

Technique

1. *News should have one or more of the following characteristics.*

    (a) Something that actually happens (e.g. calf rallies or milk yield competitions)

    (b) Unusual or extraordinary (e.g., 50 litres milk production per day)

    (c) Important (not trivial)

    (d) Near to the point of publication or audience

    (e) New, recent or timely

    (f) Something that makes the farmers interested

    - Catastrophe

    - Fight, conflict, struggle e.g, competing for a prize in a cattle show

    - New knowledge (e.g. about new type of technology).

2. *Some accepted principles of reporting*

    (a) Write the lead sentence, which tells the crux of the situation, i.e., something of importance to the reader.

    (b) Use the pyramid form of writing i.e. put the important paragraphs first, so that it won't matter even if editor cuts off the last paragraph or two.

    (c) Use the five W's and the H as guide i.e, see if you have answered the *Who, What, Why, When, Where, and How.* Get as many of these as possible in the first paragraph, as a safeguard against editor's cutting.

    (d) Write in simple language.

    (e) Avoid using your personal opinion.

    (f) Be accurate, fair and brief.

    (g) Include motivating appeals.

3. *Evaluate the effectiveness of news articles*

Advantages
1. Low cost.
2. Large coverage in short time.
3. Efficient source of timely information.
4. Carries the prestige and confidence of the printed word.
5. Reinforcing effect on other extension methods.
6. Tax payers come to know about extension activities (public relations).

Limitations
1. Of no value if people are illiterate or do not read or hear a newspaper
2. Difficult to check the results
3. Requires special training to write good articles

Radio

It is a medium for mass communication, a tool for giving information and entertainment.

Purposes
1. To reach large numbers of people quickly and inexpensively.
2. To reach people not reached by other means.
3. To stimulate participation in extension through all other media.
4. To build enthusiasm and maintain interest

Procedure or Technique
1. Determine its place in the teaching plan
2. Be clear about the purpose of your broadcast.
3. Keep the interests and needs of the audience in view
4. Select topics of current interest.
5. Time the broadcast to synchronise with the farmers leisure hours.
6. Decide what treatment to give – straight talk, interview, panel, drama etc.
7. For writing the script, follow the principles given for writing news articles.
8. Encourage people to listen to rural programmes.
9. Encourage them to write to the broadcasting stations about their likes, needs and opinions.
10. Encourage talented local people to participate in broadcasting.

Advantages
1. Can reach more people more quickly than any other means of communication.
2. Specially suited to give emergency and timely information (e.g. weather, pest outbreak etc)

3. Relatively cheap.
4. Reaches many who read little or none at all
5. Reaches people who are unable to attend extension meetings
6. A means of informing non-farm people tax payers) about agricultural matters.
7. Builds interest in other extension media
8. Possible to do other things while listening

Limitations

1. Limited number of broadcasting stations.
2. Not within reach of all farmers.
3. Recommendations may not apply to individual needs.
4. No turning back if not understood
5. Frequently loses out in competition with entertainment
6. Difficult to check on results

Television

Television is one of the important mass media for dissemination of information in the rural areas. Television has unique advantages over other mass media. While it provides words with pictures and sound effects like the movies, it scores over the latter by its high intimacy and reaches the largest number of people at the shortest possible time. The visual in it has the advantage over the radio. Television can deal with topical problems, and depict known persons who can provide the solutions. People learn through the eye, and will remember things better if they see them. Television viewing does not demand the strain and discipline needed to read the printed medium. The messages on the TV screen are pre-selected, sorted out and then presented in the simplest manner possible.

Demonstrations, "the need" in farm extension, are brought to the farmer by television. This has great value in making converts to better farm practices. Apart from the evidence by their own eyes, farmers also respond readily to what is said, especially by other farmers, and if the same point as extension people make in their interpersonal communications are highlighted, the combination is doubly effective. It is within the power of television to provide the dynamic presentation to bring ideas in a compelling way into the receptive environment of the farmers home or community.

However, the sights should not be set too high. Experience both in India and in other countries shows that it has limitations. Does television change behaviour or induce action? Many countries have come to the conclusion that the answer is a 'no'. As a mass medium, TV Programmes lead to awareness, contribute information and perhaps help to form opinion. Before the farmer thinks of taking action, he will require the televised information and impression to be reinforced by local demonstration and individual personal confirmation.

Awareness creates curiosity about a new idea in the minds of farmers, leading them to seek more information on it. Before the idea is adopted in practice, the farmer undergoes two more important stages of evaluation and trial. Television everywhere is concerned very strongly with the first stage of awareness. Apart from that, it speeds up the entire process of adoption. Television is strong in providing the stimulus, and exposing the audience to a whole range of ideas and experiences. Programmes in agriculture have an immediate effect, if the ideas put forward come along at a time when the farmer needs the most, deal with subjects of which he had no fixed idea but was groping in the dark, and will speed up changes already taken on hand by the farmer.

Exhibitions

An exhibition is a systematic display of information, actual specimens, models, posters, photographs, and charts, etc., in a logical sequence around a theme to create awareness and interest in the community. It is organized for arousing the interest of the visitors in the things displayed. It is one of the best media for reaching a large number of people, especially illiterate and semi-illiterate people. Exhibitions are used for a wide range of topics, such as planning a model village, demonstrating improved practices, different feeding methods, showing high-producing animals, new technologies and the best products of village industries.

Objectives
1. To acquaint people with better standards.
2. To create interest in a wide range of people.
3. To motivate people to adopt better practice.

Procedure
A. Planning and Preparation
1. Form a steering committee with specialists, local leaders and administrators
2. Decide on the theme and organisations to be involved.
3. Prepare a budget estimate and procure funds
4. Prepare a written programme and communicate to all concerned in time.
5. Get the site ready within the scheduled date.

6. Earmark a stall for display of exhibits to be brought by the farmers
7. Display posters at important places and publicise about the exhibition through mass media.
8. Decorate the stalls and make adequate arrangements for lighting.
9. Display the exhibits at eye-level.
10. If possible arrange, action and live exhibits
11. Train up interpreters and allot specific duties.

B. Implementation

1. Let the interpreters briefly explain the exhibits to the visitors, so that the intended message is clearly communicated.
2. Organise formal opening of the exhibition by a local leader or a prominent person
3. Arrange smooth flow of visitors
4. Organise a panel of experts to be present nearby, so that the visitors who would like to know more about problems can access to the desired information.
5. Conduct meetings, training programmes as per schedule during the day time and use the stage for entertainment during nights
6. Keep the exhibits and the premises clean
7. Judge the stalls on the basis of their quality of display, ability to draw visitors and effectiveness in communicating message
8. Conclude the exhibition as per the schedule

C. Follow up

1. Meet some visitors personally and maintain a visitor's book for feedback information.
2. Talk to the local leaders and assess success of the exhibition
3. Ensure availability of critical inputs and facilities emphasized during the exhibition
4. Look for changes in practice in the community in the future

Advantages

1. Eminently suited to teach illiterates.
2. Promotes public relations and goodwill towards extension.
3. It can fit into festive occasions and serve recreational purposes.
4. Can be used to stimulate competitive spirit.
5. Can create market for certain products.

Limitation
1. Requires a lot of funds and preparation
2. Cannot be held frequently

Score Card for Judging an Exhibit

	Points
1. Suitable subject (timely, personal) ..	10
2. Effective title (short, perusal, active verb) ..	10
3. Attracts attention (stopping power) ..	20
4. Holds interest (encourages study) ..	10
5. Conveys message (accomplishes purpose ..	30
6. General appearance (simple, balanced, orderly).	10
7. Workmanship (well constructed) ..	10

Campaign

It is an intensive teaching activity undertaken at an opportune time for a brief period; focusing attention in a concerted manner on a particular problem, with a view to stimulate the widest possible interest in a community, block or other geographical area. Campaigns are launched only after a recommended practice has been found acceptable to the people as a result of other extension methods like method or result demonstrations etc.,

Purpose

To encourage emotional participation of a large number of people and to foster a favourable psychological climate for quick and large scale adoption of an improved practice.

Procedure

1. Determine the need for a campaign.
2. Be clear about the purpose. By making sure that it fulfils the need of local people.
3. *Plan the Campaign*
    (a) Consult local leaders and organisations
    (b) Consult specialists
    (c) Ensure timely supply or men and materials
    (d) Select a suitable time for launching the campaign
    (e) Give wide publicity in advance
    (f) Build up enthusiasm of the people
    (g) Allot specific areas and items to each service personnel and local leaders.

4. *Conduct the campaign*
   (a) Ensure that campaign is carried out as per plan
   (b) Work with and through local leaders
   (c) Watch the campaign closely throughout
   (d) Avoid failures.

5. *Follow up*
   (a) Make individual and group contacts to find out reactions.
   (b) Assess extent of adoption
   (c) Find out and analyze failures.
   (d) Publicise successful items.
   (e) Give due recognition to local leaders responsible for success.

Advantages
   1. Specially suited to stimulate mass scale adoption of an improved practice in the shortest time possible.
   2. Facilitates exploitation of group psychology for introducing new practices.
   3. Successful campaigns create conducive atmosphere for popularizing other methods.
   4. Builds up community confidence.
   5. This method is of special advantage in the case of certain practices which and close association of officials and non-officials, require preparation concerted efforts and propaganda technique.

Limitations
   1. Applicable to only a few topics of common interest; but not suited to solve individual problems.
   2. Successful only when all participants co-operate in the campaign.
   3. Not useful when advocated practice involves complicated technicalities.
   4. Requires adequate preparation and close association of officials and non-officials, concerted efforts and propaganda technique.

Factors to be Considered in Selection and Combination of Extension Methods

No single "rule of thumb" can be given for the selection and use of various extension methods to ensure success in all situations. However, some guiding principles will be helpful in general. Basically the individual contact methods furnish the most direct opportunities for influencing people effectively.

All the other methods of group and mass procedures are dilutions or compromises created by the pressure of necessity. We must reach more people, teach them more often, and keep down the cost per contact. In order to get most effective results, the extension worker should (i) select the appropriate methods (ii) have a suitable combination of the selected methods and (iii) use them in proper sequence, so as to have repetition in a variety of ways.

A. Selection of Extension Teaching Methods

1. *The Audience :*

(a) Individual and collective differences in people vary greatly in their knowledge; attitudes, skills, their position in the "diffusion process", and in the "adoption categories", their educational training, age, income level, social status, religious beliefs etc. Some are progressively seeking change, others are slow to change. Some are "eye minded" while others are "ear minded". These individual and collective differences influence the teaching approach. For instance, people with little or no education, and low incomes may respond to personal visits and result demonstrations. The better educated and the more progressive elements of the population usually respond well to the methods like group meetings and discussions, exhibits and written materials. A man in the "awareness stage" cannot straight away jump to "adoption stage" but can be gradually brought to the adoption stage by using suitable methods. For "late adopters" (conservatives), direct approach may not yield so good results as approaching through the "early adopters" and "informal leaders".

(b) ***Size of Audience*** : is also a factor influencing the choice of extension methods. For instance, group discussion cannot be used effectively when the number of participants exceeds thirty; method demonstration can be used for a relatively small audience, while lecture meetings can be used for large audiences.

2. *The Teaching Objective* : (or nature of change aimed at). Do you want to bring about a change (i) in thinking or knowledge? (ii) in attitude or feeling? (iii) in action or skill? If you want merely to inform or influence a large number of people slightly, you should use mass media. If you want to influence a relatively small number of people to make maximum improvements, resort to individual contact methods. If you want to change attitudes, or arrive at a consensus of opinion, arrange

group discussion or work through village leaders. If you want to teach a skill, use the method demonstration.

3. ***The Subject Matter*** : Where the new practice is simple or familiar (i.e., similar to those already being followed) the news article, radio or circular letter will be effective, whereas complex or unfamiliar practices will require face to face contacts, written materials and audio visual aids.

4. ***The Stage of Development of Extension Organisation*** : In the initial stages of extension, result demonstrations will be necessary to gain confidence of farmers. But if extension work is already well established and the farmers have confidence in extension services, result demonstrations may not be necessary and local illustrations of adoption by village leaders will suffice.

5. ***Size of extension staff*** : in relation to the size of extension clientele: The larger the number of extension workers, the greater is the scope for direct or personal contact method.

6. ***The Availability of Communication Media*** : such as newspapers, telephones, radio etc., will also have a direct bearing to the extent of which these methods can be used.

7. ***The Relative Cost of the Method*** : (i.e., the amount expended on extension teaching in relation to the extent of practices changed) is also an important consideration in their selection and use.

8. ***Familiarity of Extension Worker*** : Familiarity of extension workers with and skill in the use of the several extension methods will also influence his choice and use of the methods.

B. Combination of Extension Teaching Methods

Extensive extension field studies over a long period of years show that people are influenced by extension education to make changes in behaviour in proportion to the number of different teaching methods with which they come in contact. As the number of methods of exposure to extension information increases, the number of farm families changing behaviour increases. Therefore, if widespread response is desired, people must be exposed to teaching effort in several different ways (i.e. repetition but in a variety of ways). It is also proved that combined use of several different methods is of the utmost importance in extension teaching. The adoption rate and percentage of practices was high when more than five methods were used as compared to single and two to five methods.

C. Using Extension Teaching Methods in Proper Sequence

To answer our teaching needs, our extension plans of work must include methods that, (a) enable our farmers to see, hear and do the thing to be learned, (b) enable us to reach large numbers of people and (c) create confidence – building situations. Our complete plan should provide not only for doing each of these three things but must be so organised that the completed plan, as a unit, does all three of these things. For instance, a personal contact is made through an office call or farm & home visit. A leader is visited. A demonstration is established. A meeting is held to discuss the demonstration. Circular letters are sent to advertise the meeting. A news story is written on the results of the demonstration as seen at the meeting. These happenings and results are broadcast over the radio. Pictures are taken and a slide story is shown at a meeting, where one method helps another, and many of them are used in combination and sequence to repeat the story.

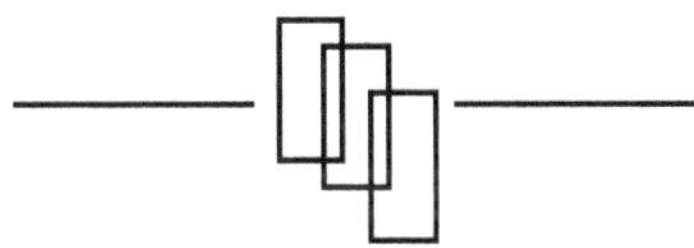

CHAPTER - 6

Audio-Visual Methods and Materials

Definitions

1. A visual aid is an instructional or communicating device in which the message can be seen but not heard.
2. An audio aid is an instructional device in which the message can be heard but not seen.
3. An audio-visual aid is an instructional device in which the message can be heard as well as seen.

In common usage, some forms of educational aids are loosely called audio-visual material. Some of these are specifically visual, some audio, and a few are true audio-visual media as illustrated in the classification given below. Strictly speaking, no extension medium is complete without talk in some form at some stage, even if it were only to introduce the material to the villagers. It must be remembered that audio-visual aids can only supplement the teacher but can never supplant him.

Purpose of Audio-Visual Aids

- Audio-visual aids are used to improve teaching, i.e., to increase the concreteness, clarity and effectiveness of the ideas and skills being transferred.
- They enable the audience to look, listen and learn (by doing);
- To learn faster, to learn more and to learn thoroughly and
- To remember longer.

Cone of Experience

It is devised by Edgar Dale in explaining the inter relationships of the various types of audio-visual materials, as well as their individual "position" in the learning process. In this cone each division represents a stage between the two extremes direct experience at the base, and pure abstraction at the apex. (The bands on the cone are not rigid divisions). (Fig. 6.1).

1. *Direct, purposeful experience :* It is the unabridged version of life itself, with three elements directness, purposefulness, and responsibility for the outcome e.g., making a piece of furniture or milking process.

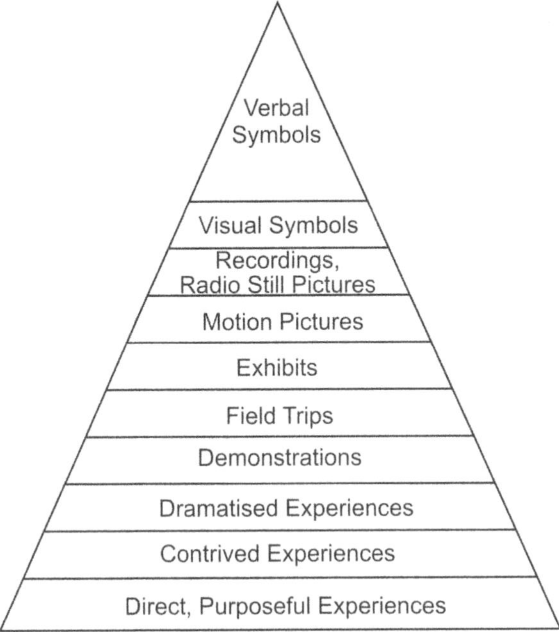

Fig. 6.1 Cone of experience.

2. *Contrived Experiences :* A contrived experience in an "editing" of reality, differing or not from the original in size, in complexity or in both e.g., models of animals, mock-ups of machinery, objects, specimens.

(a) *Model*

Is essentially a recognisable imitation or replica of the original, whether workable or not, and whether differing or not from the original in size e.g. models of compost pits, improved cattle sheds, farm machinery.

Purposes of Using Models

   (i)   To get over the disadvantages in the size of the original from instructional view point e.g. too big a size for the eye to take in, as in the case of a soil conservation project extending over many square miles, or too small for study as in the case of thrips, or aphids.

   (ii)   To make the past or the future visibly real e.g. an outdated farm implement or a new implement likely to be introduced in future.

   (iii)   To get over physical inaccessibility e.g. seasonal fruits and vegetable etc.

      Note : The models used for the above three purposes represent the external form and shape of the original object and or called 'Scale Models' or Solid Models'

(iv) To circumvent "unusable" reality – e.g. models to explain the physiology of a cow's udder or a hen's reproductive system, or hidden features of any machinery. Such models which reveal the internal structure of a real object and or constructed in such a way as to be dismantled easily are called 'Cross sectional Models'.

(b) *A Mock-up (or working model)* : Differs from the model in that it is a functional (workable) device which alters the essential elements that are being studied in the original, and concentrates on these elements only to the exclusion of others e.g. operating mock up of cylinder and piston of a diesel engine.

(c) *Specimens* : Real objects taken out of their natural settings, e.g. specimens of crop shown at a meeting or exhibition, preserved or mounted specimens of insects, plants etc.

(d) *Objects* : Pieces of reality or sample e.g. ear heads, diseased parts of plants.

3. *Dramatised Experiences :* i.e. participating in a reconstructed experience, e.g. dramas, puppet shows.

Long before any one devised lectures as a means of education; knowledge and understanding were conveyed from one generation to another through songs and ballads, dramas and dances; puppet shows and festivals. These combine entertainment with education, and have much fascination especially for village people. Although these ancient and native arts are of late, on the decline in India, they deserve to be revived and encouraged in view of their impact on the country side. In the past, they themselves have mainly been historical and mythological with religious or ethical import. Recently, some political parties have been exploiting folk dances and dramas etc., for their propaganda purposes. The harnessing of this type of audio-visual aids for rural development has till now been negligible, although their potentiality for this purpose is great. Panchayat Samithis and Panchayats will do well to discover and encourage histrionic talents and playwrights among Extension workers, school teachers, members of Youth Flubs, Mahila mandals as well as professional artists where available, and make use of dramas, puppet shows etc. as a means of attracting villagers and disseminating useful information on agriculture and other aspects of community development in an interesting way that appeals to the rural folk.

Whether one acts or observes, the dramatisation is a substitute for real experience. Contrived experience is also a substitute, but it differs from dramatised experience in one basic respect ; contrived experience retains a good deal of the original reality. A model may differ (in size texture, workability) from the original but it possess the physical appearance of the direct reality. A dramatisation does not necessarily look like the original; it is a new and different thing, a reconstruction. Time, events, speeches are all shifted and foreshortened. The characters are viewed under a special condition of a world without time.

4. Demonstrations
5. Field Trips   Already dealt in chapter 5
6. Exhibits

7. Motion Pictures of Films- silent pictures or combination of sight and sound.

Motion pictures are really a series of still pictures on a long strip of film. Each picture is flashed momentarily on the screen and the rapid succession of still pictures – (each of which shows the subjects in a slightly different position) – gives an illusion of movements. Usually 70mm and 35mm. films are used for commercial entertainment. 16 mm film for educational moves, and 8mm film for domestic pictures.

Purpose
- To present facts in an interesting way.
- To attract audience.
- To arouse interest.
- To change attitudes.
- To bring new practices to a village in a short time.
- To reach illiterate as well as literate people.

Points to Remember
- Be thoroughly familiar with the subject you plan to teach and how exactly the film supports the idea you want to get across.
- Preview the film.
- Before showing the film, explain the subject, say why it is important and stimulate viewers to look for certain things in the film.
- Have spare projection lamps on hand.
- Before audience arrives, set up the machine, have the film threaded, and test by focusing on the screen.
- The projector should be high enough to project over the heads in the audience; and the screen high enough from the floor for all to easily see the bottom of the picture. (4 feet from the floor to bottom of screen).

- Adopt "2-and-6" formula. (i.e., front row to be 2 screen widths away from the screen and hind row to be 6 screen widths away from the screen). Adopt 20° angle for beaded screen, and 30° angle for matte screen.

Advantages

1. A complete process involving motion can be shown in a short time.
2. People identify themselves with those in the picture.
3. Compels attention.
4. Heightens reality.
5. Speeds up or slows down time.
6. Brings the distant past and the distant space into learning situation.
7. Enlarges or reduces actual size of objects.

Limitations

1. Special equipment is necessary
2. The equipment is costly.
3. Some sort of power is required to operate the projectors.
4. Transportation, maintenance and storage of equipment and materials require special consideration and skill.
5. Suitable halls for showing motion pictures are not available in many places.

8. Radio and recordings – on disc, tape, or wire and still pictures (non-projected e.g., photographs, illustrations and projected e.g., slides, filmstrips.)

   Sound can be recorded in 3 ways. The mechanical process; tape recording and wire recording by the magnetic process, and movies film recording by the optical process make the disc recording. The record players, and the tape recorder are the commonly used audio aids, besides the radio.

   Tape Recorder : Is an audio equipment for recording sound on magnetic tape by electro-magnetic process. It is used to capture original sound and preserve it for later reproduction. Tape recorders are usually operated by (a) electricity, and (b) transistors. The transistorized tape recorder is inexpensive, handy and portable and can be used even in the fields by extension workers.

   The tape recorder is a versatile, popular educational aid because of the following advantages.

   1. It can be used to reproduce information in regional languages or dialects.
   2. Operation cost is low as the same tape can be used over and over again.

3. Tape recording can be immediately played back without undergoing any processing.

4. Editing is easy with tape recordings.

One Picture is worth a Thousand Words

A. Projected

Slides : The lantern slides is one of the most popular and versatile visual in extension education. It is a transparent picture (on glass or film), which is projected by focusing light through it from electric bulb, petromax or lantern. The most popular type of lantern slide used to-day is made on 35mm film, and the slides (transparencies) are of 2" x 2" size of 2 ½" x 3 ½". The correct way of inserting a slide carrier is to place it upside down and reversed.

Filmstrips : A filmstrip is a series of still photographs, diagrams, drawings or letterings on a strip of 235 mm. film. It may be of 2 types (a) Single Frame (24 x 18) (b) Double Frame (24 x 36). The number of frames in a strip may range from 30 to 60. Perforated edges of the film fit over projector sprockets. Once adjusted to project the first frame, each succeeding image will be in focus and in proper position on screen. When audience participation is desired projection can be paced at a speed suitable to the speaker, when accompanied by a carefully prepared script or talk, new ideas can be presented forcefully and dramatically.

Advantages of Filmstrips

1. Filmstrips are light, unbreakable, easily stored, and condense much information in a small package.
2. Filmstrips and filmstrip projectors are much less expensive than motion picture films and projectors.
3. The machines are simple to operate besides being relatively inexpensivee.
4. The pictures can be held on the screen for long time.
5. The village worker with camera can take good pictures of local practices and have them made into a filmstrip at very little expense.
6. The filmstrip and projector take little space and can be carried easily.
7. The villagers can participate through discussions on each picture, as presentation can be stopped without breaking sequence.
8. Filmstrips have the additional advantage that a complete process such as the Japanese method of growing paddy can be shown at one short session, step by step
9. A filmstrip, when projected can be accompanied with commentary or music played back by a tape recorder or gramophone.

Disadvantages of Filmstrip

1. Its sequence of projection is fixed and cannot be altered.
2. The surface of the filmstrip may become scratched after prolonged use, or it may be burnt if not properly handled while projecting.
3. The teacher is often dependent on filmstrips produced commercially, which may not suit his requirements.

B. Non Projected

Photographs : are exact visual recordings of things. They may be mounted or unmounted photographic prints or reproduction of photographs taken from a magazine, newspaper or book. They may be in black and white, or coloured. They may be used in personal teaching situations or as display type visuals in exhibitions or bulletin boards. They may be projected with an opaque projector. To be an effective teaching aid a photograph must; 1. tell a story 2. illustrate only one point 3. have plain and simple background 4. show the main subject prominently.

Illustrations : are non-photographic reconstruction of reality; e.g., drawings, paintings, etchings etc. These are used much in the same way as photographs.

9. *Visual symbols* : For example, flat maps, chalkboards, sketches, cartoons, posters, diagrams, charts, graphs, bulletin boards, flash cards, flannel graphs.

Chalk Board : It is most universally used of all the teaching aids. It is not itself a visual material but a vehicle for a variety of visual materials. It is one of the cheapest, most effective, most versatile and easiest to use of all the visual aids. The same black board can be used for the flannel graph. It can be used as a screen in showing slides or filmstrips, by covering it with a clean white cloth. There are two basic kinds of chalk boards viz., rigid and roll up. The latter is lighter, more compact and hence more easily portable. The rigid type is more durable and easier to use, although not easily portable due to its size and weight.

Roll up chalk boards are usually made of heavy cloth, canvas or oil cloth, coated with the chalk board paint slating. It should be placed against a smooth flat surface like a wall or up turned table, so that chalk pressure can be applied at any point on the surface to give a good impression. *Rigid chalk boards* can be made of wood, plywood, metal, fibre board or even heavy card board. Yellow chalk on dark green paint gives better visibility than white chalk on black paint.

Suggestions for using the Chalk Board

  (i)   Have it clean

 (ii)   Use clean eraser

(iii)   Write in large letters

 (iv)   Don't talk as you write

  (v)   Face group after writing and continue the discussion

 (vi)   Don't fill the board – avoid clutter

(vii)   Don't use abbreviations

(viii)   Keep drawings simple

 (ix)   Use coloured yellow chalk which is good at night

  (x)   Don't stand in front of the black board, stand to one side.

Bulletin Board : It is a simple inexpensive device that can be placed either outdoor, or indoors. A soft board that will hold pins or tacks is most suitable. A soft board that will hold pins or tacks is most suitable. It can perform basic communication functions. It can attract attention, stimulate interest, deliver a message and promote action. Items generally used on a bulletin board include photographs, cut out illustrations from publications, drawings, specimens, notices, posters and wall news papers. Since there are few village newspapers, a well planned bulletin board kept up to date can be of great help to the Gram Sevak if used for.

  (i)   Local announcement of importance to all the villagers.

 (ii)   Photographs to show local activities.

(iii)   Follow up instructions for the villagers on things demonstrated and emphasised.

(iv)   Village reminders for things to be done. When, how and by whom.

Useful Hints :

1. Avoid packing the bulletin board so full of information that nothing stands out. Better to communicate one or two ideas than to confuse the audience and communicate nothing at all.

2. Don't clutter the board with small illustrations and captions. Use fewer but larger materials.

3. Use colour for attraction and effectiveness.

4. Change material on the bulletin board regularly.

Flannel Graph : A flannel graph or khaddar graph is a visual teaching aid. Pieces of flannel felt or sandpaper, having rough surfaces, or nap, will stick to another piece of flannel stretched on a firm flat surface called a "flannel board". When you attach pieces of flannel felt or sandpaper to the back of pictures, photographs, drawings, letters etc.

these objects will also stick to the flannel board. This device is called a "flannel graph". The surface cloth may or may not be mounted on a permanent backing. Some extension workers prefer to carry with them only a piece of folded or rolled up flannel along with the symbols or parts. When they arrive where the lesson is to be given, they pin the flannel to a flat surface such as an upturned table, wall or fence. Some extension workers whose travel on foot or bicycle prefer to take only the flannel graph parts with them. When they arrive where the lesson is to be given, they borrow a blanket a piece of rough textured cotton cloth or perhaps a mosquito net. This is draped over an upturned table, bed or fence, or is attached to a wall to provide support.

The size of flannel graph to use depends on the size of the audience. A flannel graph 30 by 40 inches can be used to tell a story to about 150 people if the parts are sufficiently bold. Experience will soon tell you whether or not your flannel graph is of satisfactory size. It will be convenient to keep several different sizes to accommodate different sizes of audience.

The first step in planning your flannel graph presentation is to decide exactly what you want to tell your audience. The story should be developed in a logical, step by step sequence. It should be kept as simple as possible, covering only those parts that are important and omitting unimportant details. Then you are ready to visualise the important points in your story. This is where you decide what kind of parts are needed and what they will illustrate. If you have an artist or photographer to help you make the parts, you are fortunate. Most extension workers do not have such help and must plan and make their own parts or symbols.

Some persons make the error of standing in front of the flannel graph. This blocks the view and may irritate the audience since they are interested in looking at the parts and in seeing the story. With practice you can learn to pick up a part, apply it quickly with a firm downward pull and step to one side, continuing your story in words and pointing if necessary, to the part from the side. Wherever possible store the parts of the presentation in folders arranged and numbered for step by step presentation. This assures a smooth, logical presentation. The appeal of the flannel graph is in its action and suspense. In some ways it is like a drama. It has a story or plot. It has a background or set. It has parts that can be moved about the actors. Like a drama, the flannel graph story unfolds before your eyes. You both see and hear the story. The action of the moving parts attracts your attention. The suspense of the unfolding story holds your interest.

Flash Cards : Flash Cards are a series of illustrated cards which when flashed or presented (before a group) in proper sequence tell a complete story. They are used in the same way as filmstrips. In flash cards, however, people see the picture directly, instead of seeing it on a screen. The story is told as each card is held before the group. The story is simple and talks about one theme. e.g., Swiss method of dairy production. Flash card should: (1) Be used for groups of not over 30 people (ii) Be large enough for every one to see: at least 22 by 28 inches (iii) Be simple line drawings or photographs, or cartoons (iv) Be adapted to local conditions. (v) Have plenty of colour. It is best to limit the number of flash cards to 10 or 12 for one talk. In order to plant the most effective cards, study your talk and pick the main ideas that you want your villagers to remember.

Tips to Use Flash Cards

(i)   The story on each card must be familiar to you.

(ii)   You must use simple words and local expressions.

(iii)   You must bring in local names of people and villages.

(iv)   You must hold cards so that people can see clearly.

(v)   You must hold cards against body and not up in air. (You turn your body towards the different parts of the group to show cards to all members of the group).

(vi)   You glance down at card as you tell the story.

(vii)   You point to important objects without covering the card with your hand.

(viii) You must be enthusiastic; you must enjoy telling the story.

(ix)   You have the cards stacked in order. As one card is finished it is slid behind the other so that it will be in order when next time it is used.

Posters : The poster is an important visual aid. But like other "aids" the poster is never used alone. It must always be part of a campaign or a teaching programme. It will serve first to inspire the people. It will prove to the villagers that there is official interest in the problem treated. Lastly, as long as it remains in the village it will serve as a reminder to the villagers. A good poster arouses or urges people to immediate action and is highly suggestive. It makes them to feel a part of the work at hand. To be useful, a poster must be planned for a special job. It must be planned for the people who are supposed to do the job. The following points should be considered in making a poster.

(a) Promote one point,

(b) Support local demonstrations,

(c) Support local exhibits,

(d) Contain dramatic pictures that will stop people and make them look,

(e) Tell the story at a single glance,

(f) Have few words.

(g) Have simple words

(h) Have bold letters

(i) Must picture every day living

(j) Should be in pleasing colours

(k) Should be at least 20 by 30 inches in size

(l) Must be timely.

Generally speaking, a poster should contain three main divisions. The first part usually announces the purpose of a project. The second sets out conditions. The third recommends actions. Each of these three main divisions may be illustrated with striking art supported by brief language. Posters that are produced properly are often not effective because they are put in a poor place or not pasted. Posters should be placed where people pass or where people gather. Some posters fail to do good because they are not followed with other devices such as meeting, demonstrations, films etc.

Charts : Charts are visual symbols for summarizing, comprising, contrasting, or performing other services, in explaining subject matter. In other words, they are diagrammatic presentations of facts or ideas.

Pull Charts : Consist of written messages which are hidden by strips of thick card board or plywood. The messages can be shown to the viewer, one after, another by pulling out the concealing strips. These strips can again be restored to the concealing position after the presentation or whenever needed.

Strip Tease Charts : As is true in the case of pull charts, the appeal of the strip tease chart is in its suspense. It 'teases' the interest and imagination of the audience. The information on the chart is covered with thin paper strips to which wax, tape or other sticky substance has been applied at each end of the strip. Pins or tacks also can be used. As the speaker wishes to visually reinforce a point with words or symbols, he removes the appropriate strip of paper. It is possible to add considerable interest to the presentation by removing the paper with a dramatic flourish. The strip tease chart adds sparkle to what might otherwise be a drab presentation. It centres attention on the most important fact at any one time. The technique increases learning and aids recall.

Flow Charts : These are diagrams used to show organisational or administrative relationships. Boxes connected with lines show levels and lines of authority. You could use organisational charts to show administrative relationship in a ministry, an extension service or a university.

Bar Charts : These are used to compare quantities at different times or under different circumstances. They are composed of measured blocks spaced along a clearly marked scale. For instance, the effect of fertiliser in increasing crop yields on test plots in three successive years might be shown in a bar chart.

Table Charts : A railway time table is a familiar example.

Job Charts : e.g. Gram sevak's job chart.

Tree Chart : are used to show the development or growth of something in the shape of a tree or stream, e.g., genealogical tree.

Flip Charts : Consist of a series of individual charts which are tacked or bound together and hung on a supporting stand. These individual charts carry a series of related messages in sequence. The teacher flips them one after another, as the lesson or story progresses. To be effective, a flip chart should deal with only one broad theme and give only the salient points without too much data or details.

Over Lay Charts : Consists of a number of illustrated sheets, which can be placed one over the other conveniently and in succession. The drawing or illustration on each individual sheet forms a part of the whole picture. This enables the viewers to see not only the different parts but also see them against the total perspective when one is placed over the other. When the final over lay is placed, the ultimate product is exposed to view. Such a presentation has a dramatic effect on the viewers.

Pie Charts : These are in the shape of circles and used to show how several parts make up the whole. A pie chart might be used to show the relative proportion of different crops produced by a country. Each section of the pie should have its own colour. A colour key or code in the margin will help the audience remember what the different sections represent.

Line Charts : These are particularly useful in showing trends and relationships. A single continuous line may represent growth or expansion. Multiple lines may show the relation between market price and quantity of a farm product. A cumulative line chart may show relation trends between production costs and market price.

Pictorial Graphs : To give the viewer a vivid picture and to create a rapid association with the graphic message, cartoons and other types of illustrations may be used. Each visual symbol or "isotype" may indicate quantity, as shown when you compare the number of tractors on farms in different years.

10. *Verbal symbols* : designations that bear no physical resemblance to the objects or ideas for which they stand. These are used together with every other material on one of the experiences.

Selection of Audio-Visual Aids

Audio-visual aids are used singly or in combination, taking the following factors into consideration.

1. *The teaching objective :* i.e., the type of behaviour change you want to bring about gaining information, or changing attitudes, or learning some skill.
2. The nature of subject matter being taught.
3. *The nature of audience :* age level, educational level, interest, experience, knowledge of the subject, intelligence.
4. *The size of audience :* Flash cards can be used for a small audience only: motion picture, for a large audience.
5. *Relative cost of the various aids :* Effective aids need not necessarily be expensive.
6. *The teacher :* The extension worker's will and skill in using the several aids and his originality and skill in selection; preparation and use of aids.
7. *The availability :* An effective extension worker makes use of indigenous materials, when the teaching aid he would like to use, is not available.

Evaluation of Audio-Visual Materials
(Criteria for Selecting Audio-visual Aids)

1. Do the materials give a true picture of the idea they present?
2. Do they contribute meaningful content to the topic under study?
3. Is the material appropriate for the age, intelligence and experience of the learners?
4. Is the physical condition of the materials satisfactory?
5. Do they make learners better thinkers; critical minded?
6. Do they tend to improve human relations?
7. Is the material worth the time and effort involved?

    (a) Planning
    1. Know clearly the objectives of the presentation.
    2. Plan well in advance, this helps anticipate problems and avoid them.
    3. Anticipate size of audience as closely as possible and make sure the aids are visible and / or audible to the entire audience.

    4. Plan for the use of a variety of colorful visual aids. They help change the pace of presentation and help hold audience interest.

    5. Determine the appropriate timing for the presentations.

(b) Preparation

1. Prepare by rehearsing or previewing in order to make a smooth presentation.

2. Select as convenient and as comfortable a meeting place as possible, with acoustics and seating arrangements suited to the specific purpose.

3. Anticipate need for special lighting or for total darkness and be prepared to provide either, at the right time.

4. Make sure that all equipment is in good working order before starting the meeting.

5. Arrange the audio-visual aids in sequence and have them within easy reach.

6. Keep aids out of sight until actually required for use.

(c) Presentation

1. Motivate the audience and stress the key points they should observe during the presentation.

2. Present aids at the right moment and in proper sequence.

3. Display only one aid at a time.

4. Remove all unrelated material.

5. Stand beside the aid, not in front of it.

6. Speak facing the audience and not the aids.

(d) Evaluation

1. In the end, evaluated by providing for discussion and application, to discover and dispel misunderstanding, if any

2. Undertake follow up studies and observe results.

Advantages

1. The learner can

    (a) Learn faster

    (b) learn more

    (c) learn more thoroughly and

    (d) remember longer.

2. The teacher could organise his teaching material in a systematic order.

3. Clarify ideas being presented.

4. Impress ideas more indelibly on the mind.

5. Vitalise and make teaching more real

6. Picture experiences outside one's own environment.
7. Combat verbalism
8. Overcome the language barrier.
9. Attract and hold attention
10. Arouse and sustain interest
11. Stimulate thinking and motivate action
12. Change attitude or point of view
13. Save time because they make learning easier and faster.

Disadvantages

1. Learners may some times form mistaken or distorted impressions, unless audio-visuals are supplemented with required explanation.
2. Temptation for the teacher to narrow down his teaching to only a few big ideas, not giving the complete picture of a subject.
3. Some teachers acquire the mistaken idea that they have little to do when audio-visuals are used.
4. Possible risk of spectatorism, instead of the attitude of thoughtful enquiry.

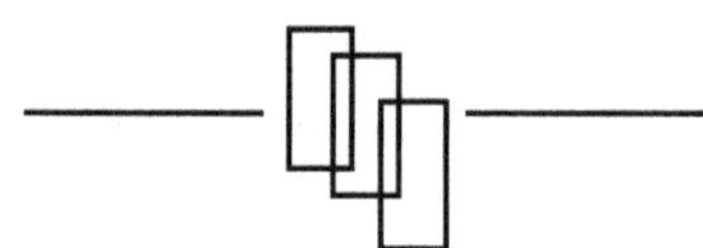

CHAPTER - 7

Communication for Development

The term communication stems from the Latin word 'Communis', meaning common. According to Rogers and Shoemaker (1971), *Communication is the process by which message are transferred from a source to receiver.* Van den Ban and Hawkins (1997) defined communication as the process of sending and receiving messages through channels which establishes common meanings between a source and a receiver. Leagans (1961) defined communication as the process by which two or more people exchange ideas, facts, feelings or impressions in ways that each gains a common understanding of the meaning, intent and use of messages. Communication, then is a conscious attempt to share information, ideas, attitudes and the like with others.

Functions of Communication

1. Information Function

The basic requirement of adapting and adjusting oneself to the environment is information. There must be some information about what is going on in the environment which concerns the people. The getting or giving of information underlies all communication functions, either directly or indirectly.

2. Command or Instructive Function

Those who are hierarchically superior, in the family, society or organisation, often initiate communication either for the purpose of informing their subordinates or for the purpose of telling them what to do, how to do, and when it. The command and instructive functions of communication are more observable in formal organisations than in informal organizations.

3. Influence or Persuasive Function

The sole purpose of communication is to influence people. Persuasive function of communication *i.e.,* to induce people, is extremely important for extension in changing the behaviour of the clientale in the desirable direction..

4. Integrative Function

A major function of communication is integration continuously offsetting any disintegration at the interpersonal or at the organisational level. This helps to maintain individual, societal or organisational stability and identity.

Communication Theories

Three kinds of investigators have made most academic studies of human communication: mathematicians, social psychologists and linguistic and symbolic anthropologists. Human communication has been divided into three principal parts: syntactics, semantics and pragmatics. These investigators and parts cut across one another, creating a matrix of three rows and three columns that formulate and answer most questions about human communication and has been evolved with three different communication theories: mathematical, social psychological and linguistic.

Mathematical Theory

The mathematical theory stems from the work of mathematicians and engineers who helped create the modern radio, telephone and television as means of electronic communication. These mathematicians and engineers also created the computers and other instruments of modern automation in communication. (Fig. 7.1).

Claude E. Shannon proved to be the key theorist, conceptualizing the theory in the late 1940s that remains central to communication study today. Published, with Warren Weaver, as a Mathematical Theory of Communication (1948), Shannon's theory was born out of his research at Bell Labs. "Shannon's initial goal was simple: to improve the transmission of information over a telegraph or telephone line affected by electrical interference, or noise. The best solution, he decided, was not to improve transmission lines but to package information more efficiently" (Horgan, 1990). Shannon's concept was quickly adopted by researchers in various disciplines and applie to computer science, physics, molecular biology and biotechnology, psychology, linguistics and communications.

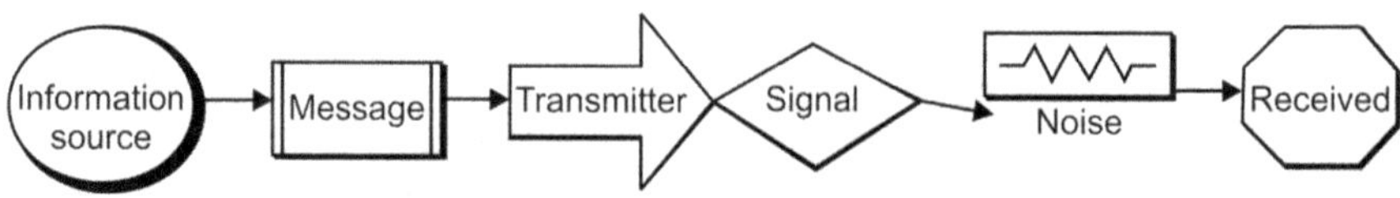

Fig. 7.1 Linear Transmission of Messages.

Mathematical Theory and Human Interactions :
A Shannon and Weaver Model

Shannon clearly designed his theory as a mathematical model that does not take human emotions and experiences into account (Rogers, 1995). However, communication scholars immediately applied the theory to human interaction. Schramm adapted the model to deal with the concern of "communication, reception, and interpretation of meaningful symbols processes at the heart of instruction" (Heinich *et al.*, 1996). The success of this interaction would then be measured by the feedback the receiver would give to the sender once the message has been transmitted. Feedback, in an instructional setting, may take the form of discussions, observations or tests. Schramm's adaptation provides a communication model that is measurable and adjustable for the production of effective communication in an instructional setting.

Social Psychological Theory

The key orientation for the social psychological theory of communication is in the analysis of human codes and networks during social interactions. Among the social psychologists some have given greater emphasis to the social aspects of communication while others have emphasized the psychological aspects.

Newcomb (1953) introduced the concept of coorientation as the basis of human communication. He developed ABX model as a helpful tool in relational analysis of dyadic pairs. This simple yet insightful model consists of two communicators, A and B, and their "orientation" toward some "object of communication", X. Any subject, behaviour, attitude, belief, event, or object, which is the focus of communication for the two participants, has the potential to be the "object of communication". Each communicator, A and B, has a simultaneous coorientation toward his or her communication partner (usually the level of attraction and feelings toward the partner) and toward the object of communication (the degree of positive or negative attitude about X). (Fig. 7.2).

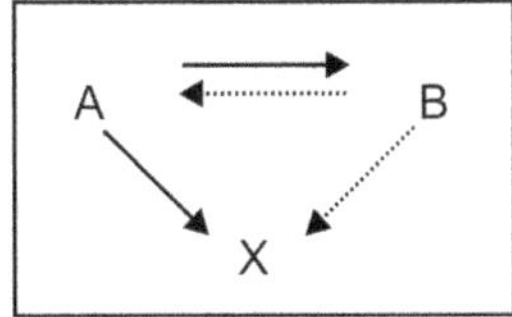

Fig. 7.2 Newcomb's A-B-X or Coorientation Model.

Newcomb sees four basic components of this relational system: A's attitude toward X; A's attraction to B; B's attitude toward X and B's attraction to A.

In this model, at the very minimum, any thorough index of a dyadic relationship should include the following two items of information: (1) each person's orientation (that is, their attitude toward the object of communication and their attraction toward their communication partner) and (2) what each person perceives their partner's orientations to be.

Linguistic Theory

This theory is primarily concerned with analysis of speech, but speech is more than just another form of signaling: it is also a form of human behaviour in general. The aim of this theory is to develop a model and a procedure for making a complete, concise and objective description of the speech of a people.

Greenberg (1966) introduced this theory with the description of verbal behaviour. The linguistic descriptions emphasize patterns and structures; they do this by isolating structural units such as phonemes and morphemes (basic elements in the sound and grammatical systems of language). Using linguistically based concepts, Greenberg describes not only what a people say or have said, but how they act.

Continuing Applications of Communication Theory

Applications of Shannon and weaver Model and are prevalent in everyday life, especially in areas where "computers are a key component, telephone systems, banking, airline reservations, scientific research, weather forecasting and hundreds of other applications of information technology (Rogers, 1995).

Researchers stress that the importance of Shannon's theory cannot be overstated: "It's like saying how much influence the inventor of the alphabet has had on literature" (Horgan, 1992). The linear model presents the measurable variables for all communication processes.

Future applications went far beyond the original focus on channel capacity. For example, Shannon's research helped to develop the idea of check digits, or, error correcting codes, (what Shannon referred to as "correcting device") in a string of binary numbers which is applied in computer technology.

With the advent of internet, the idea of effectively communicating ideas to the masses moved to the forefront of research in developing countries. Whether it is in terms of communication for effective instruction, or theory to enhance technological processes, the communication theory that was developed in the 1940s had an immeasurable impact on all forms of communication today, both human and non-human. Theories about mass communication processes need to be set forth as systematic sets of prepositions that show in straightforward terms just what is supposed to be related to what in terms of dependent and independent variables. To do this, orders of dependency between propositions can be established. If this is done, the theories can be tested empirically and their validity can be adequately assessed.

Models of Communication

According to Aristotle (1933), communication has three ingredients.

1. Speaker – the person who speaks
2. Speech – the speech that the individual produces
3. Audience – the person who listens

The Shanon and weaver (1949) the elements of communications are

1. Source
2. Transmitter
3. Signal
4. Receiver
5. Destination

Compared with the Aristotelian model, the source is the speaker, the signal is the speech and the destination is the audience, plus two added elements, a transmitter which sends out the source's message and a receiver which catches the message for the destination.

According to Berlo (1960) the model of communication consists of

1. Source
2. Encoder
3. Message
4. Channel
5. Decoder
6. Receiver

Code is a system of signals for communication. Encode means to put the message into code or cipher. Channel means the medium through which the signals move, the decoder means which converts the message in the code into ordinary language which may be easily understood.

He further elaborated that all human communication has some source, some person or group of persons with a purpose. The purpose of the source has to be expressed in the form of message. The communication encoder is responsible for taking the ideas of the source and putting them in a code, expressing the source's purpose in the form of a message. A channel is a medium, a carrier of message. For communication to occur there must be somebody at the other end, who can be called the communication receiver, the target of communication.

According to Schramm (1961), the communication process involves
1. Source
2. Encoder
3. Signal
4. Decoder
5. Destination

This model of communication is particularly relevant for the mass media. In human communication it is most important whether people can properly encode or decode the signal (message), and how they interpret it in their own situations.

Rogers and Shoemaker (1971) thought of the communication process in terms of the S-M-C-R-E model, (Fig. 7.3) the components of which are

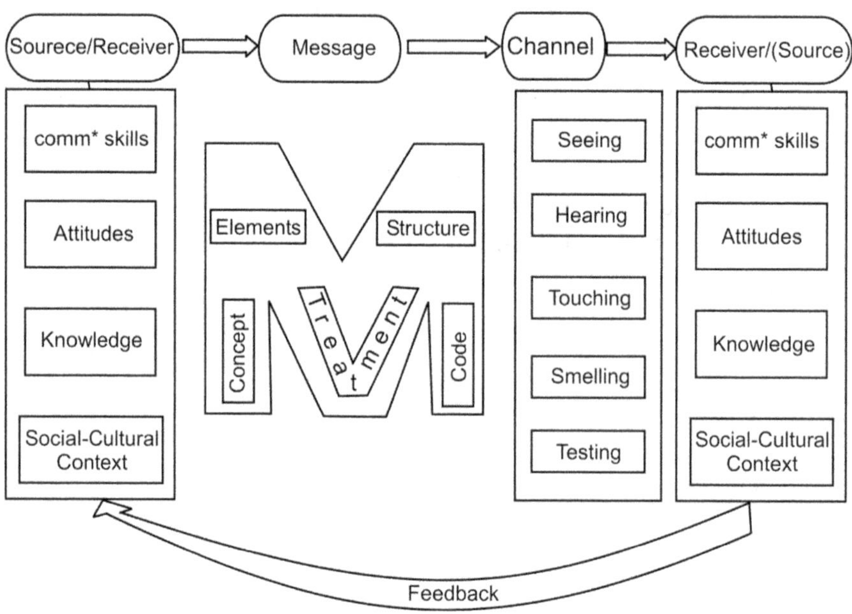

Fig. 7.3 SMCR Model of Communication

Communicator : Otherwise called as sender, is the originator of the communication. As a communicator the extension worker must be credible. He should gain the confidence of his clientele. Credibility can be improved by learning to communicate effectively.

A Good Communicator Must:
*   know his clientele, their wants and needs
*   know his message, its content and how to present it

- know effective channels of communication for his message to get across and know his own abilities and limitations
- be interested in his clientele and their welfare and how his message can help
- be interested in improving his skills in communication
- prepare his message carefully, using appropriate materials to elicit interest from the clientele
- speak clearly
- use simple words to be easily understood
- realize the mutual understanding between extension workers and the fishfarmer is mostly the worker's responsibility
- be time conscious.

Message *:* Extension workers have important information and ideas which he hopes will be received and interpreted by the clientele as he intended. Oftentimes, this is not the case due to incomplete information, poor presentation, and other reasons. To avoid these difficulties, extension workers should be prepared to reiterate the information. The purpose or objective should be clear in mind. What change in behaviour do you want to bring about? It can be a change in knowledge, attitude, skills, or in what you expect the clientele to do. The content of the message should be of interest to him. It must be related to something he understands, feels or thinks, something he can accept. The treatment of the message is important to make it acceptable and understandable to the receiver. It should be organised in terms he understands. It should conform to accept social standards. Treatment can make a message interesting or boring.

Channels : Extension methods are channels of communication. These methods may be classified as visual, spoken, or written or a combination of both. Spoken methods include field and home visits, office calls, meeting of all kinds, radio, television and telephone calls. Except for radio and television, the rest are a two-way communication. Differences of opinion can be cleared up on the spot. There are also disadvantages and obstacles to be overcome. Since an oral message is not always recorded, the receiver may remember it differently than the sender wants. Where premise statements are only spoken, the receiver has no way to refer back to what was said. In spite of its problems, spoken communication when supplemented with visual aids is the best method of extension work. Written communication is indispensable in day-to-day activities of extension. Records and reports must be prepared, kept available for use, and submitted to superiors. The clientele must be kept informed of activities and accomplishments.

Written communications have greater status and carry more authority than oral communications. Letters, bulletins, circulars, announcements of events and magazines contribute to extension in literate societies. They provide a low-cost method to disseminate information to a large number of people, but this is only a one-way communication. Few people will change their methods only because they read about it. An effective extension worker will adapt his extension methods to the subject, to the communication skills of the clientele and to the facilities available.

Receiver *:* Otherwise called as audience, who are also the acceptors. If the receiver did not accept, there would be no communication at all. The significant thing to remember is this: while some may agree on some aspect, they may also differ in thousands of ways. Often, some of these differences block communication. Differences in education mean different abilities to understand a difficult concept and other technicalities. For this reason, communication often fails because of a gap or language barrier.

Communication, in extension, may also be thought of as two-way stimulus-response (S-R) situation in which the necessary stimulus is provided by the communicator, the extension agent, in the form of a message, which produces certain response in the audience, the farmers and *vice-versa.* A favourable response by the audience reinforces learning. The communication model forwarded by John Paul Leagans (1963) has the following elements

1. Communicator
2. Message
3. Channel
4. Treatment
5. Audience
6. Audience response

The task of communication, according to him, is to provide powerful incentives for change. Success at this task requires thorough understanding of the six elements of communication, a skilful communicator sending useful messages through proper channels, effectively treated, to an appropriate audience that responds as desired.

Elements of Communication System

The elements of the extension communication system are discussed in brief. The characteristics of each of the elements, which may contribute to the success or failure of communication are furnished as per Leagans (1961).

1. *Communicator*

In the context of agriculture and rural development, extension agent is the communicator who starts the process of communication. The extension agent and mass media like radio are sometimes visualised as sources or originators of a message, which is not correct. Knowledge generates through research and as such the Research Institutes, Research Projects and Universities are the originators or sources of knowledge. The extension agent obtains the required information from research and carries it to the audience, the farmers. The extension agent is the communicator, a carrier of information. To enhance the process, extension agents may take the help of some aids, known as audio-visual aids. They also carry back the reactions of the farmers, their problems etc. as feedback information to research, for finding out solutions for the same.

The credibility of the communicator and the organization the individual represents is important for effective communication. *CREDIBILITY means trustworthiness and competence. Before the audience accepts any message it will judge whether the communicator and the organisation the individual represents, can be relied upon and is competent enough to give the information.* Studies have revealed that scientists and extension agents having status, expertise, accomplishment, authority and experience are perceived as highly credible by farmers in communicating information on agriculture and rural development. Who tells is, therefore, very important in extension communication. The characteristics of a good communicator are:

The Individual Knows

(i)   The objectives – have them specifically defined;
(ii)   The audience – their needs, interests, abilities, predispositions;
(iii) The message – its content, validity, usefulness, importance;
(iv)   Channels that will reach the audience;
(v)   Organisation and treatment of the message;
(vi)   The professional abilities and limitations.

The Individual is Interested In

(i)   The audience and its welfare;
(ii)   The message and how it can help people;
(iii) The results of communication and their evaluation;
(iv)   The communication process;
(v)   The communication channels – their proper use and limitations;
(vi)   Improvement of the communication skill.

The Individual Prepares

(i)   A plan for communication – a teaching plan;
(ii)   Communication materials and equipments;
(iii)   A plan for evaluation of results.

The Individual has Skill In

(i)   Selecting messages;
(ii)   Treating messages;
(iii)   Expressing messages – verbal and written;
(v)   The selection and use of channels;
(v)   Understanding the audience;
(vi)   Collecting evidence of results.

Poor Communicators, on the Other Hand

(i)   Fail to have ideas to present that are really useful to the audience.
(ii)   Fail to give the complete story and show its relationship to people's problems,
(iii)   Forget that time and energy are needed to absorb the material presented,
(iv)   Feel they are always clearly understood,
(v)   Refuse to adjust to 'closed' minds,
(vi)   Talk while others are not listening,
(vii)   Get far too ahead of audience understanding,
(viii)   Fail to recognize that communication is a two-way process,
(ix)   Let their own biases over-influence the presentation,
(x)   Fail to see that everyone understands questions brought up for discussion,
(xi)   Fail to provide a permissive atmosphere,
(xii)   Disregard the values, customs, prejudices and habits of the people and
(xiii)   Fail to start where people are, with respect to knowledge, skill, interest and need.

2. *Message*

The recommendations from research, the technology, constitute the content or subject matter, the message. Information which is relevant to a particular set of audiences constitutes the message, otherwise for them this is 'noise'. A good message should clearly state what to do, how to do, when to do and what would be the result. To produce desirable changes in human behaviour, the message must be motivating. *Messages which are relevant, interesting, useful, profitable, credible (latest and best, based on research findings) and complete (neither too much, nor too little) are likely to motivate the people.*

A Good Message Should be

(i)   In line with the objective to be attained;

(ii)   Clear – understandable by the audience;

(iii)   In line with the mental, social, economic and physical capabilities of the audience;

(iv)   Significant – economically, socially or aesthetically to the needs, interests and values of the audience;

(v)   Specific – no irrelevant material;

(vi)   Simply stated – covering only one point at a time;

(vii)   Accurate – scientifically sound, factual and current;

(viii) Timely – especially when seasonal factors are important and issues are current;

(ix)   Supported by factual material covering both sides of the argument;

(x)   Appropriate to the channel selected;

(xi)   Appealing and attractive to the audience – having utility and immediate use;

(xii)   Applicable – can apply recommendation to one's own particular situation;

(xiii) Adequate – combining principle and practice in effective proportion; and

(xiv)   Manageable – can be handled by the communicator and within the limits of time.

In contrast, Poor Communicators often

(i)   Fail to clearly separate the key message from the supporting content or subject-matter;

(ii)   Fail to prepare and organize their message properly;

(iii)   Use inaccurate of 'fuzzy' symbols – words, visuals or real objects – to represent the message;

(iv)   Fail to select messages that are in line with the felt needs of the audience;

(v)   Fail to present the message objectively – present the material, often biased, to support only one side of the proposition;

(vi)   Fail to view the message from the standpoint of the audience; and

(vii)   Fail to time the message properly within a presentation or within a presentation or within a total programme.

3. *Channel*

Channel of communication constitutes the medium through which information flows from a sender to one or more receivers. *Face-to-face, word-of-mouth is the simplest and yet one of the most widely used and effective means of communication, particularly for the developing countries. As society changes from traditional to modern, the emphasis shifts from oral to media system of communication.* Because of the large number of audience or receivers of information and because of physical distance of the communicator and the receivers of information, it is necessary to use different media of communication. Even in interpersonal, face-to-face, word-of-mouth communication, it becomes necessary to use some aids to make communication more effective.

In communication there are what we call filters. A filter in this sense is anything that prevents a message from getting through to the intended audience. Filters may be fear, prejudice, inability to grasp the idea, or any possible barriers. The point is that a good communicator anticipates and tries to prevent filters if he can; he is ready with every means to overcome barriers in any case. Communication failure may also occur when the idea being communicated is contrary to the accepted local customs and beliefs. This too is a filter. Recognising this danger, alternative approaches to the problem should be used. The process by which a sender can convey his message to the audience often affects the transfer of an idea. If he is sincere and respected, he is more likely to succeed in transmitting his idea to his audience.

The channels of communication may be classified in a number of ways according to different criteria.

According to Form

Spoken :   Farm and home visit, farmer's call, meetings, radio talk, etc.

Written :   Personal letter, farm publications, newspaper, etc.

According to Nature of Personnel Involved

Personal Localite *:* They are the local leaders and local people who belong to the receivers' own social system. Personal localite channels are important in a traditional social system.

Personal Cosmopolite *:* These are the channels of communication from outside the social system of the receiver. They are the extension agents of various organisations and are important in changing the farmers from traditional to modern.

Impersonal Cosmopolite *:* Here the channels of communication are from outside the social system of the receiver and at the same time no personal face-to-face contact is involved. These are mass media, which are important in areas of high urban influence, and farmers who are modern or are changing from traditional to modern.

According to Nature of Contact with the People

Individual Contact *:* The extension agent communicates w\ith the people individually, maintaining separate identity of each person. Examples are farm and home visit, farmer's call, personal letter etc.

Group Contact: The extension agent communicates with the people in groups and not as individual persons. Examples are group meeting, small group training, field day or farmers' day, study tour etc.

Mass Contact *:* The extension agent communicates with a mass of people, without taking into consideration their individual or group identity. Examples are mass meeting, campaign, exhibition, radio, television etc.

Many obstructions can enter channels. These are often referred to as 'noise', that prevents the message from being heard by or carried over clearly to the audience. 'Noise' emerges from a wide range of sources and causes following are Some of them.

(i)   Failure of channel to reach the intended audience. All people cannot or may not attend meetings, all people may not have radio or TV, or may not be tuned if they had, or many people cannot and others may not read the written materials;

(ii)   Failure on the part of the communicator to handle channels skillfully. In a meeting, who cannot hear what is said and see what is shown, do not receive the message;

(iii)   Failure to select channels appropriate to the objective of a communicator. If the objective is to show how to do a certain thing, method demonstration and TV will be appropriate, rather than radio or newspaper;

(iv)   Failure to use channels in accordance with the abilities of the audience. Written materials can not serve as useful channels of communication for an illiterate group of persons;

(v)   Failure to avoid physical distraction. Loud noise near a place of meeting or power failure at the time of projecting visuals may cause distraction of the audience;

(vi)   Failure of an audience to listen or look carefully. There is a tendency of people not to give undivided attention to the communication;

(vii) Failure to use enough channels in parallel (simultaneously). Research indicates that up to five or six channels used in combination are often necessary to get a message through to a large number of people with enough impact to influence significant changes in behaviour; and

(viii) Use too many channels in a series. An important principle of communication is that the more channels used in a series (communicating through several levels of line personnel), the less chance a communicator has for getting the message through to the intended audience.

To overcome some of the Problems of Communication, one should take the Following Factors into Account

(i)   The specific objective of the message;

(ii)   The nature of the message – degree of directness *versus* abstractness, level of difficulty, scope, timing etc.

(iii) The audience – size, need, interest, knowledge of the subject etc.;

(iv)   Channels available that will reach the audience, or parts of it;

(v)   How channels can be combined and used in parallel;

(vi)   How channels that must be used in a series can be reduced to the minimum and those used made effective;

(vii) Relative cost of channels in relation to anticipated effectiveness;

(viii) Time available to the communicator and to the audience;

(ix)   Extent of seeing, hearing or doing that is necessary to get the message through; and

(x)   Extent of cumulative effect or impact on the audience necessary to promote action.

4. *Treatment*

Treatment means the way a message is handled, or dealt with, so that the information gets across to the audience. It relates to the technique or details of procedure or manner of performance, essential to effective presentation of the message. The purpose of treatment is to make the message clear, understandable and realistic to the audience. Treatment of the message by the communicator shall depend to a great extent on choice of the channel and the nature of audience. The task cannot be reduced to a formula or recipe. Treatment is a creative task that has to be 'tailor-made' for each communication function. For example, treatment of a message will be different when it is conveyed in a meeting, or published in a folder or broadcast. Similarly, there will be difference in treatment of the message according to the level of literacy, socio-economic status and progressiveness of the audience. Designing treatment usually requires original thinking, deep insight into the principles of human behaviour and skill in creating and using refined techniques of message presentation. The following are the three categories of bases useful for varying treatment.

Matters of General Organization

(i)   Repetition or frequency of mention of ideas and concepts;

(ii)   Contrast of ideas;

(iii)   Chronological – compared to logical and psychological;

(iv)   Presenting one side compared to two sides of an issue;

(v)   Emotional compared to logical appeals;

(vi)   Starting with strong arguments compared to saving them until the end of presentation;

(vii)   Inductive compared to deductive;

(viii) Proceeding from the general to the specific and *vice-versa*; and

(ix)   Explicitly drawing conclusions compared to leaving conclusions implicit for the audience to draw.

Matters of Speaking and Acting

(i)   Limit the scope of presentation to a few basic ideas and to the time allotted – too many ideas at one time may be confusing;

(ii)   Be yourself – you can't be anyone else, strive to be clear, not clever;

(iii)   Know the facts – fuzziness means sure death to a message;

(iv)   Don't read your speech – people have more respect for a communicator who talks to the audience;

(v)   Know the audience – each audience has its own personality, be responsive to it;

(vi)   Avoid being condescending (patronizing) – do not talk or act down to people, or over their heads. Good treatment of message results in hitting the target. Never overestimate the knowledge of an audience or underestimate their intelligence;

(vii)   Decide on the dramatic effect desired – effective treatment requires sincerity, smoothness, enthusiasm, warmth, flexibility and appropriateness of voice, gestures, movements and tempo;

(viii) Use alternative communicators when appropriate, as in group discussion, panels, interviews, etc.;

(ix) Remember that audience appeal is a psychological bridge to getting a message delivered; and

(x)   Quit on time – communicators who stop when they have 'finished' are rewarded by audience goodwill.

Matters of Symbol Variation and Devices for Representing Ideas spoken words, written materials, audio-visual aids etc. belong to this category.

5. *Audience*

The audience or receiver of message is the target of communication function. An audience may consist of a single person or a number of persons. It may comprise men, women and youth. An audience may be formed according to occupation groups such as crop farmers, fruit farmers, dairymen, poultry keepers, fish farmers, home makers etc. Audience may also be categorised according to farm size such as marginal, small medium or big farmers; or according to whether they belong to scheduled caste, scheduled tribe etc. Communication, to be successful, must be target oriented. The communicator must know the target, their needs, interests, resources, facilities, constraints and even their approximate number and location. The attitude of the audience toward the message largely depends upon who gives what message through which channel; to what extent the contents of the message are in line with their preheld experiences and preexisting preferences; and how far the message is compatible with group norms and value system to which the audience belongs. In case the audience members feel that the communicator is trustworthy, dependable and find the person communicating the message through the medium of their choice, they are likely to receive the message, provided the presentation of the content appear to the audience as interesting and comprehensive.

The communicator should, therefore, be careful in selecting message which are relevant to the audience, choose channels compatible to their cultural pattern and make treatment of the message appropriate to their levels of interest and understanding. In addition to knowing the identity of an audience and some of its general characteristics, there are other somewhat more specified aspects that help to clarify the exact nature of an audience and how to reach it. Following are some of the tips

(i)   Communication channels established by the social organisation;

(ii) The system of values held by the audience – what they think is important;

(iii) Forces influencing group confirmity – custom, tradition, etc;

(iv) Individual personality factors – change proneness etc.;

(v)   Native and acquired abilities;

(vi) Educational, economic and social levels;

(vii) Pressure of occupational responsibility – how busy or concerned they are;

(viii) People's needs as they see them, and as the professional communicators see them;

(ix) Why the audience is in need of changed ways of thinking, feeling and doing; and

(x)   How the audience views the situation.

It may be noted that the audience is not a passive recipient of message. The individuals are rather selective in receiving, processing and interpreting messages.

Selective Exposure *:* People expose themselves to messages selectively. There is a tendency for individuals to expose themselves relatively more to those items of communication that are in agreement with their ideas, beliefs, values, etc.

Selective Perception *:* Regardless of exposure to communication, an individual's perception of a certain event, issue, person or place could be influenced by one's latent beliefs, attitudes, wants, needs or other factors. Thus, two individuals exposed to the same message could go away with different perceptions about it.

Selective Retention *:* All information is not retained by the individuals. People generally tend to retain that information in which they have some interest and which they consider to be important. Research showed that even recall of information is influenced by factors such as individual's needs, wants, moods, perceptions and so on. The social categories to which people belong, their individual characteristics and social relationships greatly influence their acquisition and utilisation of information.

6. *Audience Response*

Response of the audience is the ultimate objective of any communication function. Response of an audience to messages received may be in the form of some kind of action, mental or physical. Until the desired action results, extension communication does not achieve its most essential objective.

The possible kinds of response to messages received are almost infinite. The following gives an idea of possible variety in response that may result when a useful message is received by a typical village audience.

(i) *Understanding Versus Knowledge:* People usually do not act on facts alone, but only when understanding of facts is gained. Understanding is attained only when one is able to attach meaning to facts, see the relationship of facts to each other and to the problem. Communication must promote understanding;

(ii) *Acceptance Versus Rejection:* Audience may tend either way. Communication should lead to understanding and acceptance of the idea;

(iii) *Remembering Versus Forgetting :* When opportunity for action is not immediately available or action is delayed, the message may be forgotten. Transmitting the right message to the right people at the right time is often a crucial factor in successful communication;

(iv) *Mental Versus Physical Action :* Changes in the minds of the people must always precede changes in the action by hands. People should not only understand and accept the message but must also act on it; and

(v) *Right Versus Wrong :* The intent of a communication is to promote desirable action by an audience as determined by the communicator and expressed in his objectives. Consequently, resulting action in line with the intended objectives is assumed to be 'right' action. But the problem is more complex. Unfortunately, 'noise' often plays mischief at this point. For a variety of reasons, people often fail to behave precisely according to instructions, even when they understand and accept them. Assume, for example, that a message giving five steps in seed treatment has been transmitted to a group of cultivators. Assume further that the cultivators understood, accepted and acted on the message. But the results were disastrous. This was because the cultivators, contrary to instruction, decided among themselves that if the use of one ounce of the chemical in treating a maund of grain (as instructed) was good, two ounces would be better. Individually and in groups, human beings have their own ideas about how to act.

The many kinds of possible behaviour are mentioned here only to illustrate the range of response that may result when a message is received by an audience. So, communicators engaged in promoting rural development must be concerned not only with process but with product; not only with means but with ends attained. Progress requires refutation of the status quo and building a more desirable situation in its place. This requires action by people who need to make changes in what they think; feel and do. The statement is axiomatic that continuation of the means in programmes of change is justified only by the product they produce. The ultimate question that may be asked about the elements of communication, therefore is; who communicates what, to whom, for what purpose, by which media, with what results?

FeedBack

Extension communication is never complete without feedback information. *Feedback means carrying some significant responses of the audience back to the communicator.* Communication work is not an end in itself. The extension agent should know what has happened to the audience after the message has reached them.

Characteristics of Feedback

    (i)   Feedback is source oriented,

    (ii)   Feedback varies in different communication situations,

    (iii)   Feedback affects the sources or communicator,

    (iv)   Feedback exerts control over future messages,

    (v)   Feedback affects communication fidelity, and

    (vi)   Feedback maintains the stability and equilibrium of a communication system.

Feedback should be a continuous process as the audience and communicators are neither always the same persons, nor they are interacting in the same situation. The extension agent shall take steps to analyse the responses of the audience, which may be positive, negative, or no response. If there has been no response or negative response to a message, the extension agent shall find out reasons for the same. If it pertains to research, the problem should be referred to as feedback information to research, to find out solutions for the same.

If the problem does not relate to research, the extension agent shall find out whether the message has been relevant to the audience, or whether the channel, treatment, and audio-visual aids have been appropriately used. If not, corrective steps should be taken without any loss of time. For a season-bound programme, if nothing can be done in that particular season, the extension agent shall take appropriate steps next season, so that the mistakes are not repeated. If there has been a favourable response to the message by the audience, the extension agent shall find out what next is to be done to reinforce the learning already made by the farmers. At this stage, supply of critical inputs and services including credit are important. Adequate and correct feedback are essential for purposeful communication. Feedback information provides the communicator an opportunity to take corrective steps in communication work, helps in identifying subsequent activities and acts as a pathfinder for need-based research.

Critical Factors in Communication

Each act of communication has at least three phases – expression, interpretation and response. If the expression is not clear, the interpretation accurate and the response logical, one's effort to communicate will not succeed. As it is difficult to control how an audience shall interpret the message and respond to it, a powerful communication effort by the extension agent must be constantly exerted. The critical factors in extension communication according to Leagans (1961) are as follows :

1. Communication is limited by one's concept of the communication process.
2. Communication is a two-way process, involving interaction between the communicator and the receivers.
3. One must have ideas before one can communicate with others.
4. The system of symbols used to represent ideas, objects or concepts must be relevant, accurate and skillfully used.
5. Cultural values and the social organisation are determinants of communication.
6. The environment created by the communicator influences one's effectivenss.
7. To make sense, the communication effort must be organised according to some specific form or pattern.
8. Cooperation, participation and involvement are essential to communication.
9. The standards of correctness, effectiveness, good taste and social responsibility of communication influence its success.
10. Evaluation is necessary to improve communication.

All communication is not verbal or vocal, some messages are transmitted non-verbally too. Non-verbal or 'second-order messages' provide for receivers the context in which specific verbal and vocal cues are interpreted. Non-verbal cues are messages about messages. Because some of these cues are culture-specific, they are hard to decode without knowledge of the cultural context in which they are used.

Concepts Relating to Communication

Frame of Reference

Each person has a stored experience of beliefs and values as an individual and also a member of a society. This provides the background of stimulation which influences a person's behaviour in a particular situation and is called the individual's frame of reference. The functionally interrelated external and internal factors operating at a given time constitute the frame of reference of the ensuing reaction.

A message received by an individual is interpreted in terms of the frame of reference of the individual. The message which challenges these beliefs and values may be rejected or misinterpreted. This tendency on the part of the receiver obstructs communication, in case the receiver and the sender do not have a common frame of reference.

Perception

Gibson (1959) defined perception as the process by which an individual maintains contact with the environment. Kollat, Blackwell and Engel (1970) explained perception as the process whereby an individual receives stimuli through the various senses and interprets them. Perception of the same solution may differ from individual to individual due to differences in their experiences and ways of looking into it. The expectations, needs and ways of thinking influence how an individual interprets what is observed.

Perception is selective and we perceive what we want to perceive. Our perceptions are organised and we tend to structure our sensory experiences in ways which make sense to us. Perception is influenced by the environment in which communication takes place. It is not the intrinsic quality or attribute of an object, individual or message, but how people individually and collective perceive them is important for extension.

Communication Fidelity

Fidelity is the faithful performance of communication process by all its elements Communicator, message, channel and receiver. Noise and fidelity are two sides of the same coin. Eliminating noise increases fidelity, the production of noise reduces fidelity. The basic concern related to noise and fidelity is the isolation of those factors within each of the ingredients of communication which determine the effectiveness of communication.

The communication fidelity can be explained as the extent of desirable changes in receivers' behaviour as a result of communication. The desirable changes are in receivers' knowledge, attitude and action. The objective of any communicative effort is to have communication fidelity as high as possible. The communication fidelity can be explained as the extent of desirable changes in receivers' behaviour as a result of communication. The desirable changes are in receivers' knowledge, attitude and action. The objective of any communicative effort is to have communication fidelity as high as possible.

Communication Gap

Gap means the difference between what was intended and what has been obtained or achieved. Communication gap refers to the difference between what was communicated by the extension agent and what has actually been received by the audience. Desirable action by the audience cannot take place if there is a large communication gap. The nature of communication gap may be of two types, the message does not reach the target and the message fails to produce the desired impact, even if reaches the target. The following steps must be well taken care for reducing the communication gap.

Where the Message does not Reach the Target

   (i)   Communication must be made available,

   (ii)   Communication must be need based

   (iii) Communication must be in time, and

   (iv)   Use more than one channel of communication (minimum three channels comprising both mass media and interpersonal may be used simultaneously).

Where the Message fails to Produce the Desired Impact

   (i)   Use credible (trustworthy and competent) channels of communication,

   (ii)   Repeat the message at least thrice at suitable intervals, in different time slots. In repeating the message some variation may be introduced in the format, keeping the central theme intact. This shall help in sustaining the audience interest,

   (iii) Take precaution against distortion of message (repeat and use printed media),

   (iv)   Increase understandability of message,

(v) Give complete information,

(vi) Help in maintaining the equilibrium (new technology may create some disequilibrium in farm and home for which adjustments must be made), and

(vii) Give new ideas to create and sustain audience interest.

Time Lag

Lag means delay. While communication reduces time lag, the communication process itself may involve some time lag. There may be delay in getting the relevant information in the form of a message and treat the message according to channel requirement and needs of the audience. There may be delay in organizing extension programmes. Some time may be spent in contacting the channels and the channels themselves may require some time to attend to the message in view of their preoccupations. The communicator has to remain alert and take in to consideration all possible delays that may occur during the communication process. By computing this time lag, the communicator should plan and initiate the communication action well in advance so that the intended message reaches the audience in time.

Empathy

Empathy is the ability on the part of one person to understand the other person's integral frame of mind and reference and accept the same. This acceptance does not mean agreement. Empathy is also defined as the ability of an individual to project oneself into the role of another person, to be able to appreciate the feelings, thinking and actions of another person. An extension agent who is empathic shall be able to understand and appreciate the farmers' situations and communicate with them effectively. Similarly, an empathic farmer shall be able to communicate with the outsiders to get the desired information. Empathy is an indispensable skill for people moving out of the traditional settings.

Homophily – Heterophily

According to Rogers and Shoemaker (1971) one of the most obvious and fundamental principles of human communication is that the transfer of ideas most frequently occurs between a source and receiver who are alike, similar, homophilous. As defined by them Homophily is the degree to which pairs of individuals who interact are similar in certain attributes, such as beliefs, values, education, social status and the like. On the other hand, Heterophily is the opposite of homophily, and is defined as the degree to which pairs of individuals who interact are different in certain attributes.

When communicator and receiver share common meanings, attitudes and beliefs and a mutual language, communication between them is likely to be effective. Most individuals enjoy the comfort of interacting with others who are quite similar. Heterophilic interaction, on the other hand, may produce a sort of disagreement because the receivers are exposed to messages that may be inconsistent with their existing beliefs and create an uncomfortable psychological state. Differences in technical competence, social status, attitudes and beliefs, all contribute to heterophily in language and meaning, thereby leading to messages that go unheeded.

To maintain a homophilic status, along with modernisation, the farmers shall seek more and more specialised information and assistance. For this purpose they shall try to build up and maintained contact with research and extension personnel higher up in the administrative hierarchy.

Propaganda, Publicity, Persuasion

Propaganda is deliberate manipulation of people's beliefs, values and behaviour through words, gestures, images, thoughts, music, etc. Propaganda aims at propagating aims beliefs and values of the propagandist and presents only the communicator's side of arguments without considering the arguments of the receiver's side.

Publicity is based on truth and propaganda often suppresses the truth. One sided communication giving view-points of only the message-source, ignoring the view-points of receivers of the message, may sound live propaganda despite the message being based on truth. Propaganda is often authoritative in approach in influencing the people

On the contrary, persuasion is more democratic in influencing the audience to bring about change in their attitude and behaviour. In persuading people, the extension agent provides lots of arguments in favour of acceptance of the recommendations and provides evidences of gain.

Development Communication

Development communication is a communication which is purposive, pragmatic, goal directed and audience oriented. It has 'popular participation' as an essential component. Development communication is used to inform and motivate all levels and sectors of a poor or developing country, to use new skills and equipment in accordance with their needs. Without this social aspect, development will remain detached from the intended population because of the cultural gaps which exist in the society.

Mere acquisition of information will not result in development. For example, a communication programme cannot supply fertilisers, provide loans or establish industries required for economic development. As such, communication programmes should be so timed that these facilities are available to the audience almost simulataneously.

Management of Information System

According to Koontz and Weihrich (1988) Management of Information System is a formal system to gather, integrate, compare, analyze and disperse information internal and external to the enterprise in a timely, effective and efficient manner. The management information system has to be tailored to specific needs and may include routine information, such as monthly reports; information that points out exceptions, specially at critical points and information necessary to predict the future.

Electronic equipments permit fast and economical processing of huge amounts of data. The computer can, with proper programming, process data toward logical conclusions, classify them and make them readily available for a manager's use. In fact, data do not become information until they are processed into a usable form that informs.

Rural Journalism

In presenting information to the rural audience, whether through print media or through electronic media, the following principles may be followed. There shall improve comprehension of the information by the audience.

1. *Use Simple Language :* Explain the technical terms in short and simple sentences, using common words which have concrete meanings. Abstract ideas and 'jargon' should not be used. Avoid using text book language.

2. *Structure and Arrange Arguments clearly* : Present ideas in a logical order, clearly distinguishing between the main and the side issues. Presentation must be clear, with the central theme remaining visible so that the whole message can be reviewed easily. Separate key points or sections of the message by use of careful layout and typography.

3. *Make Main Points Briefly* : Restrict arguments to the main issues, clearly directed towards achieving stated goals without unnecessary use of words.

4. *Keep important Information at the Top* : Organize the write-up like an inverted pyramid, keeping the most important information at the top,

so that if some portion from the bottom is deleted during editing, it won't affect the write-up much.

5. *Use lively pictures and Photographs* : The pictures and photographs should be simple, bold, with good contrast of light and shade, so that the message intended to be conveyed is clearly brought out.

6. *Prepare a Stimulating Write-up* : The presentation should be interesting, inspiring, personal and sufficiently diversified to sustain audience interest. Accuracy, Brevity and Clarity (ABC) are fundamental to a good presentation of information.

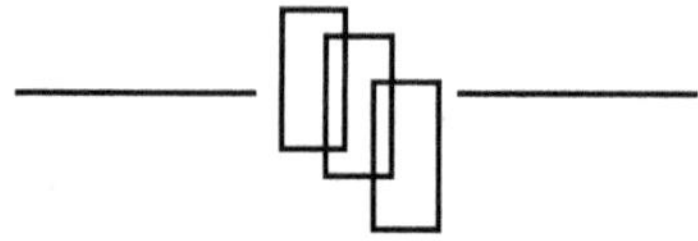

CHAPTER - 8

Extension Management

Management is universal in the modern world. Every organisation whether formal or informal is required to make decisions, coordinate its activities, handle people, evaluate performances and direct the people towards goal or group objectives. Management in extension is more important as extension is becoming more and more specialised and the scale of operation is ever increasing. Because of the increasing importance of management and new challenges it faces; many researchers in various disciplins have concentrated their attention on this part of the subject. In the past management developed only in industrial and business sectors. But now it involves public sectors, information societies and even entrepreneurial organisations.

Management : According to Mary Parker Follett (1933), management is the art of getting things done through people. According to Joseph L. Massie (2000), management is the process by which a cooperative group directs actions towards common goals.

Management Process : A series of operations done by a manager over a period of time is called management process.

Concept of management : Harbinson and Myers (1961) offered a three-fold concept for the viewpoint of management, such as:

(a) As viewed by the economist, management is one of the factors of production i.e. land, labour and capital or an economic resource.

(b) As viewed by the specialist, organisation and administration, management is a system of authority.

(c) As viewed by sociologist, management is a social class and social status system.

The total concept of management requires an understanding of the meaning of liberal education and its relationship to management. A liberal point of view is not merely a sum of a number of narrow approaches, but it's the freedom to choose from the widest range of possibilities by discovering new possibilities and recalling possibilities, which are previously developed but forgotten.

Management Analysis : Although academic writers and theorists contributed notably little to the study of management until the 1950s, the past three or four decades have seen a veritable deluge of writing from the academic halls. The variety of approaches to management analysis and number of differing views have resulted in much confusion as to what management is, what management theory and science is, and how managerial events should be analyzed. There are various approaches to management analysis, such as:

Empirical or case Approach : Studies experience through cases and then identifies the successes and failures.

Interpersonal Behaviour Approach : Focus on interpersonal behaviour, human relations, leadership and motivation based on individual psychology.

Group Behaviour Approach : Emphasis on behaviour of people in groups. Based on sociology and social psychology. The study of large groups is often called as organisational behaviour.

Cooperative Social System Approach : Concerned with both interpersonal and group behavioural aspects leading to a system of cooperation.

Decision Theory Approach : Focus on the making of decisions, persons or groups making decisions and the decision making process. Some theorists use decision making as a springboard to study all enterprise activities, where the boundaries of the study are no longer clearly defined.

Systems Approach : Systems concepts have broad applicability. Systems have boundaries, but they also interact with the external environment, i.e. organizations are open systems, which recognizes importance of studying interrelatedness of planning, organizsing and controlling in an organisation as well as the many subsystems.

Mathematical Approach : Managing is seen as a mathematical process, concepts, expressed in mathematical symbols and models. Looks at management as a purely logical process, expressed in mathematical symbols and relationships. Many aspects in managing can not be modeled. Mathematics is a useful tool, but hardly an approach to management.

Situational Analysis : Managerial practice depends on circumstances i.e. a contingency or a situation. Contingency theory recognizses the influence of given solutions on organisational behaviour patterns.

Managerial Roles Approach : original study consisted of observations of five chief executives. On the basis of this study, ten managerial roles were identified and grouped into Interpersonal roles, informational roles and decision roles.

Mckinsey's 7-S Approach : The seven S's are strategy, structure, system, style, staff, shared values and skills.

Operational Approach : Draws together concepts, principles, techniques and knowledge from other fields and managerial approaches. The attempt is to develop science and theory with practical application, which distinguishes between the managerial and non-managerial knowledge. The operational approach recognizses that there is central core of knowledge about managing pertinent only to the field of management. In addition, this approach draws on and absorbs knowledge from other fields, including systems theory, decision theory, theories of motivation and leadership, individual and group behaviour, social systems, cooperation and communications and from the application of mathematical analyses and concepts and the nature of operational approach can be seen in Fig. 8.1.

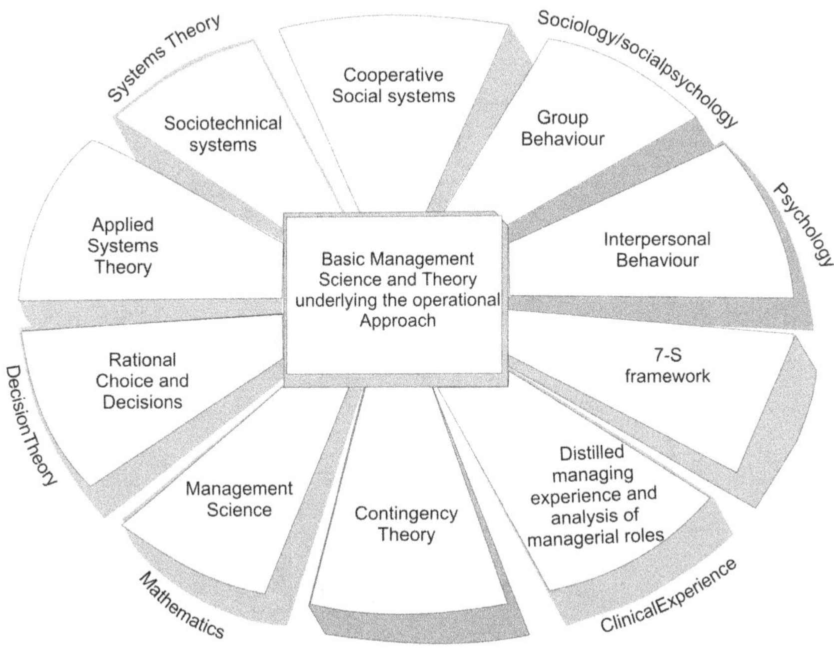

Fig. 8.1 Operational Approach of Management.

Systems Approach to Operational Management

An organised enterprise does not, of course, exist in a vaccum. Rather, it is dependent on its external environment; it is a part of larger systems such as the industry to which it belongs, the economic system and society. Thus, the enterprise receives inputs, transforms them and exports the outputs to the environment or reenergises the system. The model of the systems approach to management is the foundation for organising managerial knowledge (Fig 8.2).

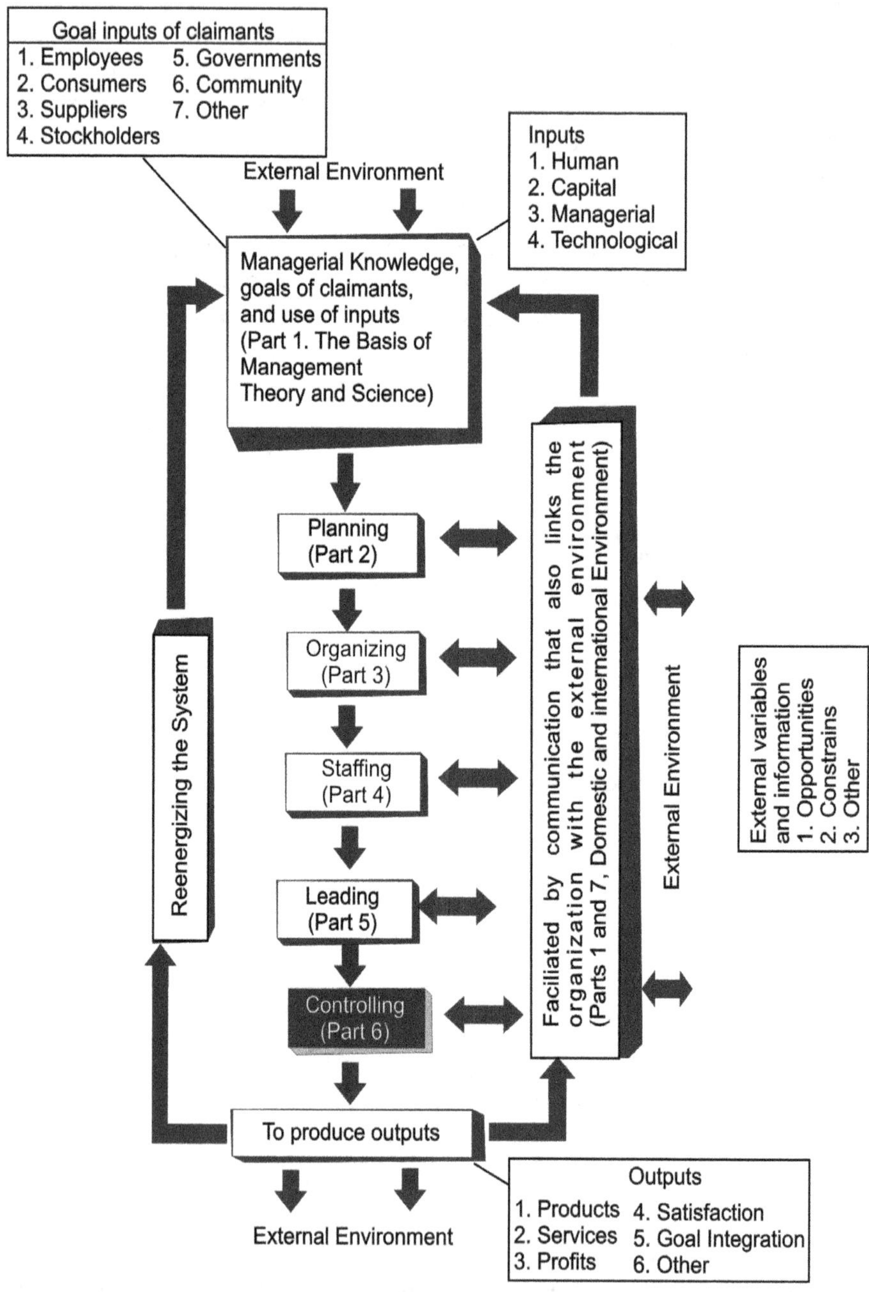

Fig. 8.2 Systems approach of management

Functions of Extension Management Process

Series of operations done by a manager over a period of time called as management process. Firstly, there has to be a horizon or an ambit within which management must perform, and while this ambit may be large or small but it ought to be properly defined. The management functions are ordered in a sequence just to avoid the overlapping of functions, such as:

A. *Planning* : First element /function in management which sets the model points in management, under this the following functions are carried out:

(i)   Crystallisation of objectives

(ii)   Collection and synthesis of information

(iii) Developing alternative courses of action

(iv)   Comparing the alternatives in terms of objectives, feasibility and consequences.

(v)   Selecting optimum course of action

(vi)   Establishing policies, procedures, programmes, schedules, systems, standards and budgets.

B. *Organization* : According to Louis Allen, (1958) the task of organization has got four basic elements.

(i) **Work load** : Identifying the work that must be accomplished to attain the given objectives.

(ii) *Departmentation* : Identifying and grouping of similar activities on a logical basis so that a team of operators can be organised in order to attain the objectives of enterprise may be called as departmentation. Factors deciding the departmentation are: specialisation, coordination, control, economy, adequate attention, human consideration and local concerns.

Forms of Departmentation

- *Functional Departmentation* : It refers to the grouping of the organisational activities on the basis of major functions i.e. each function is performed by separate department.

- *Product Departmentation* : On the basis of products, i.e. each product would have separate department.

- *Process Departmentation* : On the basis of processing steps. For example, production, procurement, processing and marketing units in any industry will have separate departments.

- *Territorial Departmentation* : On the basis of regions/geographic areas. For example, North, South, East and West regions.

(iii) *Delegation* : According to Louis Allen, delegation means entrustment of responsibility and authority to others and creation of accountability for performance.

Characteristics of Delegation

- Delegation is an authorisation to a manager to act in a certain way independently.
- Delegation may be in a written or unwritten form or may be general or specific.
- Authority once delegated to a subordinate is withdrawn, reduced, or even can enhanced.
- A superior delegates authority to subordinates; at the same time   still retains the authority.
- A manager cannot delegate any authority, which the individual does not possess.

(iv) *Relationship of Structure :* The organization has got a structural framework for carrying the managerial functions. There are three organizational structures.

- *Line organisational Structure* : Refers to the backbone of the hierarchy
- *Functional organisational structure* : Refers to the execution aspects of the authority
- *Staff organisational Structure* : Here the organization is basically structured on line authority relationship.

C. *Staffing* : Staffing is defined as filling and keeping filled the positions in the organisation structure or it is the process of flow of human resources into, within and out of the enterprise. It involves: identifying work force requirement, inventorying the people available, recruitment, selection, placing, promoting, appraising and training of both current job holders and new recruitees to accomplish their tasks effectively and efficiently. Managers often say that people are their most important asset. Staffing will determine the success or failure of an enterprise and it is considered as a crucial function of managers. Managers often overlook the fact, that it's their job to fill the positions in their organization and keep them filled with qualified people.

Functions of Staffing

- Anticipating manpower needs
- Developing job description and specification
- Deciding appropriate source of recruitment
- Arranging the selection procedures

- Deciding manpower development activities
- Determining suitable training programmes and techniques.

D. *Directing* : According to Koontz and O'Donnel, direction is a complex function that includes all those activities, which are designed to encourage subordinates to work effectively and efficiently in both short and long runs.

Characteristics of Directing

- It is a continuous process
- It is a managerial function performed by all the managers at all levels.
- It initiates the actions
- It activates subordinates to go according to plan.
- It provides opportunities to superiors to do important work, which subordinates cannot perform
- It follows hierarchically from top to bottom

Principles of Directing

- Principle of leadership
- Principle of integration of group and organisational goals
- Principle of getting maximum individual contribution
- Principle of unity of command
- Principle of effective communication
- Principle of efficient control
- Principle of efficiency
- Principle of follow through

Direction has got Three Dimensions

(i) ***Motivation:*** It is a process of stimulating people to accomplish desired goals. Motives are of two types; (a) Primary motive: Biological in nature e.g. food, clothing, shelter etc.

(ii) *(b) **Secondary:*** Social in nature e.g. needs, status etc.

(iii) Leadership: It is the art or process of influencing the people, so that they contribute willingly and enthusiastically towards group goals.

(iv) ***Communication:*** It is defined as interchange of thought or information to bring about mutual understanding and confidence. The purposes of communication are: information, instruction, persuasion and integration.

E. *Coordination* : It is the essence of managership, for achieving harmony among individual efforts toward the accomplishment of group goals. Each of the managerial functions is an exercise contributing to coordination.

Purposes of Coordination
- To harmonise conflicts
- To emphasize human relations
- To give total accomplishment
- Key to other functions
- Retention of good personnel

Characteristics of Coordination
- It is a continuous and never ending process
- It refers to group effort but not individual
- It emphasises on unity of efforts and united actions
- Always aims at improvement

Principles of Coordination
Principle of direct contact

Principle of continuous process

Principle of reciprocal relation of all factors

Coordination can be started at any stage of management process

Forms of Coordination
- Vertical coordination or Horizontal coordination
- Internal coordination or external coordination
- Procedural coordination or substantive coordination

Limitations of Coordination
- Uncertainty of the future
- Confused and conflicting ideas and objectives
- Lack of administrative skills and techniques
- Lack of orderly method of developing and adopting new ideas and programmes
- Incompleteness of human knowledge, wisdom with regard to man and life.

F. *Controlling* : It is the process of checking the correct performance against predetermined goals with a view to ensure adequate progress and satisfactory performance or it implies the measurement of accomplishment against the standards and the correction of deviations to assure attainment of objectives according to plans. The basic control process, involves three steps:

(a) *Establishment of Standards* : Standards are by definition simply criteria of performance. They are the selected points in an entire planning programme at which measures of performance are made so that managers can receive signals about how things are going and thus do not have to watch every step in the execution of plans. There are many kinds of standards. Among the best are verifiable goals or objectives.

(b) *Measurement of Performance*: Although measurement is not always practicable, the measurement of performance against standards should ideally be done on a forward looking basis so that deviations may be detected in advance of their occurrence and avoided by appropriate actions. If standards are appropriately drawn and if means are available for determining exactly what subordinates are doing, appraisal of actual or expected performance is fairly easy. But there are many activities for which it is difficult to develop accurate standards, and there are many activities that are hard to measure. It may be quite simple to establish labour-hour standards for the production of a mass produced item, and it may be equally simple to measure performance against these standards, but if the item is custom made, the appraisal of performance may be a formidable task because standards are difficult to set.

(c) *Correction of Deviation* : Standards should reflect the various positions in an organisation structure. If performance is measured accordingly, it is easier to correct deviations. Managers know exactly where, in the assignment of individual or group duties, the corrective measures must be applied. Correction of deviations is the point at which control can be seen as a part of the whole system of management and can be related to the other managerial functions. Managers may correct deviations by redrawing their plans or by modifying their goals. Or they may correct deviations by exercising their organising function through reassignment or clarification of duties. They may correct also, by additional staffing, by better selection and training of subordinates or by re-staffing measures.

Purpose of Controlling
- To measure progress
- To uncover deviation
- To indicate corrective action

Characteristic of Controlling
- It is a continuous and never ending process

- It is forward looking
- It is a dynamic process
- It does not curtail the rights of the individuals

Forms of Controlling

- Direct control or Indirect control
- Physical control or Financial control
- Policy control or Quality control or Inventory control or Overall control

Control as a Feedback System : Managerial control is essentially the same basic control process as that found in physical, biological and social systems. Many systems control themselves through information feedback, which shows deviation from standards and initiates changes. In other words, systems use some of their energy to feedback information that compares performance with a standard and initiates corrective action. Management control is usually perceived as a feedback system similar to that which operates in the usual household thermostat. (Fig. 8.3).

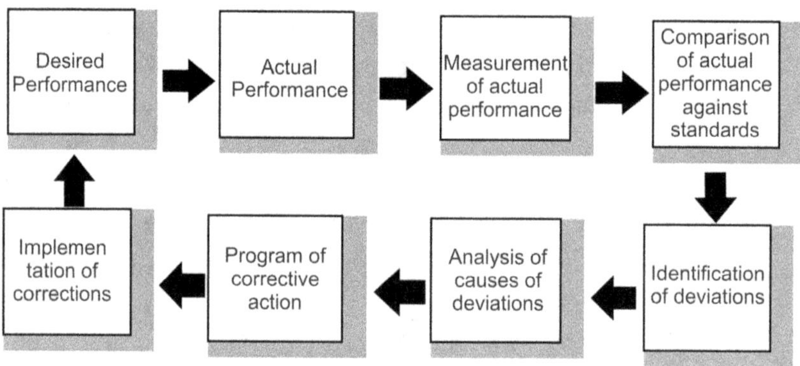

Fig. 8.3 Feedback loop of management control

G. ***Decision Making*** : Whatever the manager performs, the manager does through making decisions. A manager is oriented towards making decisions rather than towards performing the actions personally the actions are carried out by other subordinates. Decision-making can be defined as a course of action consciously chosen from available alternatives for the purpose of achieving a desired goal.

Characteristics of Decision-Making

- It involves choices
- It is a mental process at the conscious level i.e., emotional, non-rational etc.

- It is influenced by sub-conscious factors
- It is purposive in nature i.e. to facilitate the attainment of some objectives
- Initiatives or facts or experiences or authority are the elements in decision-making

Decision making by Individuals : Human beings make decisions that affect their own actions. Managers are chiefly concerned with making decisions that will influence the actions of others. Thus, the environment of the decision makers and the role that they assume affects the decision-making process of management. (Fig. 8.4).

However, for convenience, the decision-making process may be described in five steps:

- A good decision depends on the maker's being consciously aware of the factors that set the stage for the decision.
- A good decision is dependent upon the recognition of the right problem i.e. the use of concept of the limiting factor.
- Search for and analysis of available alternatives and their probable consequences is the step most subject to logical and systematic treatment.
- Even in the best-designed framework of alternatives, the crucial step involves: selection of the solution.
- A decision must be accepted by the organisation.

Fig. 8.4 Decision making process

Decision Making by Group : In a cooperative endeavour, one person may appear to have made the decision but in fact may have performed only one step in the process. Committees usually make decisions in a social environment. A committee may be defined as any group interacting in regard to a common, explicit purpose with formal authority delegated from an appointing executive.

Purpose of Committees

- For fact finding, investigation and collection of information
- To avoid the appearance of arbitrary decision and to secure support for a position
- To make a decision
- To negotiate between conflicting positions taken by opposing interests
- To stimulate human beings to think creatively, to generate ideas and to reinforce thoughts
- To provide representation for important elements of an organization.
- To coordinate different parts and sub-groups of an organisation.
- To train inexperienced personnel through participation in groups.

Advantages of Committees

- A decision can be approached from different viewpoints by individual specialists
- Coordination of activities of separate departments can be attained
- Motivation of individual members to carry out a decision may be increased
- It provide a means by which executives can be trained in decision making
- It permits representation of different interest groups
- Group discussion is one of the methods of creative thinking

Disadvantages of Committees

- Considering the values of the time of each individual member are costly
- The length of time required to make a decision is more
- Group action may lead to compromise and indecision
- A superior line may dominate the decision making process
- Group decisions may be reached by a method in which no one is held responsible for a decision. 'Buck passing' may result

Organization

It is the structure or framework on which the management of the enterprise rests; or organization is a social system that is deliberately established for achieving certain predetermined goals and is characterised by prescribed roles, authority, structure and formally established system of rules and regulations to govern the behaviour of its members (Rogers and Shoemaker, 1971).

Characteristics of Organisation

1. It has clearly defined limits (organization may seek new goals to justify their existence).
2. Membership is formal, voluntary and motivated by individual interest.
3. Membership may involve restrictive qualifications and certain minimum requirements.
4. Membership grants certain rights, principles and benefits.
5. It has its own administrative structure with roles and functions clearly defined and prescribed.
6. Organisations have operative principles, procedures and goals.
7. The rules and regulations of an organization for maintaining conduct and behaviour of members.
8. Number of organisations increases with natural calamities and other conditions causing crisis in society.
9. Organisation enable a group of persons sharing a common interest in society to associate with one another.
10. Organisations may influence social decisions and stimulate social change.
11. Organisations may act as testing ground for new programmes or projects.

The development and growth of organisations generally follows a pattern involving four stages. They are:

(i) *Period of Cultivation :* Idea is stimulated any interested person, interest is cultivated and enthusiasm generated.

(ii) *Period of Formation :* After sufficient stimulation and cultivation of interest among people a meeting of all interested persons is held to structure/ form the organisation.

(iii) *Period of Normal Functioning :* In this period, the organisation operates in accordance with prescribed rules and regulations to achieve its purpose.

(iv) *Period of Decline and Passing :* When an organization no longer meets the interests, purpose and objectives for which it was created. The bond holding together members weakens.

Organisational Theories

A. *Need-Want-Satisfaction Theory* : The felt needs give rise to wants or goals sought, which cause tensions, which give rise to actions toward achieving goals, which finally result in satisfying actions. Needs do cause behaviour, but needs also may result from behaviour. Satisfying one need may lead to a desire to satisfy more needs. For example, a person's need for accomplishment may be made keener by the satisfaction gained from achieving a desired goal or it may be dulled by failure. (Fig. 8.5).

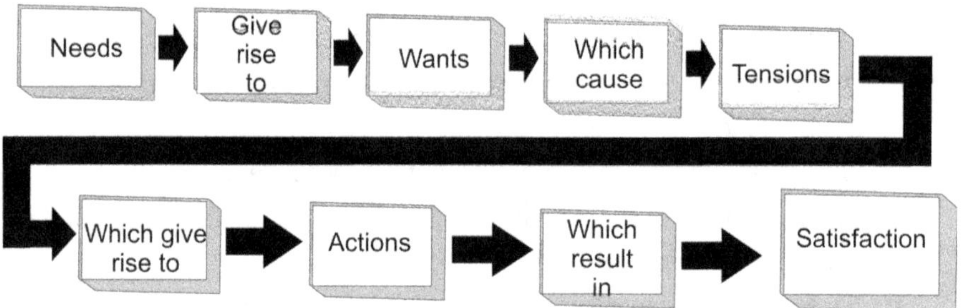

Fig. 8.5 Need-want-satisfaction chain.

B. *Hierarchy of Needs Theory or Motivational Theory* : One of the most widely mentioned theories of motivation is the hierarchy of needs theory put forth by psychologist Abraham Maslow (Fig. 8.6). The basic human needs placed by Maslow in an ascending order of importance, such as:

(i) *Psychological Needs :* These are the basic needs for sustaining human life itself, such as food, shelter, clothing and sleep.

(ii) *Security or Safety Needs :* These are the needs to be free of physical danger and of the fear of losing a job, property, food or shelter.

(iii) *Affiliation or Acceptance Needs :* Since people are social beings, they need to belong to be accepted by others.

(iv) *Esteem Needs :* According to Maslow, once people begin to satisfy their need to belong, they tend to want to be held in esteem both by themselves and by others. This kind of need produces such satisfactions as power, prestige, status and self-confidence.

(v) *Need for Self-Actualisation :* Maslow regards this as the highest need in his hierarchy. It is the desire to become what one is capable of becoming to maximize one's potential and to accomplish something.

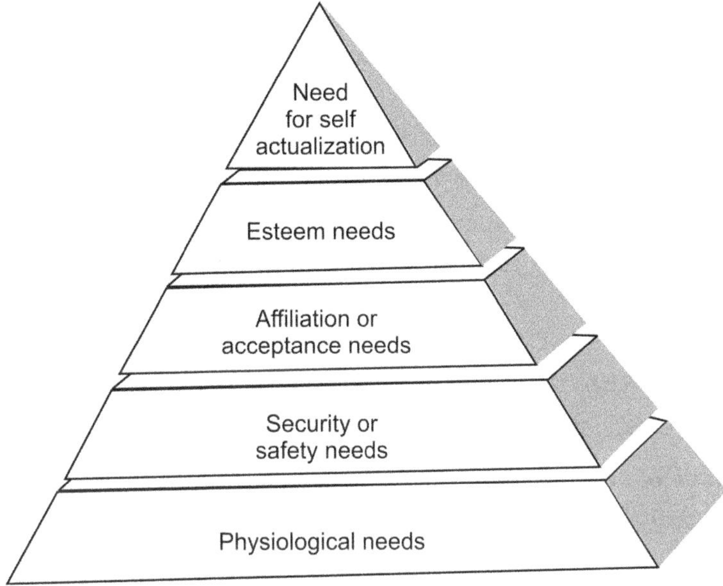

Fig. 8.6 Maslow's hierarchy of needs

C. *Herzberg's Theory* : Maslow's need approach has been considerably modified by Frederick Herzberg and he purports to find a two-factor theory of motivation such as; Motivation-Hygiene theory and Motivation-Maintenance theory. In the first group of factors (Dissatisfiers) will not motivate people in an organisation; yet they must be present, or dissatisfaction will arise. While the second group factors (Satisfiers) are the motivators, which include achievement, recognition, challenging work, advancement and growth in job.

D. *Vroom's Expectancy Theory* : People will be motivated to do things to reach a goal. If they believe in the worth of that goal and if they can see that what they do will help those individuals in achievement of the goal.

$$\text{Force} = \text{Valence} \times \text{Expectancy}$$

Where, force is the strength of motivation; Valence is the strength of a person's performance for an outcome and expectancy is the probability that an particular action will lead to a desired outcome.

E. *Skinner's Reinforcement Theory* : Psychologist B.F. Skinner developed an interesting but controversial technique for motivation, this approach is called positive reinforcement or behavioural modification, holds that individuals can be motivated by proper design of their work environment and praise for their performance and that punishment for poor performance produces negative results.

Principles of Organization

1. It should reflect to a virile awareness of the dynamic goal of the enterprise.
2. It should reflect to adequate decentralisation of powers.
3. It should reflect to unity of command.
4. It should reflect to line of responsibility.
5. It should reflect to interrelationship of human elements within an enterprise.
6. The channels of communication should be free, easy and logical.
7. The job chart should avoid overlapping of functions and authority.
8. The organisational tree like structure has properties of a living organism.

Functions of Organization

1. Divide the work.
2. Design the job structure.
3. Define the responsibilities.
4. Delegate the authority.
5. Establishing structural relationship to secure coordination.

Table 8.1 Difference between formal and informal organizations

S.No.	Characteristics	Formal organization	Informal organization
1	General Structure	Official	Unofficial
2.	Major concept	Authority	Power and Politics
3.	Primary focus	Position	Person
4.	Source of leader power	Delegated by management	Delegated by the group
5.	Guidelines of behaviour	Rules	Norms
6.	Sources of control	Rewards and Penalties	Sanctions

Social Organisation

It is a condition an a society, in which the various institutions function in accordance with their implied purpose. Without a general basic issue, a society can't continue to exist. All the societies have ethos i.e., shared values, objectives, preferences etc., and the society can exist only when a great number of people consider great number of things from the same point of view and when they hold same opinion on any subject. Following are the few organizations that execute work in rural areas.

Family Organization : Joint family is the common type of organization, most commonly seen in India.

Caste System : It's the basis for formation of various organisations in rural India.

Religion Organization : Undoubtedly India is a religious country; each and every aspect of rural life is heavily loaded with religious thoughts and forms organisations accordingly.

Marriage Organization : It is still considered to be a sacrament by villagers and paves the way for formation of organizations in rural society.

Panchayats : It's the organisation at village level, provides media through which every individual had the opportunity to express himself in the progresses of the community.

Educational Organization : The significance of education in modern society can't be over-estimated and provides an opportunity to form organizations.

Occupational Organization : People who followed their parental occupation or caste occupations very strictly are now giving them up for the sake of better income and accordingly involving themselves in various organisations.

Youth Organization : Youth are the hope of nation, future citizens of the country, creative forces of the society and shoulders the aspiration of the country. Therefore they need to be organised on better lines so as to utilize their potentialities effectively and efficiently.

Women's Organisation : It is felt necessary that women need to be organized for their development.

Informal Organization : The informal organisation is a network of personal and social relations not established by formal authority but arising spontaneously as people associate with one another. Managers cannot create or abolish the informal organisation and here, power attaches to a person and the power is earned or given permissively by group members. Much emphasis was given to people and their relationships.

Characteristics of Informal Organization
- It is a loosly defined structure.
- It has less well defined goals.
- It has less conciosusness in membership.
- It has shorter life span.
- It is smaller in size.
- It has no clearly identifiable beginning.
- It is flexible.

Rural Organizations

Rural organisations (member-run and financed rural cooperatives, agricultural producer and rural workers associations, rural credit unions, women and youth associations and other self help groups) are important forms of rural social capital that empower collective self-help action that makes rural development happen. Acting through their own organisations, small-scale rural producers and workers and the poor access inputs, markets and government services more efficiently aimed at improving their livelihoods and undertake other self-help action to improve their communities. Government and NGO agencies also reach more people and deliver services more effectively. But there are socio-political advantages as well. Through groups, rural people elect their own leaders, mobilise their own resources to improve their livelihoods and their communities and learn the value of cooperation. This reduces the risk of conflict and contributes to the improved local governance and the growth of more stable and democratic institutions serving the interests of rural people.

The Sustainable Development Department assists the governments' own efforts to promote decentralisation, foster farmers' and rural people's organisations and strengthen their capacity to participate, at local, regional and national levels, in defining and implementing rural development policies and programmes, as well as in the sustainable management of natural resources. Assistance, in the form of policy advice, guidelines and training materials, is provided in the three major areas:

- Rural cooperative development: policy advice and training on cooperative reform, cooperative business management, cooperative finance and business information systems.
- Self-help group development: assistance and training on group formation, enterprise management, savings and inter-group associations.
- Promotion of partnerships with rural organizations, NGOs and government to improve information exchange and cooperation for achieving sustainable rural development and food security.

Administration

The word administratoin derived from a latin word 'Ad minister', which means to serve. The management of public affairs of an institution is called as administration. Most of the modern, large-scale organization or institutions have an administrative form called 'Bureaucracy'.

"The Basic Features of Administration are Authority and Power"

Concept Relationship

First view : Perceived as management and administration are different by Sheldom, Schulture, Spriegel and Tred.

Second view : Perceived as management included in administration by Brech.

Thrid view : Perceived as management and administration are same by Henry Fayol.

Functional Elements of Administration

Administrators, broadly speaking, engage in a common set of functions to meet the organisation's goals. The idea of a set of standard administrative functions carries back to Luther H. Gulick, who in 1937 established the acronym POSDCoRB (pronounced "poz dee korb") which stood for planning, organizing, staffing, directing, coordinating, reporting, and budgeting.

Planning is deciding in advance what to do, how to do it, when to do it, and who should do it. It bridges the gap from where the organization is to where it wants to be. The planning function involves establishing goals and arranging them in logical order. Administrators engage in both short-range and long-range planning. Planning has both symbolic and functional value. The resulting plan provides standing information to members/employees of the organization, and it convinces stake holders to buy into the organization's goals.

Organising involves identifying responsibilities to be performed, grouping responsibilities into departments or divisions, and specifying organisational relationships. The purpose is to achieve coordinated effort among all the elements in the organization. Organizsing must take into account delegation of authority and responsibility and span of control within supervisory units.

Staffing means filling job positions with the right people at the right time. It involves determining staffing needs, writing job descriptions, and recruiting and screening people to fill the positions.

Directing is leading people (see Leadership) in a manner that achieves the goals of the organisation. This involves proper allocation of resources and providing an effective support system. Directing requires exceptional interpersonal skills and the ability to motivate people. One of the crucial issues in directing is to find the correct balance between emphasis on staff needs and emphasis on production.

Coordination this function's purpose is to interrelate various parts of work and eliminating the overlapping of functions and authorities to aviod conflict.

Reporting this function's purpose is to keep informed both the superiors and subordinates about the organisation, its activites and functions.

Budgeting excepted from the above list, incorporates most of the administrative functions, beginning with the implementation of a budget plan through the application of budget controls.

Table 8.2 Comparision between management and administration.

S.No.	Management	Administration
1.	It is the art of getting things done through people	It is the structure or framework on which the management of the enterprise rests
2.	Relatively narrower term	Broader term
3.	Specific in nature i.e., objectives are specific	General in nature i.e. with long term objectives
4.	Flexible	Rigid
5.	Not so many rules	Rules bound
6.	Major concern is input-output relationship (Man, Money and Material)	Major concern is task achieving capabilities
7.	It is used during execution of function	It is used for the execution of function.

Principles of Good Administration

1. Definite and clear cut responsibilities should be assigned to each executive.
2. Each administrative head should have a single responsible head.
3. Responsibility should always be coupled with corresponding authority.
4. No executive should receive orders from more than one source.
5. No order should be given directly to subordinate by passing their immediate supervisors.
6. Closely related individuals should be grouped together for better efficiency in coordination.
7. People under a supervisor should be limited for effective supervision.
8. Criticism of subordinates should be made privately but not in the presence of employees of lower or equal ranks.
9. Rewards must be based on objective assessment and given in presence of all the employees.
10. Authority should ensure that, the employees act in accordance with the plan of the organization.
11. All administrators should understand and follow the art and science of human relations in the management of people.
12. No executive or employee should expect to be an assistant to others of equal or higher rank.

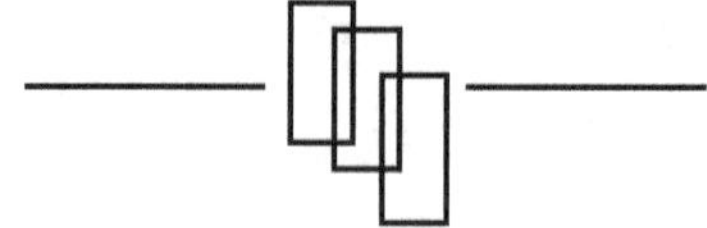

CHAPTER - 9

Adoption and Diffusion of Innovations

The major goal of extension is to get new and sustainable technologies adopted by the rural community. An understanding of adoption and diffusion processes will help the extension workers to accelerate the process. This chapter has been developed mainly on the basis of definitions, principles and concepts of Everett M. Rogers (1983) and Rogers and Shoemaker (1971).

Adoption: It is the mental process through which an individual passes from first hearing about an innovation to final adoption, or it is a decision to make full use of an innovation as the best course of action available.

Diffusion: It is the process by which an innovation is communicated through certain channels over time among the members of a social system.

Innovation: It is an idea, practice or object that is perceived as new by an individual or other unit of adoption.

Adoption Process

Adoption is essentially a decision making process (Johnson and Haver, 1955), which may be divided into a sequence of stages with a distinct type of activity occurring during each stage. The sequential stages are as follow:

1. Observing the problem
2. Making analysis of it
3. Deciding the available courses of action
4. Taking one course at a time and
5. Accepting the consequences of the decision

The adoption process has five stages, which has received world-wide attention and they are:

Awareness Stage: At the awareness stage the individual is exposed to the innovation but lacks complete information about it. The individual is aware of the innovation, but is not yet motivated to seek further information. The primary function of the awareness stage is to initiate the sequence of later stages that lead to eventual adoption of the innovation.

Interest Stage : At the interest stage the individual becomes interested in the new idea and seeks additional information about it. The individual favours the innovation in a general way, but he has not yet judged its utility in terms of his own situation. The function of the interest stage is mainly to increase the individual's information about the innovation. The cognitive 'knowing' component of behaviour is involved at the interest stage. The individual is more psychologically involved with the innovation at the interest stage than at the awareness stage. Previously, the individual listened to or read about the innovation; at the interest stage he actively seeks information about the idea. His personality and value, as well as the norms of his social system or groups may affect where he seeks information, as well as how he interprets this information about the innovation.

Evaluation Stage : At the evaluation stage the individual mentally applies the innovation to his present and anticipated future situation and then decides whether or not to try it. A sort of 'mental trial' occurs at the evaluation stage. If the individual feels the advantages of the innovation outweigh the disadvantages, he will decide to try the innovation. The trial itself, however, is conceptually distinct from the decision to try the new idea. The evaluation is probably least distinct of the five adoption stages, and one of the most difficult from which to question respondents. The innovation carries a subjective risk to the individual. He is unsure of the results, and for this reason, a reinforcement effect is needed at the evaluation stage to convince the individual that his thinking is on the right path. Information and advice from peers is likely to be sought at this point.

Trial Stage : At the trial stage the individual uses the information on a small scale in order to determine its utility in his own situation. The main function of the trial stage is to demonstrate the new idea in the individual's own situation and determine its usefulness for possible complete adoption. It is thus a validity test or "dry run"; the decision to use the ideas on a trial basis was made at the evaluation stage. The individual may seek specific information about the method of using the innovation at the trial stage.

Adoption Stage : At the adoption stage the individual decides to continue the full use of the innovation. The main function of the adoption stage is consideration of the trial results and the decision to ratify sustained use of the innovation. Adoption implies continued use of the innovation in the future.

These then are the stages in the mental process of accepting new ideas and practices. Individuals may go through these stages at different rates depending upon the practice itself. The complexity of the practice seems to be a major factor in determining the rate and manner with which people go through these mental stages. An innovation may be rejected at any stage in the adoption process. The individual may decide at the evaluation stage that the innovation will not apply to his situation and mentally reject it or the innovation may be rejected at the trial stage, where the individual decides that the rewards expected from adoption will not outweigh the cost and effort of doing so. Rejection may occur for less rational reasons.

Factors Affecting Adoption Process

The relative speed with which a new idea is adopted depends partially upon the characteristics of the new idea. Some factors affecting the rate of adoption are:

1. ***Cost and economic return*** *:* New practices that are high in cost generally tend to be adopted more slowly than do the less costly ones. However, regardless of cost, practices which produce high returns for money invested tend to be adopted more rapidly than those which yield lower returns. Also, practices producing quick returns on investment tend to be more rapidly adopted than those which produce deferred returns or returns over a long period of time.

2. ***Complexity*** *:* New ideas that are relatively simple to understand and use will generally be accepted more quickly than more complex ideas. For example, increased fertilizer application is likely to be more readily accepted than an innovation in fertilizer application methods.

3. ***Visibility*** *:* Practices also vary in the extent to which their operation and results are easily seen or demonstrated. The more visible the practice and its results, the more rapid is its adoption.

4. ***Divisibility*** *:* Practices such as fertiliser application, different fertiliser analysis, feed additives, weed sprays, or seed varieties may be tried on a sample basis and the results compared with those from previous practice. A practice that can be tried on a limited basis will generally be adopted more rapidly than one that cannot.

5. ***Compatibility*** *:* A new idea or practice that is consistent with existing beliefs will be accepted more rapidly than one that is not.

Innovation Decision Process

Rogers defines the innovation-decision process as the process through which an individual or other decision making unit passes from first knowledge of an innovation to, forming an attitude towards the innovation to a decision to adopt or reject the new idea. This process consists of a series of actions and choices over time through which an individual or an organisation evaluates a new idea and decides whether or not to incorporate the new idea into the ongoing practice. The innovation-decision process is conceptualized to have five stages, such as:

1. ***Knowledge*** *:* It occurs when an individual or other decision-making unit is exposed to an innovation's existence and gains some understanding of how it functions. Knowledge seeking is initiated by an individual and is greatly influenced by one's predispositions. A need can motivate an individual to seek information about an innovation and the knowledge of an innovation may develop the need.

2. ***Persuasion*** *:* It occurs when an individual or other decision-making unit form a favourable or unfavourable attitude towards the innovation. At this stage the individual becomes more psychologically involved with the innovation and actively seeks information about it. The individual perceives the attributes the of innovation, which is conditioned by one's personality and social system norms and develops a general idea about the innovation. There may be two levels of attitudes, a specific attitude towards the innovation and a general attitude towards change. A previous positive experience helps the process and a previous negative experience develops resistance to future new ideas.

3. ***Decision*** *:* It occurs when an individual or other decision-making unit engages in activities, which lead to a choice to adopt or reject the innovation. Considering the relative advantage, risks involved and many other related factors like availability of markets, need of the family etc. The individual makes a decision to adopt or reject the innovation.

4. ***Implementation*** *:* It occurs when an individual or other decision-making unit puts an innovation into use. At this stage the individual is generally concerned with where to get the innovation, how to use it and what operational problems will be faced and how these could be solved. Implementation may involve changes in management of the enterprise or modification in the innovation, to suit more closely to the specific needs of the particular person who adopts it.

5. ***Confirmation*** *:* It occurs when an individual or other decision-making unit seeks reinforcement of an innovation-decision already made, but may reverse this previous decision if exposed to conflicting messages about the innovation. The decision to adopt or reject an innovation is not a terminal act. Throughout the confirmation stage, the individual seeks to avoid a state of internal disequilibrium or dissonance and to accomplish it by changing one's knowledge, attitude or actions.

   Adoption : It is a decision to accept an innovation

   Rejection : It is a decision not to adopt an innovation

Discontinuance : It is a decision to reject an innovation after having previously adopted it.

Over Adoption : It is a decision to continue to adopt an innovation, rather vigorously.

Attributes of Innovation

An innovation has some qualities or characteristics. But it is not the intrinsic quality, but the quality of the innovation as people see them which is important for extension. The perceived attributes of innovation, which are basic to extension, are as follows:

Relative advantage : The degree to which an innovation is perceived as better than the idea it supersedes. Multiple use of an innovation may be a form of relative advantage. The advantage of location for specific enterprises in specific areas, may provide some relative advantage. The innovations, which have more relative advantage, are likely to be adopted quickly.

Compatibility : The degree to which an innovation is perceived as being consistent with existing values, past experiences and needs of potential adopters. It has two dimensions; situational compatibility and cultural compatibility. When a new breed suits the agro-climatic condition of the farmer, it indicates situational compatibility. When a breed of livestock advocated to the farmers is in agreement with their beliefs and values, it is the cultural compatibility of an innovation that is essential for adoption.

Complexity : The degree to which an innovation is perceived as difficult to understand and use. An innovation should as far as possible be less complex for the farmers to understand and use. However, complexity of an innovation may not deter its adoption, provided it has more relative advantage, but complex technologies often require complementary adoption. Because of their complicated nature, require the consistent training and communication support for the clientele, for their adoption and continued use.

Trialability : The degree to which an innovation may be experimented on a limited basis. Earlier adopters appear to be more concerned about the trialability of an innovation than later adopters.

Observability : The degree to which the results of an innovation are visible. The visibility of an innovation facilitates its diffusion in the social system. The problem of lack of observability may however, be overcome by strengthening extension efforts like training, communication etc., which can increase one's vision and reasoning.

Predictability : It refers to the degree of certainty of receiving expected benefits from the adoption of an innovation. Farmers are reluctant to adopt any technology which introduces a higher level of uncertainty into the operation of the farm enterprise.

It may be generalized that all the attributes of an innovation, as perceived by the members of social system are positively related to its rate of adoption expect complexity of an innovation, which is negatively related to its rate of adoption.

Consequences of Innovation

Consequences are the changes that occur to an individual or to a social system as a result of the adoption or rejection of an innovation. There are three categories of consequences:

1. ***Desirable vs Undesirable :*** Consequences, depending on whether the effects of an innovation in a social system are functional or dysfunctional.

2. ***Direct vs Indirect :*** Consequences, depending on whether the changes to an individual or to a social system occur in immediate response to an innovation or as a second order result of the results of direct consequences of an innovation.

3. ***Anticipated vs Unanticipated :*** Consequences, depending on whether the changes are recognised and intended by the members of a social system or not.

Adopter Categories

In any community, the readiness to accept new ideas and to put them into practice varies from farmer to farmer. We can classify them according to their readiness to accept new ideas. The main purpose of studying this classification is to understand how people can accept the new ideas that are being taught to them and be able to select leaders who can help program their objective and follow-up actions (Fig. 9.1).

Innovators : These are farmers who are eager to accept new ideas. Usually there are only a few people in this class in the normal farming community. In village societies, innovators are often looked on with suspicion and jealousy. Yet, they are important to the success of an extension program for they can be persuaded to try new methods and thereby create awareness in the communities where they live. However, the extension worker should exercise tact and caution and avoid individual preferences publicly which could result in rejection of the idea as a whole.

1. Innovators are venturesome, desire for the rash, the daring, and the risky,

2. Innovators must control of substantial financial resources to absorb possible loss from an unprofitable innovation.

3. Innovators has the ability to understand and apply complex technical knowledge, and

4. Innovators has the ability to cope with a high degree of uncertainty about an innovation.

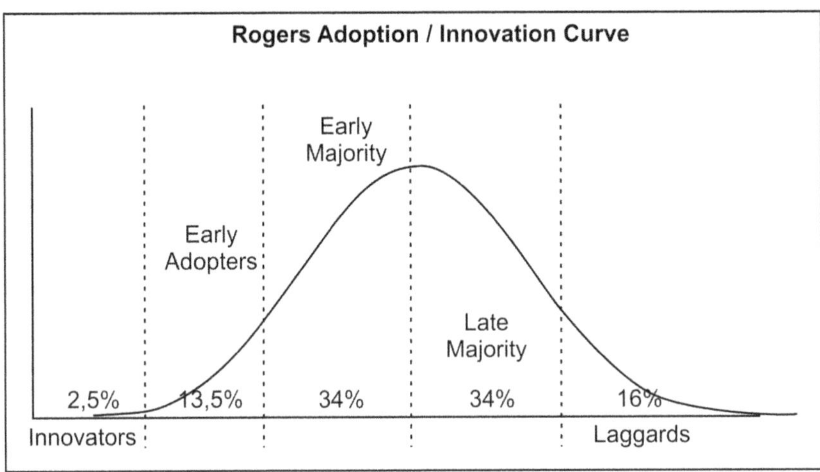

Fig. 9.1 Adopter Categories on the Basis of Innovativeness.

Early Adopters : They are literate, localite and have a large scale enterprise, high income and maintain good contact with cosmopolite sources of information and want to see the idea tried and proven first under local conditions. They express early interest but must be convinced by result of demonstrations of the direct benefit to them. Usually this group of people includes local leaders and others who are respected in the community. Their support is vital to the process of acceptance by the community.

1. Early adopters are integrated part of the local social system,
2. They possess the greatest degree of opinion leadership in most systems,
3. They serve as role models for other members or society,
4. They are respected by peers, and
5. They are the most successful.

If farmers in an area are not aware of the new idea, the innovators and early adopters must help demonstrate the accepted results for the benefit of the community.

Early Majority : They adopt new ideas just before the average members of the community. They are neither very early nor late to adopt an innovation.

They differ in the speed of adoption. Some may never adopt the idea at all and may continue to oppose it. The early majority is more likely to be influenced by the opinions of local leaders and neighbors than by the extension workers and are those from whom the members of the community seeks information about innovations.

1. Early majority interact frequently with peers,
2. They seldom hold positions of opinion leadership,

3. One-third of the members of a system, making the early majority the largest category and

4. Deliberate before adopting a new idea

Late Majority : They are cautious and skeptical and adopt new ideas just after the average members of the community. They adopt mainly because people have already adopted the innovation and getting the benefit out of it. They have low level of education, participation and depend mostly on localite sources of information.

1. They comprise one-third of the members of a system,

2. Experience pressure from peers,

3. They possess economic necessity,

Laggards : They are traditional and last to adopt any innovation. By the time the laggards finally adopt new idea, it may already have been superseded by more recent ideas, which the innovators are already using.

1. They possess no opinion leadership,

2. They are isolates,

3. Always their point of reference in the past,

4. They are suspicious towards innovations,

Rejection and Discontinuance

Of course, as Rogers points out, an innovation may be rejected during any stage of the adoption process. Rogers defines rejection as a decision not to adopt an innovation. Rejection is not to be confused from discontinuance. Discontinuance is a rejection that occurs after adoption of the innovation.

Rogers synopses many of the significant research findings on discontinuance. Many discountenances occur over a relatively short time period and few of the discountenances were caused by supersede of a superior innovation replacing a previously adopted idea. The relatively later adopters had twice as many discountenances as the earlier adopters. Previous researchers had assumed that later adopters were relatively less innovative because they did not adopt or were relatively slow to adopt innovations. This evidence suggests the later adopters may adopt, but then discontinue at a later point in time. Rogers identifies two types of discontinuance

1. Disenchantment discontinuance - A decision to reject an idea as a result of dissatisfaction with it's performance

2. Replacement discontinuance - A decision to reject an idea in order to adopt a better idea.

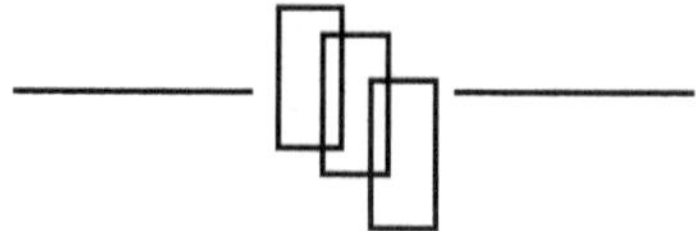

CHAPTER - 10

Programme Planning and Evaluation

Definitions

The educational efforts of any organization should be to teach persons how to think and not what to think. It is the function of the educational system to teach people to determine accurately their own needs and the solutions of their problems, to help them acquire knowledge and to inspire them to action. But it should clearly be understood that action must be their own and made of their own knowledge and convictions. The ultimate objective of extension teaching is to promote the physical, mental, spiritual and social growth of the individual farmer and his family members. This can be done by helping them in analysing their problems, in finding amicable solutions to them and in bringing about active participation in carrying out the plans. In order to obtain their participation, it is necessary that they should be involved in the preparation of the plan.

Programme : Programme is a procedure of working with the people in an effort to recognise the problems and to determine the possible solutions / objectives and goals.

Plan of work : It is an outline of activities so arranged as to enable efficient execution of the entire programme.

Project : It is an outline of procedure and pertains only to some phase of extension work.

Calendar of work : It means plan of work arranged chronologically.

Programme Planning Process

The first step in any systematic attempt to promote rural development is to prepare useful programmes based on peoples, needs. The development of such programmes, which harmonise with the local needs as the people see them and with the national interests with which the country as a whole is concerned, is an important responsibility of extension personnel at all levels national, state, district, block and village. Programme planning is the process of making decisions about the direction and intensity of extension-education efforts of extension-service to bring about social, economic and technological changes or it is a process of making decisions that will carry into the future.

'If we could know where we are now and where we ought to go then we could better judge what to do and how to do it'............

- Abraham Lincoln

The controlling objective of the National Extension Service is to develop the rural people and the first step in any systematic attempt to promote rural development is to prepare a useful programme.

Basis for Sound Extension Programme

- Without education: Unfelt needs cannot be converted into felt needs
- Without felt need: Voluntary action does not occur
- Without voluntary action: Objectives are seldom attained

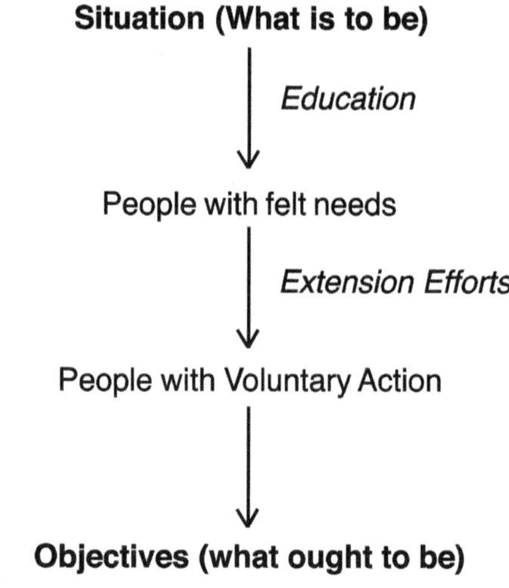

Fig. 10.1 Basis for Sound Extension Programme.

Principle of Extension Programme Planning

The planning of an extension programme is done on the basis of certain well recognised principles which should be clearly understood and followed by extension workers. The main principles are:

(i)   The programme-planning should be based upon a careful analysis of a factual situation.

(ii)   In a good programme-planning, problems for action are selected on the basis of recognised needs.

(iii)   A good programme-planning determines objectives and solutions which are feasible and offer satisfaction.

(iv)   The programme should be permanent and flexible; to meet a long-term situation, short-time changes, and emergencies.

(v)   A sound programme should have both balance and emphasis.

(vi)   A good programme has a definite plan of work.

(vii)   Programme-planning is a continuous process.

(viii)   Programme-planning is a co-ordinating process.

(ix)   Programme-planning should be educational and directed towards bringing about improvement in the ability of the people to solve their own problems individually and collectively.

(x)   A good programme-planning provides for the evaluation of results.

Steps in Programme Planning Process

Programme planning is a continuous series of activities or operations leading to the development of a definite plan of action to accomplish particular objectives. In this process they agree and feel that the goals and experiences may help them in reaching the objectives. The process has eight steps, out of which the first four steps are included in programme planning and the rest four steps are grouped under programme action. The programme planning process is a continuous one and it is better if it starts from the first step and moves to the last step. Each step has its own importance; therefore it is necessary the steps should not be overlooked while preparing and implementing the programme. In preparing, executing and evaluating the extension programme, there are nine steps. (Fig. 10.2).

1. *Collection of Facts :* Sound plans are based on availability of relevant and reliable facts. This includes facts about the village people, physical conditions, existing farm and home practices, trends and outlook. Besides, other facts about customs, traditions, rural institutions, peoples' organisations operating in the area, etc. should be collected. The tools and techniques for collecting data include systematic observations, a questionnaire, interviews and surveys, existing governmental records, census reports, reports of the Planning Commission, Central Bureau of Statistics, and the past experiences of people.

2. *Analysis of the Situation :* After collecting facts, they are analyzed and interpreted to find out the problems and needs of the people.

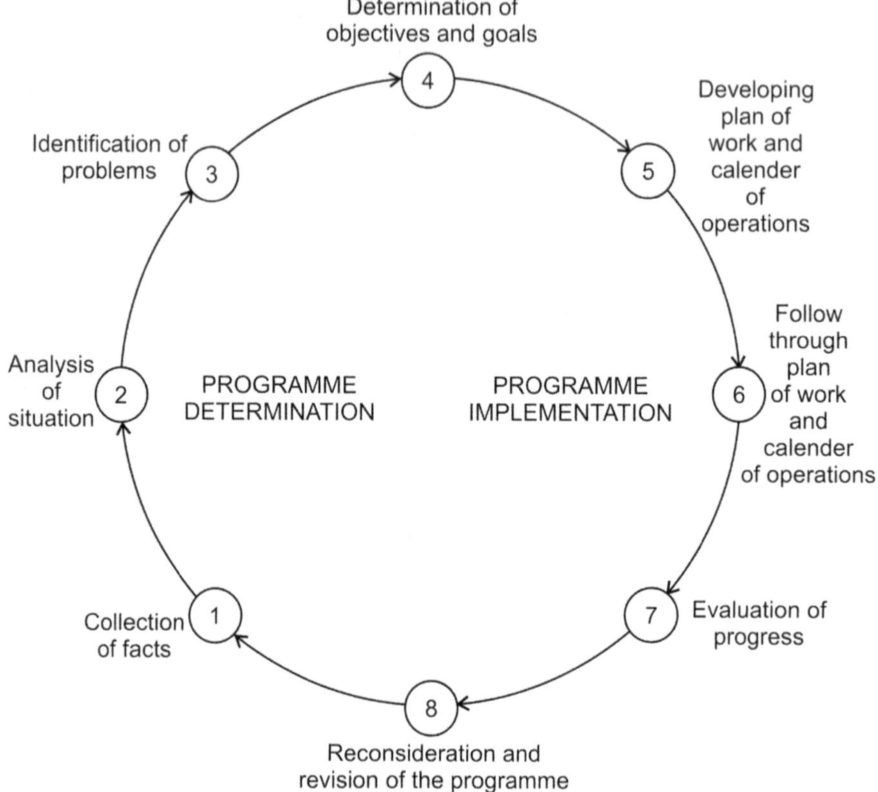

Fig. 10.2 Steps in programmes planning process.

3. ***Identification of Problems :*** As a result of the analysis of facts the important gaps between 'what is' and 'what should be' are identified and the problems leading to such a situation are located. These gaps represent the people's needs.

   - Problems which can be solved by villagers
   - Problems which can be solved by involuntary community co-operation
   - Problems which can be solved with the assistance from outside sources

4. ***Determination of Objectives :*** Once the needs and problems of the people have been identified, they are stated in terms of objectives and goals. The objectives represent a forecast of the changes in the behaviour of the people and the situation to be brought about. The objectives may be long-term as well as short-term, and must be stated clearly.

5. ***Developing the Plan of Work :*** In order to achieve the stated objectives and goals, the means and methods attaining each objective are selected and

the action plan, i.e. the calendar of activities is developed. It includes the technical content, who should do what, and the time-limit within which the work will be completed. The plan of work may be seasonal, short-term, annual or long-term.

6. ***Execution of the Plan of Work*** : Once the action plan has been developed, arrangement for supplying the necessary inputs, credits, teaching aids, extension literature etc. has to be made and the specific action has to be initiated. The execution of the plan of work is to be done through extension methods for stimulating individuals and groups to think, act and participate effectively. People should be involved at every step to ensure the success of the programme.

7. ***Evaluation*** : It is done to measure the degree of success of the programme in terms of the objectives and goals set forth. This is basically done to determine the changes in the behaviour of the people as a result of the extension programme. The evaluation is done not only of the physical achievements but also of the methods and techniques used and of the other steps in the programme-planning process, so that the strong and weak points may be identified and necessary changes made.

8. ***Reconsideration*** : The systematic and periodic evaluation of the programme will reveal the weak and strong points of the programme. Based on these points the programme is reconsidered and the necessary adjustments and changes are made in order to make it more meaningful and sound. Programmes which are accepted by the people should be expanded to the neighbouring areas and in some are not accepted by the people, research should be conducted on those areas and the reasons for their failure be ascertained.

Programme-planning is not the end-product of extension activities but it is an educational tool for helping people to identify their own problems and make timely and judicious decisions. From the above-mentioned cycle, it is clear that the planning of an extension programme comprises a logical series of consecutive steps. The procedure followed for planning extension programmes, including rural development programme. It should be noted that evaluation, decision, planning and action takes place continuously, in varying degree throughout all steps of programme planning process.

Importance of Programme Planning in Extension Work

- To ensure careful consideration of what is to be done and why.
- To establish objectives with which the progress can be measured and evaluated
- To furnish a guide against which new proposals are to be judged
- To present in written form a statement for public use.

- To have means of choosing the important one from identical problems and the permanent one from temporary changes.
- To develop both felt and unfelt needs.
- To give continuity during changes of personnel.
- To justify appropriations by public bodies.
- To avoid waste of time and money and promote general efficiency.
- To aid in the development of leadership.
- To coordinate the efforts of the different people working for rural development.

Characteristics of a Good Extension Programme

An extension programme is a carefully prepared statement written in a form that clearly sets forth the significant changes that are needed in the behaviour of people and in the conditions in which they live, to be obtained over a period of time. A good programme should have the following characteristics:

1. It should be suitable for use by the staff, planning group and other individuals or groups concerned with programme.
2. It should state the primary facts that should clearly reveal the situation on major subjects or problem areas.
3. It should state the important problems or needs identified by the staff.
4. It should state both long-term and short-term objectives for each major subject that is to be focused in programme execution over a period of time.
5. It should state the objectives of the programme clearly and meaningfully.
6. It should specify the subject matter related to each objective that is highly significant to people, socially and economically.
7. It should include a summary of the programme suited to public distribution.
8. It should be circulated by appropriate means so that all can understand its nature and objectives.
9. It should be used as a basis for developing an annual plan of work.

Programme Evaluation

The term evaluation is derived from a Latin word 'Valere' which means 'Determination of the value' i.e. 'judging the value of something'. It may be formal or informal. The formal evaluation may be defined as a process of systematic appraisal by which we determine the worth, value or meaning of something. This something in extension may be a programme, method or situation. Extension evaluation is the process of determining how well the desired behavioural changes have taken place as a result of extension educational efforts.

Principles of Evaluation

- It should give a basis for adjusting the programme like: How to judge the effectiveness of planning procedures? What modifications in the planning procedure are desirable? Who should take what responsibility? How does the present programme contribute to any long-term programme?
- It should help in determining the degree to which we set out to do.
- It should serve as a check on extension teaching methods.
- It should provide information about the people with whom we work.
- It should serve to appraise the effectiveness of organisational, administrative and supervisory procedures.
- It should provide report to the public, which in turn provides evidence to the society.
- It should give satisfaction to the local leaders and extension workers in creating a sense of accomplishment.

Important Elements of Evaluation

1. Observation or collection of information
2. Applying some standards or criteria to the observations made
3. Finally, forming some judgments or drawing some conclusions or making some decisions.

Degrees of Evaluation

Casual Everyday Evaluation : It is an informal evaluation, where we make value judgments everyday. In this evaluation, simple observations are important and we must be careful to distinguish what is actually present from what we think or see. For example, "best show I have ever seen; one of the worst foods I have ever taken". But it has certain limitations such as:

- Personal ideas used instead of standard measurements.
- Intuition and personal bias cannot be eliminated.
- No systematic plan for arriving at conclusion.
- May have only part of the information.

Self-checking Evaluation : It is an informal evaluation, where conscious attempts are made to apply principles of evaluation. For example, checking on ordinary observations, talking with others, getting other people's judgments.

Do-it-yourself Evaluation : It is an informal evaluation, which involves more careful planning, apply principles of evaluation and more systematically done. They usually require surveys and score cards.

Extension Studies: It is formal evaluation, which is more complex and involves more scientific approach.

Scientific Research: It is an formal evaluation, where experimental studies were carried out scientifically to determine cause and effect relationships. It must be factual, analytical, reliable, objective and impartial.

Uses of Evaluation

1. It helps us to determine the degree to which we are accomplishing that which we set to do.
2. It gives a basis for adjusting the programme planning.
3. It provides evidence to the community of the value of the programme.
4. It serves as a check on extension teaching methods.
5. It serves to appraise the effectiveness of organisational, administrative and supervisory procedures.
6. It provides information about the people with whom we work.
7. It provides a report to the public.
8. It gives satisfaction to the local leaders and extension workers in creating a sense of accomplishment.

Steps Involved in Evaluation

The following five simple questions in evaluation indicate the five main steps in the process.

What information do you want to get and why?

Where, when and how will you get the information?

Who will collect the information?

How will it be analysed?

What does it mean?

The above five questions can be expanded into following steps as a continuing plan of action.

1. Need for the evaluation
2. Purpose of the evaluation
3. Questions to be answered by the evaluation
4. Sources of information
5. Collecting the information
6. Selecting or constructing the tools for data collection
7. Analysis and tabulation of the data
8. Interpretation, reporting and applying the findings

Important Elements for Evaluating the Extension Work

1. *Statement of Objectives :* State the objectives of an activity to be evaluated in terms of behaviour changes in the people who are undergoing learning. For example, farmers to learn which breed is best adapted to their conditions.

2. *Source of Evidence :* Only those individuals whom you try to reach can provide proof for the success or failure. For example, those livestock farmers, who attended the demonstration meeting.

3. *Representative Sample :* Those individuals who actually provide the evidence of success must be representative of all whom you tried to approach. For example, Select the farmers from the list of dairy farmers attending the demonstration and see that each selected farmer answers the questionnaire, this is to avoid sending the questionnaire to all the dairy farmers.

4. *Appropriate Methods :* The teaching methods of obtaining evidence must be appropriate to the kinds of information being collected. For example, behaviour change is to be evaluated i.e. recorded observation of what an individual does in his/her community before and after the teaching.

5. *Reliable Questions :* Frame the questions carefully so as to obtain reliable and unbiased data. For example, while asking questions about the farmer's field tours i.e., Did you see any new breeds during your tour? If 'yes' what are they?

6. *Plan to use Results :* Decide how you will analyse and use your evaluation results before evaluation is done. For example, is the percentage of adoption of artificial insemination high or low; expected or unexpected.

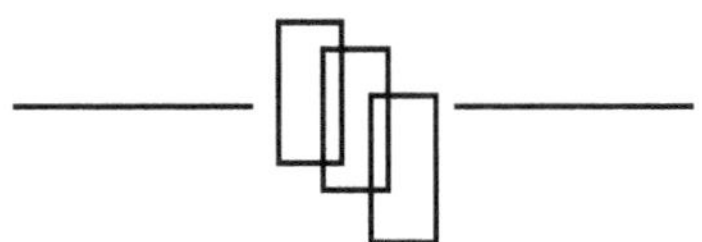

CHAPTER - 11

Leaders and Leadership

During the early stone age, man was ignorant of the art of cultivation and knew only hunting. During this stage man much like an animal, wandering in groups from one place to other and season after season, in search of food and water. Later these groups settled down at a common space of land. Then social history of man and village community started where, village administration was initiated. It was handled by some persons and then leadership developed to regulate the characteristics of man and their culture. Everyone perhaps exercises special influence over few persons, but we cannot call this as leadership. There must be both special influence and a number of people involved. It is the interaction between specific traits of one person and other traits of many persons. Leadership is a phenomenon, since every person has not only leadership traits but also the followership traits.

Leader : is a person who affectively influences a group to cooperate in achieving the goals or a person who create most effective change in group performance.

Leadership : It is an activity in which effort is made to influence the people to cooperate in achieving a goal viewed by the group as desirable. It is the direct face-to-face contact between leader and follower and is the personal and social control on the follower.

Types of Leaders

A. *Depending upon the Chief Interest :* There are several types of leaders such as political, military, business, religious etc.

B. *Whyte has Classified them into Four Categories :*

Operational Leaders: those persons who actually initiate action within the group, regardless of whether or not they hold an elected office.

Popularity Leadership: means the popular person was elected to a position of leadership because he/she was well liked by the members. Such an individual may or may not be the actual leader of the group. If such persons who hold elective positions do very little about initiating action for the group, and are mere figureheads or ornamental leaders they are called nominal leaders.

Assumed Representative Leadership : refers to a person selected to work with a committee or other leaders because the latter have assumed that he/she represents another group they desire to work with; he/she may or may not be a leader of the group.

Prominent Talent Leadership : artists and musicians who have exhibited an outstanding ability and accomplishment in their respective fields. It may include the experts and intellectual leader.

C. Depending upon the Trainings Received :

Professional Leader : is one who has received specific specialised training in the field in which he/she works full time as an occupation, and is paid for the work (e.g.Gram sevaks, extension workers)

Lay Leader : is one who may or may not received special training, is not paid for the work and generally works part-time with local organizations. (e.g. Gram sahayaks). Lay leaders are also called volunteer leaders or local leaders or natural leaders, who may be either formal leaders or informal leaders, depending upon whether they are regular office bearers of organized groups or not.

D. Depending upon Behavioural Nature : This is the most significant classification from the view point of modern research as well as practical application of the results of research. It is Designat them into the following three types:

Autocratic Leader : operates as if he/she cannot trust people such a leader thinks the subordinates are never doing what they should do and because the employees are paid to work, therefore they must work. If the leader is a benevolent autocrat, he/she may tend to view employees as children and encourage them to come to him with all their problems, no matter what the nature or the magnitude of the problem.

Democratic Leader : shares with the group members decision making and the planning of activities. Participation is encouraged. H/she works to develop a feeling of responsibilities on the part of every member of the group. He/she attempts to understand the position and feelings of the employees.

Laissez-Faire Leader : believes that if you leave workers alone, the work will be done. He/she seems to have no confidence in himself/herself. If at all possible he/she puts off the decision, often rationalises and tends to withdraw from the work group. The results of this kind of leadership are: low morale and low productivity within the work group; employees are restless and lack the incentive for teamwork; another leader arises and problems of administration, supervision and coordination are multiplied.

Roles of Leaders

1. *Group Spokesman* : The leader has the responsibility of speaking for the group and representing the group's interests and position faithfully and accurately. That means the leader is fully aware of the group's consensus of opinion and how it may or may not coincide with his individual thinking. Whenever a leader is uncertain regarding the freedom he/she has as the group spokesman, he she can ask the group to define more precisely what responsibility and privilege they are delegating to him/her.

2. *Group Harmoniser* : All social groups usually have both uniformities and differences of opinion. To maintain harmony emphasis must be placed upon the uniformities among members rather than upon individual differences. The leader is responsible for pointing out to the group, when potential conflict situations arise, that the common purpose is sufficiently worthy of cooperation that the and differences be resolved peacefully. But it should be remembered that the role of the group harmoniser is to promote harmony in line with the basic purpose of the group, not to promote harmony simply for harmony's sake. The leader should have courage of conviction so that if he/she cannot conscientiously work with the group in view of the conflict between his/her conviction and group's objectives and must let the members select another leader.

3. *Group Planner :* Generally persons are chosen for leadership positions because it is assumed that they know a little more about the problems confronting the group and their possible solutions than do the other members of the group. The authority of the leader rests upon his/her mastery of the field in which he/she is working. The group expects its leader to have new ideas for initiating activities, and to meet this expectations, the leader must be able to plan. Effective planning is closely related to the social insight one has into the structure and functioning of the group. This leadership role does not require that the leader should do all the planning, or that the leader should plan for the group. When the leader makes all the plans and announces to the group what they are, the group's reaction is that this is the leader's plan, so let him/her be responsible for its success. On the other hand if the members have helped in making the plans, then they will consider them and the group will ensure that those plans are carried out. It is admitted that it takes less time for the leader to make the plans than for the group to discuss the situation and reach a decision. On the other hand, it generally takes less time to carry out a group decision than a leader's decision.

4. ***Group Executive :*** Most groups have established methods of conducting business and achieving consensus of opinion on issues that come up before them. The leader is the one who presides when the group is conducting business. As a group executive the leader is responsible for seeing that the business of the organization is carried out on democratic principles. The leader has the role of seeing that the policies of the group agreed upon are put into action.

5. ***Group Educator :*** In many groups the person elected as a leader is one who has had more training and experience than most of the members of the group. It may be rightly assumed that the leader knows more about the work of the particular group than do most of the members. The leader should raise the level of understanding of the work of the groups and in order to do this he/she must share with the followers the knowledge and experience, such sharing of experience and insight is teaching. Group leadership depends in a large part upon teaching. Because the good teacher is not a dictator, it is possible that the best way to play group planner is by playing the role of group teacher. Learning takes time and patience. The leader needs to learn the level of understanding of followers and in this way, the followers can come to understand and appreciate the positions and views of the leader because they have been able to share with the knowledge and experience.

6. ***Group Symbol :*** All social groups have implicit or explicit norms or ideals. Persons accepted as leaders are those who have adopted these norms and live by them. The group expects its leadership to embody the ideals of the group. If a person cannot accept the ideals of the group and consistently make an effort to achieve them, leader should decline to accept the role of leadership for the organization.

7. ***Group Chairman :*** This leadership role is related to that of group executive, there is reason to indicate it as a separate role. In the recent years there has been an increased interest in group discussion., in view of this, leaders have been given training along this line.

8. ***Group Supervisor :*** Professional leaders, in addition to serving as leaders of social groups also devote a portion of their time to working with lay leaders and group organizations This role is not to take over the work of the lay leaders, but rather to serve in the capacity of advising them.

Qualities or Traits of a Leader

Following are the qualities or traits, desirable for effective leadership

1. Physical fitness

2. Intelligence
3. Sense of purpose
4. Social insight i.e. sensitivity to other person's position
5. Communication abilities
6. Love for the people
7. Democracy
8. Imitativeness
9. Enthusiasm
10. Authority
11. Decisiveness i.e. ability to make good decisions
12. Integrity
13. Teaching ability and
14. Conviction and faith

Principles of Leadership

- Leadership behaviour may be learned, because one is not born to the purpose
- Leadership in a group will not concentrate on one or few persons but it is widely spreaded throughout the group
- Leadership must be accepted by the other members of the group
- Leadership skills may be learned and imposed upon by almost every one
- Leadership is more about having a pleasing personality
- A leader in one group may not be a leader in another group
- A leader for one group in one situation may not be the leader for the same group in different situation

Methods of Identification of Leaders

Identification of responsible leaders has been stated as the starting point in working with rural livestock farmers/people. Most of these leaders operate in informal face-to-face setting and act as key communicators and decision makers at different levels of village atmosphere. Through these leaders, active participation of rural people can be brought about. Identification of these key persons is the first step in implementing as well as formulation of rural development programmes. The functional leaders are of two types such as, professional leaders and lay leaders and the methods of identifying these leaders are as follows :

Methods for identifying Lay leaders

- Discussion method
- Workshop method
- Group observation method
- Sociometry method
- Key informants
- Self designating technique
- Election method
- Seniority and past experience

Discussion method *:* In this method, person who has the ability to express his ideas and shows his willingness to participate in a programme is considered. Those persons with sound knowledge who actively participants in the discussion can be identified. Discussion gives encouragement to locate potential key communicators. Hence through this method, an individual with sound knowledge and experience can be located and identified as a leader.

Workshop method *:* In this method a large group breaks into smaller groups and the responsibility rests on each small group and through this the leader can be identified/emerges out in each group. The extension worker can spot the leader who shows the readiness to accept responsibilities for carrying out any programme.

Group observation method *:* The extension worker can select the key communicator by constant observation of a group in action without the awareness of the members of the group.

Sociometry method *:* This is the most useful method in identifying a lay leader in a village. The extension worker goes to a rural community in a village with an interview schedule to find out the persons who can guide the rural community, if the extension worker wants to introduce any improvement in the latest scientific livestock farming. He goes into the given area and asks the farmers through the interview schedule to indicate whom they consult on the matters of livestock farming for needed help. After a few such interviews he can locate the influential person or the opinion leader in the village. This method is popularly known as Sociometry test. For example, When 'F' is interviewed he may indicate that he goes to 'D' for seeking advice on livestock farming. Similarly the farmers A, B, C, and H, who are interviewed also indicate that they will go the 'D' to take advice on livestock farming. The extension worker, then can prevail that 'D' is the farmer who is considered as the influential person to take up any scientific livestock farming practices because other in the village will also be influenced by his advice and motivation.

Sociometric tests are suitable for identifying lay leaders. It is necessary that persons involved in Sociometric test should be known to one another. This method is time consuming and costly. (Fig. 11.1).

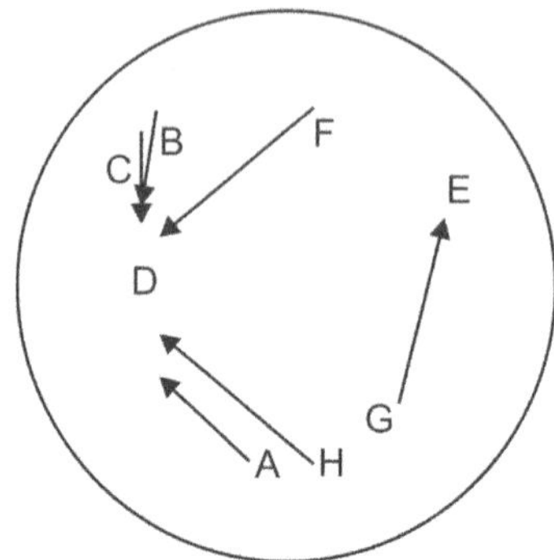

Fig. 11.1 Sociogram.

Key informants : Asking the key informants to indicate the leaders in that particular area. This method is cheap and time saving when compared to Sociometry method.

Self-designating technique : This method consists of asking a respondent a series of questions that determine the degree to which he perceives himself to be a leader.

Election method : Electing the right people for right job. Extension worker should assist the people in electing the suitable person.

Seniority and past experience : Persons with knowledge, seniority and past experience are considered to be potential leaders. Individuals with due knowledge and experience in livestock farming can be considered as potential key communicators.

Methods for Identifying Professional Leaders

- Interview method
- Battery test
- Performance test

Interview method : This is most widely used method of identifying professional leaders. Persons are identified on the basis of knowledge, past experience and professional records. A large amount of information can be acquired through interview method. But the difficulty is that, the ability of an individual has to be judged only during a brief period of time.

Battery tests : This method is based on the evaluation of the ability and aptitude of a persons to solve problems with in the field of specialization in which a person wants to be utilized for the programme. The use of such type of test along with the interview method provides a better selection for identifying leaders.

Performance test : In this test a group of 7 to 8 persons are given a common task to perform and depending upon their performance the leader is identified. It is of two types

1. *Leader less group test :* In which a group of 7-8 persons are given a common task to perform and it is left to the group to select the leaders themselves.

2. *Leader appointed group test :* This is to appoint a person as a group leader and then observe how well he directs the activities of the members of his group.

Training of Leaders

Training is the planned and systematic process by which an individual's efficiency and effectiveness in the given context of a job can be maximized. The very idea of training leaders implies that we assume leadership as something that can be taught and learned. This incidentally brings us to the two general theories like:

The Biological Theory : assumes that certain persons are born to be leaders and that these persons will be leaders, regardless of the circumstances in which they find themselves. According to this theory, great leaders have children who also become great leaders. While it is true that some families have contributed several leaders over a period of generations, but it cannot be generalized.

The Socio-Psychological Theory : This theory emphasizes that; it is the situation and not the person, alone; which makes the individual as leader. The leader is a product of his/her own times, his/her background, experience and education plus the situation in which he/she finds himself/herself, makes him/her a leader.

Objectives of Training

1. To understand and interpret group behaviour, social learning and cultural differences.
2. To learn methods of identifying and analyzing problems
3. To develop competence in group processes, cooperative thinking, exchange and analysis of ideas, facts and teaching process.
4. To acquire technical skills, necessary to carry out a job.

Training of Professional Leaders

1. *Background Courses in Educational Institutions :* Besides a general college education, a person preparing for professional leadership should take additional courses in psychology and sociology, because he/she needs a broad background of the social science approach.

2. *Induction Training :* Apprenticeship experience under the direction of a trained and experienced leader in the field will enable the new professional leader to develop his/her abilities for successful leadership.

3. *In-service Training :* After leaders have received their background and apprenticeship training and have worked for a period of time as leaders, a small number of them at a time may be brought together periodically for constantly improving their efficiency by focusing attention upon the problems they have faced in the field and the ways they have solved them. Such training programmes will facilitate exchange of information, which is highly beneficial to leaders working within the same field of work. In-service training has become increasingly important in view of fast changing technology.

Training of Lay Leaders

It is difficult to separate out methods, which apply only to formal or informal training. The methods overlap in many situations. Formal methods of leadership training are those which are structured to achieve specific goals and are usually set by someone seeking to train and develop leadership in others. Informal methods are not structured, but are those, which individual utilise, in personal leadership.

Informal Methods :

Observation : Noticing how others have performed or performing.

Reading : Studying printed material often found in the form of leader handbooks, newsletters, circulars, bulletins etc.

Talking : Speaking with other leaders in the same or related field of interest and also with members to determine consensus.

Formal Methods

Lecture : It is probably the most common method, through this local leaders under training are given enough material for thought, but little opportunity for self-expression. This method is effective, but should be supplemented by other methods.

Apprenticeship : Here, the leaders see someone operating with a view to learning some of the activities and ways of handling problems in the field of leadership. This serves as an instrument for the local leaders to acquire a better understanding of the job.

Direct Assistance from Experts : This will come in the form of advice.

Giving Responsibility to Leaders : Giving every one a job through which self-confidence may be attained by achievement in activities useful to the group is essential for development of leadership.

Advantages of using Local Leaders in Extension Work

1. Local leaders virtually play the role of extension teachers
2. Cost of extension is reduced, as local leaders are not paid for their work
3. Local leaders themselves become better trained, because of the experience they gain in teaching and influencing others.
4. People accept new ideas more readily from a local leader, who has practically tried it.
5. The frequent association of local leaders with officials enhances his prestige and personal following, which in turn increases the possibility of extensive adoption of new practices.

Limitations of using Local Leaders in Extension Work

1. Person selected as leader may not have expected following or may not be willing to devote required time to work or may be a poor teacher.
2. Considerable time is required to locate and train local leaders.
3. Local leaders may try to use prestige connected with position for personal advantage.
4. The more difficult task of arousing interest on the those not interested in extension is too often let to the inexperienced local leader.
5. Reorganization given to local leaders may sometimes jeopardize their position and adversely affect their influence among the people.

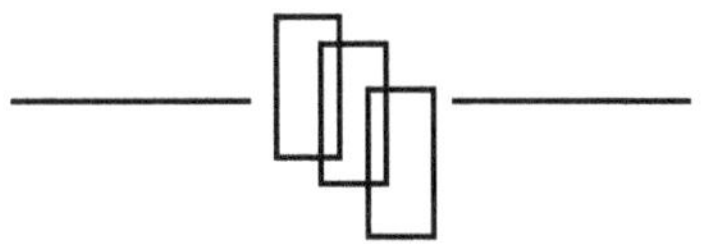

CHAPTER - 12

Livestock Development Programmes in India

Introduction

The Planning Commission was set up in March, 1950 by a Resolution of the Government of India which defined the scope of its work and in July, 1951 the Planning Commission presented a draft outline of a plan of development for the period of five years from April, 1951 to March, 1956. The Plan included a number of development projects, which had been already taken in hand, as well as others, which had not yet been begun. While the execution of development schemes which had been included in the plan after consultation with the Central Ministries and the State Governments was not to be affected

1st five year plan	:	1951-1956
2nd five year plan	:	1956-1961
3rd five year plan	:	1961-1966

Though original IV plan was drafted in 1966, it was abandoned on account of economic disturbances like two years of drought, devaluation of rupee and inflationary conditions. Instead three annual plans were implemented between 1966 - 69.

4th five year plan	:	1969-74
5th five year plan	:	1974-1979
6th five year plan	:	1980-85
7th five year plan	:	1985-90 Seventh plan got extended up to 1992
8th five year plan	:	1992-97
9th five year plan	:	1997-2002
10th five year plan	:	2002-2007

Livestock is an integral part of agriculture and has profound influence on sustainability. Apart from generating higher incomes, livestock generates employment and produces organic manure. A systematic management of livestock will have a much more positive impact on land productivity in the mixed farming context. The complementary and supplementary roles of cattle and crops under a mixed farming system through draught power for agricultural operations and rural transport, organic manure for maintaining soil fertility and fuel. Draught power provided by livestock for agricultural operations and transport, supply of dung as organic manure and the utilisation of green fodder and crop by-products for milk production are some of the links between these sectors (George, 1996).

Quantity and quality of livestock influence the soil fertility management both directly and indirectly. It contributes directly by influencing the availability of organic manure. It contributes indirectly through its influence on incomes of the households. Increased income through livestock increases the capacity of the household to invest on productivity enhancing methods through purchase of off-farm inputs. Livestock economy is changing very rapidly. The growth of draught animal stock has slowed down and milch animal stock is growing relatively fast. The proportion of cross breeds among milch animals is growing rapidly (Saxena and Sardana, 1997). Within the broad framework of policy, the Indian government has undertaken a wide variety of programs in agriculture to build up the physical and information infrastructures necessary for sustained development. There are programs for the betterment of India's rural population; research, education, and extension programmes; irrigation development schemes; plans to increase the supply of agricultural inputs, such as seeds, fertilisers, and pesticides; plans to change the institutional framework of land ownership; plans to improve agricultural financing; better marketing techniques; and plans to improve technology. These programs are administered, financed, and run by the central government and by the state governments, and both levels encourage private-sector development through direct or indirect programs.

Livestock plays a dominant role in Indian rural economy. It's of great use for cultivation, transport and the production of milk, meat, egg etc. In our country merely 35% of the national income is accounted by agriculture and contribution of livestock to agriculture income is nearly 25%. Considering the enormity of livestock in India, the percentage contribution is very meagre, this may be due to poor genetic potentiality of the stock available, inadequate fodder facilities and poor standards of managemental practices, which made the planners look into the livestock sector to extract the maximum out of it by introducing various developmental programmes.

At the time of independence, there were civil veterinary departments as different from remount veterinary corps. Provision of veterinary services was the sole activity of such departments. As the five year plans progressed, the role of livestock in rural economy was increasingly realised and was made to include a component of livestock production and development both in the curricula of veterinary colleges as well as in the mandate of civil veterinary departments. Gradually the complexion of these departments has changed so much, that they ultimately become the departments of animal husbandry with full-fledged wings to take care of veterinary care and animal husbandry. Apart from organizational changes in the set-up several programmes have been developed for livestock development and livestock as a component of rural development. The earlier plans emphasis on growth via investment in agriculture, allied activities and redistribution of income and wealth, while as in later plans the emphasis is an the programmes to tackle poverty directly through area oriented to beneficiary oriented programmes.

Dairy Development Programmes

Milk has emerged as the second largest agricultural commodity next to the rice production (1988). India ranks first in world's milk production (1996). India's milk production is 106.9 million tones (2005).

Key Village Scheme

The key village scheme was started during 1951-52, where each centre consists of three or four villages having altogether about 500 breedable population of cows over three years of age. In these areas breeding will be strictly controlled and confined to three or four superior bulls specially marked out and maintained by the farmers for the purpose. The unapproved bulls will be removed or castrated. Other essential features of cattle development, namely, maintenance of records of pedigrees and milk production, feeding and disease control, will receive full attention at every centre. The technique of artificial insemination was utilised to accelerate progress and reduce the requirements of bulls and a beginning was made by the sanctioning of 94 artificial insemination centres and about 150 artificial insemination centres at the rate of one per four key villages are proposed to be established during the entire period. The targets for the establishment of key villages, artificial insemination centres and bull rearing farms under the key village scheme during the course of four years are given below:

1. The scheme is expected to produce about 60,000 bulls per year.
2. Each Community Project will have an artificial insemination centre at a convenient place with four key villages attached thereto.
3. The improvement of common grazing grounds, the growing of fodder crops in suitable rotations, the preservations of surplus monsoon grass, the use of hitherto untapped fodder resources, etc.
4. Demonstrations under village conditions are essential

Artificial Insemination

In the year 1951, the successful application of artificial insemination brought about a revolution in livestock development in the whole developed world. India has the unique distinction among all developing countries of putting this technology to an extensive use for several decades. Initially dry ice (solid CO_2) was used as a refrigerent, later liquid nitrogen was introduced in the year 1965 then it was felt necessary to establish a centrally frozen semen bank and it was established in Karnataka in the year 1969 for processing and distribution of frozen semen. The financial and technical assistance was obtained from Danish government and the same technology was extended by Bharatiya Agro industries.

Fodder Development Programme

For the propogation of good varieties of fodder and to serve as demonstration centers. The aim of the programme is to augment the fodder production in the country and to execute the same; eight regional centers were started in various parts of the country.

1. Kalyani in West Bengal
2. Srinagar in Jammu and Kashmir
3. Suratgarh in Rajasthan
4. Hissar in Haryana
5. Gandhinagar in Gujurat
6. Alamadi in Tamilnadu
7. Hyderabad in Andhra Pradesh
8. Hissarghata in Karnataka

Main Objectives of these Regional Centers

- To demonstrate successful cultivation of good quality fodder crops suitable to that region
- To provide fodder i.e. through supply of fodder mini-kits comprising seeds, fertilizers and saplings
- To produce fodder at the regional centers as well as on willing farmers' fields or can be distributed the help of voluntary organization
    - (a) *Fodder Minikit Demonstration Programme :* This programme is implemented through ICDP, with the objective of supplying fodder minikits of various fodder seeds in Rabi Season and Kharif Season to the farmers of the State through ICDP subcentres and Veterinary dispensaries hospitals. The seeds are obtained from Central Ministry of Agriculture.
    - (b) *Intensive Fodder Development Programmes in Departmental farms:* This includes intensive fodder cultivation activities and provision for providing irrigation facilities with a view to augment the fodder production in departmental farm. It is envisaged to extend the fodder cultivation activity in the department to the unutilized areas in the farms. The fodder / grass thus produced can be distributed among the farmers for taking up fodder production in an intensive manner. Fodder slip can also be made available to the farmers.

Gosadan Scheme

The problems of unproductive cattle and its impact on the economy of the country has already been referred to. Though the availability of fodder and feed will increase to a certain extent as targets of additional production, particularly food grains, are realised, the benefit derived would not be of much consequence as the per capita increase would be insignificant. Malnutrition is to a great extent responsible for the disproportionately large number of dry animals. The census returns show that out of a total of 48 million cows over three years of age as many as 28 million are dry. The removal of useless cattle to areas of natural grazing or tracts where fodder supply is not being utilised has, therefore, been accorded a high priority in the Plan for livestock development. The Plan in this connection provides for establishing 160 Gosadans and all old and useless cattle will be segregated and sent to Gosadans located in wasteland, in forests and other out-of-the-way places where cattle grazing facilities exist which have not hitherto been utilised. At the later stages, the livestock will be castrated and remains of the dead animals such as hides, skins, horns, hoofs etc., will be fully utilized by setting up a small tannery at each centre.

Each Gosadan will maintain about 2,000 cattle. In 1952-53, therefore, the Gosadans will remove from the improvement areas 70,000 animals, in 1953-54: 220,000 and in 1954-55: 320,000. In removing cattle, preference will have to be given to areas of key villages. Surplus cattle from the key villages will, therefore, be sent to the Gosadans in the first instance. These measures by the States will, however, touch only the fringe of the problem. It is considered that this movement should receive wide public support especially from charitable institutions like Goshalas.

Aims and Objectives of Goshala

- To provide an old age home to old cows, old bulls, non-producing animals
- To ensure a gradual formation of an organic farm manure
- To ensure that all animals are well looked after, respected and loved until the last days of their life
- To extend this concept throughout the country
- To save the animals from slaughter
- To feed the hungry animals and
- To dedicate the entire concept to natural living, organic farming and green earth

Operation Flood

It is one of the world's largest rural development programmes, launched in 1970, and considered as world's biggest dairy development programme in terms of its coverage and longevity Operation Flood has helped dairy farmers direct their own development, placing control of the resources they create in their own hands. Operation Flood owes its origin to the late prime minister; Mr. Lal Bahadur Shastri, the Kaira district cooperative milk producers' union limited popularly known as AMUL (Anand Milk Union Limited). The prime minister was so convinced with the AMUL model of dairy development, that he asked Kurein to prepare a programme for replicating the model through out India. National Milk Grid links milk producers throughout India with consumers in over 700 towns and cities, reducing seasonal and regional price variations while ensuring that the producer gets a major share of the consumers' rupee. The bedrock of Operation Flood has been village milk producers' cooperatives, which procure milk and provide inputs and services, making modern management and technology available to members. Operation Flood's objectives included:

- To increase milk production
- To augment rural incomes through milk production and
- To ensure fair prices for consumers

Operation Flood was Implemented in Three Phases

Phase-I (1970-1980) : During this pahse the organization was financed by the sale of skimmed milk powder and butter oil gifted by the European Union then EEC through the World Food Programme. NDDB planned the programme and negotiated the details of EEC assistance. During its first phase, Operation Flood linked 18 of India's premier milksheds with consumers in India's four major metropolitan cities: Delhi, Mumbai, Calcutta and Chennai.

Objectives :

- To increase the capacity of milk processing facilities
- To change urban market from traditional to modern supplies, by capturing metro cities.
- To make provisions for the resettlement of city based animals
- To develop long distance milk transport and storage facilities
- To develop Anand milk procurement system
- To improve dairy farming system

Phase-II (1981-1985) : In this phase, the milk sheds were increased from 18 to 136; 290 urban markets expanded the outlets for milk. By the end of 1985, a self-sustaining system of 43,000 village cooperatives covering 4.25 million milk producers had become a reality. Domestic milk powder production increased from 22,000 tons in the pre-project year to 140,000 tons by 1989, all of the increase coming from dairies set up under Operation Flood. In this way World Bank loan helped to promote self-reliance. Direct marketing of milk by producers' cooperatives increased by several million litres a day.

Objectives :

- To cover 10 million milk producer families
- To create National Milk Herd of 14 million crossbred cattle
- To stengthen milk supply and demand centers
- To construct base structure of national dairy industry
- To increase the per capita consumption of milk to 144gms/day

Phase-III (1986-1996) : Enabled dairy cooperatives to expand and strengthen the infrastructure required to procure and market increasing volumes of milk. Veterinary first-aid health care services, feed and artificial insemination services for cooperative members were extended, along with intensified member education and has consolidated India's dairy cooperative movement, adding 30,000 new dairy cooperatives to the 42,000 existing societies organized during Phase II. Milk sheds peaked to 173 in 1988-89 with the numbers of women members and Women's Dairy Cooperative Societies increasing significantly. Phase III gave increased emphasis to research and development in animal health and animal nutrition. Innovations like vaccine for Theileriosis, bypass protein feed and urea-molasses mineral blocks, all contributed to the enhanced productivity of milch animals.

Objectives :

- To increase the coverage of milk producers
- To promote the anand pattern in the country
- To increase National Milk Grid
- To better utilisation of inputs
- To develop dairy cooperatives by improving health, environmental sanitation, nutrition etc.

The significance of operation flood is that despite several constraints, it is a successful development programme. More than 50 lakh farm families containing 50% landless and marginal; 22% small; 16% medium and 12% large farmers are getting benefit through this programme. The financial assistance of the world bank and commodity aid in the form of skimmed milk powder and butter oil has helped to fund the programme imaginatively. Although the operation flood programme has made significant positive impact on several counts and helped to enhance milk production in the country it has received baseless criticism as follows:

- Operation flood has increased India's dependency on inputs
- Operation flood is not good for urban poor but only for rural rich.

- Food aid has depressed India's production and producer's price.
- Too much hardware, too little rise in the yield

Intensive Cattle Development Project

While key village scheme was in progress it was realised that there is heavy demand for milk. In 1964, the scientists panel expressed the need for concerted efforts on large scale for rapid increase in milk production in India, including these aspirations and recommendations the ICDP was launched during 1964-65. Under ICDP it was envisaged to cover one lakh breedable population, so that it will have significant impact on milk production and efforts were made to attain the following objectives:

1. To provide controlled breeding facilities through artificial insemination for all the breedable female bovine population in the milk shed area of dairy plants with a view to increase milk production.
2. To provide balanced diet through comprehensive package programme of fodder production, setting up of feed mixing plants for distribution of feeds to milk producers through the dairy plants.
3. To protect the animals belonging to milk shed areas, against the diseases like Rinder pest, Hemorrhagic septicemia etc. to ensure safe enterprise of the farmers.
4. To provide marketing facilities for the milk produced by farmers through establishment of cooperative societies
5. To provide credit facilities for the purchase of milch animals and
6. To provide dairy extension services to make farmers more interested in scientific practices of livestock farming and to increase production incentives in milk shed areas by holding cattle shows, calf rallies and providing subsidy to the farmers

Specification of Project

- Each ICDP should cover one lakh breedable cows and buffaloes population
- It should be located in breeding tracts of indigenous breeds of cattle in milk shed areas.
- It should be linked up either with fluid milk marketing schemes or milk products manufacturing projects
- The project should be located in areas where good potential conditions existed to ensure satisfactory response to cattle improvement and milk production enhancement efforts.

Reasons for Failure of ICDP (as oriented by Anand P. Gupta)

- Considerable time lag in providing organisational structure, contents of the programme and inputs as detailed in the model plan
- Officers other than animal health specialities could hardly be associated with planning of allied activities such as: fodder production, conservation etc.
- Local conditions and changes were not sufficiently considered in planning the programme
- Dilution of inputs or cuts in financial assistance, while transferring from center to state
- 30% to 40% of the semen produced was not utilised in most of the projects
- Lack of training to stockman
- Lack of periodical evaluation of the project
- Required attention was not given to village level workers, who are most important in any livestock developmental programmes.
- No incentives were given to the best personnel for their performance in activities
- The proclaimed aim of the project is to increase milk production and its supply to dairy plants, thus the project might have named as Intensive Milk Development Project instead of Intensive Cattle Development Project

Intensive Dairy Development Programme

The programme was initiated in the year 1993-94, with the following objectives:

- Development of Milch Cattle.
- Increase in Milk Production by providing Technical Input Services.
- Procurement, processing and marketing of milk in a cost effective manner.
- Ensure remunerative prices to the Milk Producers.
- Generate additional employment opportunities.
- Improve Social, nutritional and economic status of residents of comparatively more disadvantaged areas.

Implementing Agency : State Dairy Development Departments/ Directorates of Dairy Development/ State level Milk Cooperative Federations.

Target Group/Beneficiaries : Small, marginal and landless labourers with special attention to the farmers belonging to SC/ST categories.

Technology Mission

This dairy development programme was launched in June, 1988 and started functioning from June, 1989. It aimed to accelerate the pace of rural income through dairy development, with the following objectives:

- Extending the dairy cooperative structure to 275 districts or 60% of the country.
- Increasing milk production from 44 MT in 1986-87 to 70 MT by 2000AD.
- Increasing per capita availability of milk from 158 ml in 1986-87 to 196 ml by 2000AD.
- Improving average yield of milk per animal within the project area i.e., in cows from 390 kgs to 800 kgs and in buffaloes from 910 kgs to 1100 kgs by 2000AD

Milk and Milk Products Order (MMPO)

This programme has been started by the Government of India during 1992 under the liberalisation policies. It empowers those dairy plants exceeding its utilisation of 10,000 liters per day to register with government for its modernization, product manufacturing and to collect milk in specified area.

Piggery Development Programmes

Regional pig breeding stations and model piggery units were started with the exotic pig breeds in various states over India during the second five year plan period. The regional pig breeding stations supply pure breeding stocks of exotic pigs to the farmers for further multiplication and surplus stock for slaughter and local sales.

Piggery products provide cheaper source of animal proteins and are important for improving the nutritional requirements. During the Second plan, 13 pig breeding units were set up, with a view to utilise breeding materials from these units 28 piggery development blocks were also established. In addition, 2 regional pig breeding-cum-bacon factories were established at Aligarh in Uttar Pradesh and at Haringhata in West Bengal. During the Third plan 2 more regional breeding-cum-bacon factories, 12 piggery units, 140 piggery development blocks were set up. Intensive development in this industry can make a material contribution towards raising the economic levels of several groups among the weaker sections of the village community. Out of the total slaughter rate; the percentage of meat produced from different sources are cattle and buffaloes 44, sheep and goat 27, poultry 14 and others 15. However, the slaughter rate in relation to the population are 1.45% in case of cattle, 3.45% in buffaloes, 32.5% in sheep, 35.6% in goats and 26.21% in case of pigs. In Fourth and Fifth plans, Seven bacon factories, four pork processing plants 23 pig breeding farms were established.

Objectives of Piggery Development Programmes
- To improve the socioeconomic status of the sizeable section of weaker section of the rural community
- To make available of clean and wholesome pork and pork products
- To substitute the mutton and chevon, which has increased in price
- To make available plentiful of meat at cheaper rate

Constraints of Piggery Development Programmes
- Piggery development did not take roots, since the distribution was not made in rural areas
- No remunerative market for pigs raised by the farmers
- Absence of proper animal health cover
- Lack of supply of balanced feed at economic prices
- Wrong selection of farmers and locations
- Lack of proper system of purchase of pigs
- Lack of credit facilities and market outlets
- Absence of training programmes
- Lack of scientific knowledge regarding proper upkeep of pig rearing

Poultry Development Programmes

Poultry keeping is one of the best tools available for an integrated rural development and to bring about socio-economic transformation of small entrepreneurs. No other branch of animal husbandry made such rapid strides in their development, as achieved by poultry farming. Commercial poultry development in India is barely 30 years old, although poultry raising dates back to pre-historic times. The first major step towards poultry development was taken during 1939 with the establishment of poultry research division at IVRI, Izatnagar in Uttar Pradesh.

Five Year Plans and Poultry Development

Even in First plan recognition was given to poultry as a vital tool for the socio economic uplift of the large majority in rural areas. During this plan initially 50 poultry extension centers and later on increased to 269 centres along with the establishment of five regional farms equipped with superior stock. In Second plan training programmes were organised to poultry breeders. During Third plan, Poultry farming emerged out as a vital commercial enterprise. Development of deep litter system, multiplication of exotic breeds and organization of poultry development projects, were initialised in this plan. Further, this plan provided 60 state poultry farms, 3 regional poultry farms and 50 extension cum development centers with commercial hatcheries were set up. In Fourth plan, emphasis was given to breeding better stock and popularising the latest scientific practices. Credit facilities and insurance policies were introduced for the farmers. During Fifth plan, attempts were made to improve the quality of inputs and to establish proper marketing facilities. Three regional poultry breeding farms, 3 laying testing units, 61 intensive poultry production-cum-marketing centers were established. Further 3 central farms, 14 state farms and 55 Intensive egg and poultry production-cum-marketing centers were expanded.

In Sixth and Seventh plans, all aspects of the poultry industry had been developed. Broiler farming emerged as a new wing and it occupies the pride place. Poultry layer strains like HH-260 and BH-78 and broiler strains like IBK-80 and IBB-80 were evolved. National Egg Coordination Committee (NECC) was established. This period can aptly called as 'Decades of poultry'. During Eighth plan, availability of quality chicks was encouraged through providing incentives, institutional finance and appropriate training and guidance. Strengthening of processing, marketing and storage of infrastructure facilities. In this period, National Poultry Development Board was evolved with the objectives: (i) Promote poultry [production activities, (ii) Assist the state government in organizing poultry development on cooperative basis, (iii) Provide support price for eggs through market intervention, (iv) Protect interests of producers and consumers, (v) Assist in delivery of processing and marketing inputs, (vi) Provide consultancing services to both public and private services, (vii) Conduct market surveys on various aspects of poultry production, processing, marketing and export and (viii) Liaise with national and international agencies in the field of poultry development.

Factors Responsible for Poultry Development in India
- Poultry industry gives quick returns to the farmers
- Introduction of many exotic breeds belong to layers and broilers and their availability throughout the state
- The availability of compound feeds, feed additives, medicines, vaccines etc
- Availability of knowledge regarding feeding, health coverage, scientific management from the animal husbandry departments
- Financial inputs, credit facilities and availability of insurance coverage
- Establishment of national egg coordination committee, poultry development corporation, private cooperative society.
- Poultry farming provides additional jobs, supplementary income to the weaker sections
- Flexible regulations for the import of essential inputs

Constraints in Poultry Development Programme

The major constraints limiting the growth of poultry sector in India are:
- Regional imbalance due to lack of entrepreneurship
- High concentration of poultry production in certain pockets with inadequate bio-security measures
- Inadequate infrastructural facilities such as diagnostic laboratories, specialised training centers.
- Lack of reliable data on poultry production, population and marketing intelligence
- High investment cost with decreasing profit margin
- High cost of insurance and inordinate delay in settling the claims
- Unorganised marketing network and lack of efforts to develop rural markets
- Lack of remunerative prices and
- Technology transfer, knowledge and adoption constraints

Suggestions to Overcome the Constraints

The following suggestions are given to overcome the constraints:
- Planned reduction in regional imbalances by developing entrepreneurship in all pockets of the country
- Developing regional centers to provide disease diagnostics, feed testing laboratories, training centers, promotion of consumption of poultry products
- Enhanced support of research and development on cost effective housing and feeds
- Developing effective bio-security measures
- Building up of the sound database on population, production and marketing intelligence
- Establishing integrated poultry complexes having facilities of rearing, production, health coverage and marketing
- Developing adequate storage and marketing facilities
- Development of industry friendly insurance policy/mortality fund and
- Improving the monitoring and supervision system in financial institutions

Sheep Development Programmes

Sheep development activity was undertaken as early as the early 19th century by the East India Company, which imported exotic breeds for cross-breeding with the indigenous breeds. Subsequently, with the establishment of the Imperial (now Indian) Council of Agricultural Research, research and development programmes were taken up on a regional basis; they included selective breeding within the indigenous breeds and cross-breeding them with exotic fine-wool breeds, and covered almost all the important sheep-rearing States. Major emphasis was however placed on sheep development after the country attained independence and initiated its Five-Year Development Plans.

During the Third five year Plan, a large number of sheep and wool extension centers were established, and a wool grading and marketing programme was initiated in Rajasthan. In 1962, realizing the importance of sheep in the agrarian economy, the central government established CSWRI (Central Sheep and Wool Research Institute, Avikanagar) and its regional stations, undertaken fundamental and applied research in sheep production and wool utilization and to provide post-graduate training in sheep and wool sciences. During the Fourth Plan, a large sheep-breeding farm was established in collaboration with the Australian Government, at Hissar, for pure-breeding Corriedale sheep. Corriedale stud rams are being distributed from this farm to a number of States for cross-breeding to improve wool and mutton production. Seven more such farms have been established in Jammu & Kashmir, Uttar Pradesh, Madhya Pradesh, Bihar, Andhra Pradesh and Karnataka, to produce exotic pure-bred or cross-bred rams. During the Fifth Plan, a large number of breeding farms were planned to be established in the Central and State sectors to produce genetically superior breeding stocks. It was also planned to reorganise and strengthen the existing sheep-breeding farms in the States as well as to expand and reorganize sheep and wool extension centres, and to set up scientific sheep shearing and wool-grading programmes.

The National Commission on Agriculture (NCA, 1976) reviewed the previous sheep and goat development activities and made recommendations on the approach as well as on organization with a view to implementation of various development programmes. In addition to genetic improvement, NCA laid emphasis on the provision of proper health protection, development of feed and fodder resources through silvi-pasture, and organization and extension activities for the transfer of improved sheep production technology to the farmers, and organising the marketing of live animals and wool. The breeding strategy is different for different regions of the country. In the north temperate and northwestern regions, it involves breeding for apparel wool through cross-breeding indigenous breeds with exotic fine-wool breeds. For the northwestern and central peninsular regions and Bihar, selection among better carpet-wool breeds and crossing extremely coarse and hairy indigenous breeds with exotic fine-wool and dual-purpose breeds to improve carpet-wool production and quality and mutton production has been recommended. For improving mutton production in the southern peninsular region, the strategy contemplates selection within better indigenous breeds such as Nellore and Mandya, and upgrading of inferior breeds with these two breeds. So far, there has been very little systematic emphasis on goat development. Some State governments have been distributing bucks of superior indigenous breeds, mostly Jamnapari and Beetal, or stationing them in veterinary dispensaries for natural service. The Sixth Plan evvisages the establishment of large goat-breeding farms for the production of studs of important breeds as well as breeding bucks of exotic dairy breeds to be used for cross-breeding for improving milk production. There is some emphasis on improving Pashmina production in the Ladakh area of Jammu & Kashmir, whose government has a Pashmina goat farm for the production of studs.

The State Animal Husbandry/ Sheep and Wool Departments are responsible for development programmes in their respective States; they maintain large exotic and native farms for the production of rams and are equipped to provide health protection and advice on improved management practices. Some also organize the marketing of wool and live animals. Most of the sheep development activity is now organised through intensive sheep development projects and sheep and wool extension centres. To a limited extent, under the Drought-Prone Area Programme (DPAP), sheep-breeders' cooperative societies are organized to carry out development work.

Objectives of Sheep Development Programmes

   (i)   To improve the socio economic condition of the rural population in the area of operation

   (ii)   To provide adequate employement for the youth in villages

   (iii) Provision of adequate infrastructure facilities i.e. through pasture development, exotic breeding etc.

   (iv) To bring improvement in economic traits such as mutton production and course wool production

   (v)   Distribution of improved rams to selected breeders

   (vi) Improvement of facilities in terms of health cover, extension services and research

   (vii) Improvement of marketing system of sheep

   (viii) Improvement of selected slaughter houses for hygienic mutton production

Constraints in Sheep Development Programmes

The following are found to be the reasons for the slow growth rate of sheep population and their production

- Most of the beneficiaries are selected wrongly and the unit cost was found to be inadequate

- Interference of middlemen
- Untimely assistance and diversion of the loan amount
- Absence of sufficient grazing lands
- Inadequate infrastructure in production and marketing of sheep and their byproducts
- Lack of organisation in this sector
- Lack of modernisation of abattoirs
- Lack of appropriate carcass utilisation centers

Suggestions to Overcome the Constraints

To monitor the sheep development programmes effectively emphasis should be given to the following aspects:

- Beneficiaries must be selected without any deviation from the norms prescribed by the financing agency
- Persons monitoring the various development programmes, should be given adequate training
- Unit cost should be reviewed and revised based on prevailing market price
- Rules and regulations of the financing agencies must be flexible and should not be rigid in repayment
- Frequent inspection of officials may prevent the misuse of the assistance
- Increasing demand for pasturelands may be met by converting the waste lands into pasturelands
- For effective implementation, district level monitoring cell should be started

Rabbit Development Programmes

The rearing of rabbits, especially angora rabbits in India, largely for the wool obtained from them, it has been enabled by the import in the 1960s and 1970s of rabbit stock from German, British and Russian. The Central Sheep and Wool Research Institute, Kullu region of Himachal Pradesh and its sub stations i.e. smaller enterprises like Punjab, Haryana, Kumaon and Garhwal hills of Uttar Pradesh and Sikkim have played an important part in the propagation and dissemination of rabbit stock and in providing technical back up to a large number of small entrepreneurs, as well as to backyard rabbitaries. Apart from this Central Wool Development Board has also established at Jodhpur. Under Indian conditions, wool production from angora rabbits has averaged 600-700 grams. The wool development programme will seek to promote production and processing of rabbit wool as an employment and income generating enterprise in rural areas of four regions of India, such as Himachal Pradesh, Uttar Pradesh, Sikkim and Darjeeling and cluster of beneficiaries around these centres will be provided with necessary financial and other assistance. The development objective of the wool programme is to generate employment, reduce poverty and improve standard of life of the farmers in particular in the rural areas.

The Integrated angora rabbit development project was established in 2000-2001, which aims at encouraging rabbit farming in remote hilly areas thereby increasing the production of angora wool resulting in saving of foreign exchange and generation of employment. The project will be implemented through state government organizations and NGOs, it intends to cover 50 new families.

Constraints of Rabbit Development Programmes

Following are number of factors have constrained the growth of rabbit sector:

- Lack of adequate and high yielding germplasm
- Lack of technical knowledge, which is reflected in low productivity, high mortality and poor quality wool production
- High cost of inputs
- Lack of veterinary services and health coverage and
- Lack of appropriate and standard feed

Suggestions to Overcome the Constraints

- Beneficiaries must be selected without any deviation from the norms prescribed by the financing agency
- Persons monitoring the various development programmes, should be given adequate training with regard to scientific management, balanced feeding, health coverage etc.
- Enough input supplies, credit facilities and insurance coverage may provided.

Rules and regulations of the financing agencies must be flexible and should not be rigid in repayment

Various Other Livestock Development Programmes

The central and state governments have started financing sheep industry through various sheep development programmes such as:

1. *Special livestock Breeding Programme*

   During 6th Five Year Plan, special livestock breeding programe was operating as a part of IRDP and during 1986-87, the programme was transferred to animal husbandry sector. There may not be any misunderstanding that the target group of SLBP should be below the poverty line as in case of IRDP. Here, the beneficiary could simply be from the category of small farmers, marginal farmers and agricultural labourers, where 30% of the beneficiaries should come from SCs and STs; 40% of the beneficiaries should come from women belonging to different livestock rearing sectors. In 1987-88, farmers' training component has been added, which provides short trainings of 7-10 days. It has even includes milk recording, registrations and cataloging of good quality animals.

2. *Special Animal Husbandry Programme*

It was started in the year 1978-79, to provide subsidiary income to weaker sections of society.

3. *Special Central Assistance-Sheep Production Programme*

This was introduced in the year 1983-84, where 50% of the beneficiaries must be women and with an objective; to improve input facilities and specific management of stock.

4. *Intensive health cover for sheep*

It was started in 1982 with the following objectives:

(i) Periodical deworming of sheep population, so as to augment mutton production

(ii) To help in increasing the income of economically weaker sections of the society through sheep rearing.

5. *Sheep rearing Programme*

It was started in the year 1987-88 with the financial help of STEP (Support to training and Employment Programme) by government of India, to improve the socio-economic status of rural women.

Apart from this, the assistance were given through other programmes like IRDP, DPAP, ITDP, SFDA, MFAL etc.

Veterinary Services for Cattle Development

This Scheme envisages control of livestock diseases of economic and national importance, Foot & Mouth control and to strengthen epidemiology unit of the state. Vaccination against diseases like Foot & Mouth, Rabies, Hemorrhagic Septicemia, Anthrax will be conducted. Containment vaccination will be taken up to cover the new entry and dropouts in southern region. Zero surveillance will be maintained for foot and mouth as well as for Rinderpest disease. The outlay will be utilised for State share towards staff cost, cost of Vaccination, equipments, salary of staff, office expenses etc. Poultry and Duck diseases will be controlled through vaccination to develop export potential. Collection of Data on Epidemiology of diseases for dissemination of information on Monthly / Seasonal basis will be taken up. Network with neighboring states will be linked to monitor outbreaks of that region and to initiate appropriate preventive measures in our state. Collection, compilation and analysis of epidemiological data will be undertaken. The staff cost, office expenses etc of the epidemiological wing will be met from this provision. Replacement of condemned vehicles, which are used for this project will be undertaken from this outlay. Required software on Epidemiological data will be purchased and installed. Screening of Brucella and Tuberculosis will be undertaken. Facilities will be developed for Rabies Tissue Culture Vaccine production. Project on Control of Rabies will be initiated. Strengthening of diagnostic laboratories and vaccine production centre will be taken up. Networking of District Veterinary Centres , regional laboratories, epidemiological unit, Directorate, district offices, will be undertaken initially. Present status – Second round of systematic vaccination has been completed in the Northern region. Containment vaccination by collecting 50 % cost of vaccination is being carried out in the southern region. In the event of outbreaks free vaccinations are carried out.

Importance and Status of Livestock in India

India has huge bovine population of 209 million cattle and 92 million buffaloes accounting for about 51% of Asia and 19% of World's population. There are 30 breeds of cattle out of which 26 are recognised and there are 13 breeds of buffaloes, out of which 7 are recognised. while, buffaloes constitute less than 40% of bovine population but account for more than half of the total milk production. Dairy sector in India provides regular employment to 9.8 million people in principle status and 8.6 million people in subsidiary status, which together constitutes 5% of the total work force. Whereas, the share of livestock to agriculture accounts to 26% i.e. 6% of the total GDP.

India has 57,494 thousands of sheep population. There are 40 breeds of sheep out of which 21 are recognised. The income from sheep rearing is estimated to be Rs. 140 crores per annum, which is based on the annual production of 132 million kgs of mutton, 44 million kgs of wool, 37 million pieces of skin and 20 million tones of manure. India has 122721 thousands of goat population. There are 20 goat breeds, out of which 19 are recognized. The income from goat rearing is estimated to be 244 crores per annum which is based on the annual production of 458 thousand metric tons of meat, 85 tousand metric tons of manure, 1500 thousand metric tons of milk and various others like skin and Pashmina.

Strengths

- Economic symbiosis of crop and livestock production
- For small and marginal farmers, cattle is perhaps the only tangible asset and mainstay for their socio-economic security
- Provides draught power, manure and cash income
- Dairy animals produce milk by converting the crop residues and by-products which other wise be wasted
- Largest number of buffaloes in the world
- India has abundant labour force which includes rural women force engaged in rearing of dairy animals
- Well established cooperative network in dairy sector, which has its roots in more than 270 districts to serve to improve the marketing efficiency

Weaknesses

- Raise in the non descriptive cows to 80% and buffaloes to 50%
- Low productivity
- Chronic storage of feed and fodder
- Poor nutritive value of available feeds
- Low fertility
- India has a bad reputation of being simultaneously the largest producers as well as the largest waster of several agricultural and livestock products

Rural Development Programmes In India

The term rural development connotes overall development of rural areas, with a view to improve the quality of life of rural people. In this sense, it is a comprehensive and multidimensional concept and encompasses the development of agriculture and allied activities, village and cottage industries and crafts, socio-economic infrastructure, community services and facilities and above all the human resources in the rural areas. Since time immemorial, India has been, still continues to be and will remain in the foreseeable future, a land of village communities. As a matter of fact, village was the basic unit of administration as far back as the vedic age. With more than 700 million of its people living in rural areas, and with the rural sector contributing about 29 per cent of its gross domestic product, no strategy of socio-economic development for India that neglects rural people and rural areas can be successful. The rural character of the economy and the need for regeneration of rural life, was stressed by Mahatma Gandhi. India is to be found not in its few cities but in its 7,00,000 villages. Rural development is therefore, an absolute and urgent necessity in India now and will continue to be so in future. It is the *sine qua non* of development of India.

Integrated Rural Development Programme (IRDP) : IRDP launched on October 2nd. 1980 all over the Country, initially it is operation in 8 blocks and later, all the 15 Blocks have been covered under the Scheme. The I.R.D.P. continues to be a major poverty alleviation programme in the field of Rural Development. The objective of IRDP is to enable identified rural poor families to cross the poverty line by providing productive assets and inputs to the target groups. The assets, which could be in primary, secondary or tertiary sector are provided through financial assistance in the form of subsidy by the government and term credit advanced by financial institutions. The programme is implemented in all the blocks in the country as a centrally sponsored scheme funded on 50:50 basis by the Centre and State. It is stipulated that at least 50 per cent of the assisted families should belong to Scheduled Caste and Scheduled Tribe categories. It is also required that at least 40 per cent of those assisted should be women under this programme. The Scheme is merged with another Scheme named Swarnajayanti Grama Swarojagar Yojana (SGSY) since 01.04.1999.

Training Of Rural Youth For Self Employment (TRYSEM) : It is a supporting component of the IRDP, started as a centrally sponsored scheme on 15 th. August,1979. It aims at providing technical and entrepreneurial skills to rural un-employed youths in the age group of 18-35 years from the families below the poverty line to enable them to take up income generating schemes. This scheme is no more in operation.It is merged with S.G.S.Y. since 01.04.1999.

Development of Women and Children in Rural Areas (DWCRA) : Was launched as a sub-component of IRDP and a centrally sponsored scheme of the Department of Rural Development with UNICEF cooperation to strengthen the women's component of poverty alleviation programmes.It is directed at raising the income levels of women of poor households so as to enable their organised participation in social development towards economic self reliance.The DWCRA's primary thrust is on the formation of groups of 15 to 20 women from poor household at the village level for delivery of services like credit and skill training, cash and infrastructure support for self employment. Through the strategy of group formation, the programme aims to improve women's access to basic services of health, education, child care, nutrition and sanitation. It is merged with S.G.S.Y. since 01.04.1999.

Supply Of Improved Tool-Kits To Rural Artisans (SITRA) : The programme is implemented as a part of IRDP. At the district level, the DRDA is the nodal agency. The scheme is formulated and circulated to all the State Governments on 20th. July,1992. Under this programme any suitable improved hand tool is to be provided. All the prudential rural artisans will be able to enhance the quality of the product to increase their production and their income and lead a better quality of life. No more in operation, it is merged with S.G.S.Y. since 01.04.1999.

Indira Awas Yojana (IAY) *:* Indira Awas Yojana (I.A.Y.) which was launched during 1985-86 as a sub-scheme of Rural Landless Employment Gurentee Programme (R.L.E.G.P.) has continued as part of J.R.Y. since its launch on April,1989. However from 01.01.1996, I.A.Y. has been made a separate scheme. The objective of I.A.Y. then was to provide dwelling units, free of cost to the members of Scheduled Caste / Scheduled Tribes and freed Bonded Labourers living below the poverty line. From 1993-94, the scheme has been extended to non-S.C./S.T. rural poor also. Indira Awas Yojana is a centrally sponsored scheme funded on cost sharing basis between the Government of India and the State Govt. in the ratio of 75:25. The cost of I.A.Y. houses have been enhanced from Rs.14,000/- to Rs.20,000/- in hilly and difficult areas.

Jawahar Rojagar Yojana (JRY) *:* Alleviation of rural poverty has been one of the main objective of the development programes. Since independence various schemes of employment generation were taken up from time to time in the country. The Eighth plan has also stressed the need for having a larger focus on the programmes aimed at giving self-employment and wage employment to the poorer section of the community. During the first four years of the Seventh Five Year Plan, two Wage-employment Programme viz., N.R.E.P.(National Rural Employment Programme) and Rural Landless Employment Guarantee Programme (R.L.E.G.P.) were in operation in the country. From 01.04.1989 i.e. last year of the Seventh Five Year Plan, these programmes were merged in to a single wage employment programme known as Jawahar Rojagar Yojana (J.R.Y.). The primary objectives of J.R.Y. are generation of additional gainful employment for the un-employed and under-employed men and women in rural areas. The secondary objectives of this programme are creation of sustainable employment by strengthening the rural economic infrastructure.

Million Wells Scheme (MWS) *:* Earlier it was a sub-scheme of JRY, is funded by the Centre and states in the ratio of 80:20. The objective of the MWS is to provide open irrigation wells free of cost to poor, small and marginal farmers belonging to SCs/STs and freed bonded labour.

Employement Assurance Scheme (EAS) *:* The Employment Assurance Scheme (E.A.S.) aims at providing wage employment in unskilled manual works to the rural poor who are in need of employment and seeking it. The secondary objective is to create economic infrastructure and community assets for sustained employment and development. The Employment Assurance Scheme for generating employment opportunities to the rural poor on an assured basis has been launched from 2nd. October 1993. The scheme is the single wage employment programme implemented at the district/block level through out the country. The scheme is operative in all the 12 Blocks of this district. A maximum of two adults per family are provided 100 days employment on an assured basis, who need and seek wage employment during the lean agriculture season. The ressources under the scheme would be shared between the Centre and the State in the ratio of 75:25 respectively. Men and women over 18 years of age and below 60 years of age normally residing in the village are covered.

National Social Assistance Programme (NSAP) : Recognizes the responsibility of the Central and state governments for providing social assistance to poor house-holds in case of maternity, old age and death of bread earner. NSAP is a centrally sponsored programme with 100 per cent central funding to the States/UTs that provides benefits under its three components viz., (i) National Old Age Pension Scheme (NOAPS); (ii) National Family Benefit Scheme (NFBS); and (iii) National Maternity Benefit Scheme (NMBS). On the basis of suggestions made by the Central Advisory Committee on NSAP, the Government has since approved changes relating to enhancement in the rate of benefits for NFBS and NMBS.

Drought Prone Area Programme (DPAP) : The Drought Prone Area Programme (DPAP) aims to mitigate the adverse effect of drought on the production of crops and livestock ,productivity of land, water and human resources. It strives to encourage restoration of ecological balance and seeks to improve the economic and social condition of the poor and the disadvantaged sections of the rural community. Now DPAP is a people's programme with Government assistance. There is a specific arrangement for maintenance of assets and social audit by Panchayati Raj institutions. Development of all categories of land belonging to Gram Panchayat, Government and individuals fall within the limits of the selected watersheds for development. Allocation is to be shared equally by the Centre and State Government on 50:50 basis Watershed Committees is to contribute for maintenance of the assets created. Utilization of 50 % of allocation under the Employment Assurance Scheme(EAS) is for the Watershed Development funds are directly released for sanction of projects and release of funds to Watershed Committees and Project Implementing Agencies. Village community including self help groups undertake area development by planning and implementation of projects on watershed basis through Watershed Associations and Watershed Committees constituted from among themselves. The Government supplements their work by creating social awareness imparting trainings and providing technical support through the Project Implementation Agencies.

Swarnajayanti Grama Swarojagar Yojana (SGSY) : The objective of Swarnajayanti Grama Swarojagar Yojana(S.G.S.Y.) is to provide sustainable income to the rural poor. The programme aims at establishing a large number of Micro-enterprises in the rural areas building upon the potential of the rural poor. It is envisaged that every family assisted under SGSY will be brought above the poverty line in a period of three years. This scheme is launched on 1st April,1999, the programme replaces the earlier Self Employment and allied programmes IRDP,TRYSEM,DWCRA,SITRA, and MWS, which are no longer in operation. The programme covers families under below poverty line in rural areas of the country within this target group, special safe guard have been provided by reserving 50% of benefits for SC/STs, 40% for women and 3% for physically handicapped persons subject to availability of funds. It is proposed to cover 30% of the rural poor in each block in the next five year. S.G.S.Y. is a credit cum subsidy programme. It covers all aspects of self-employment such as organization of the poor into self-help groups training, credit technology, infrastructure and marketing. SGSY is a centrally sponsored scheme and funding shared by the Central and State Government in the ratio of 75:25.

Jawahar Gram Samridhi Yojana (JGSY) : Jawahar Gram Samridhi Yojana (JGSY) is the restructured streamlined and comprehensive version of erstwhile Jawahar Rojagar Yojana, designed to improve the quality of life of the poor, JGSY has been launched on 1st. April,1999. The primary objectives of the JGSY are creation of demand driven community village infrastructure including durable assets at the village level and assets to enable the rural poor to increase the opportunity for sustained employment. The secondary objective is the generation of supplementary employment for the unemployed poor in the rural areas. The wage employment under the programme shall be given to Below Poverty line (B.P.L) families. JGSY is being implemented entirely at the village Panchayat level. Village Panchayat is the sole authority for preparation of the Annual Action Plan and its implementation. The programme will be implemented entirely as a centrally sponsored scheme on cost sharing basis between the Centre and the State Government in the ratio of 75:25.

Swarna Jayanti Shahari Rozgar Yojana (SJSRY) : It came into operation from 1.12.1997, sub-summing the earlier urban poverty alleviation programmes viz., Nehru Rozgar Yojana (NRY), Urban Basic Services Programme (UBSP) and Prime Minister's Integrated Urban Poverty Eradication Programme (PMIUPEP). The scheme aims to provide gainful employment to the urban unemployed or underemployed poor by encouraging the setting up of self-employment ventures or provision of wage employment. It is being funded on a 75:25 basis between Centre and the states. It comprises two special schemes i.e. The Urban Self-Employment Programme (USEP) and the Urban Wage

Employment Programme (UWEP). The scheme gives a special impetus to empowering and uplifting the poor women and launches a special programme, namely, Development of Women and Children in urban areas under which groups of urban poor women setting up self-employment ventures are eligible for subsidy up to 50% of the project cost.

Prime Minister's Rozgar Yojana (PMRY) : This programme provides self-employment to educated unemployed youth had been designed to provide employment to more than a million persons by setting up of seven lakh micro enterprises in Eighth Plan.

High-Yielding Varieties Programme : A new dimension which was created in the community development programme has been the Agricultural Production Programme, known as the High-Yielding Varieties Programme of wheat and paddy and hybrids of maize, sorghum and bajra evolved and introduced in the country since 1965-66. To start with, 100 districts were selected for this purpose, but later on it spread to other parts of the country also.

Multiple-Cropping Programme : Yet another strategy recently introduced in the country is of multiple cropping which aims at maximizing production per unit of land and per unit of time by taking three or four crops from the same piece of land in a year. This has been made possible because of the new short-duration, high-yielding varieties and improved agricultural technology.

Small Farmer's Development Agency (SFDA) : The aims of the SFDA are to identify the problems of the small farmers, to prepare appropriate programmes to overcome them and ensure the availability of inputs and credit. In all, about 50 SFDA projects have been established through out the country under the Fourth Five-Year Plan.

Marginal Farmers and Agricultural Labourers Projects : The principle objective of this scheme is to assist the marginal cultivators in making the maximum productive use of their smallholdings by undertaking horticulture, animal husbandry and dairying. Efforts are also to be directed towards bringing in larger incomes by channelling credits, improved inputs and improved practices. Under this scheme, 41 projects were scheduled to be established throughout the country in the Fourth Five-Year Plan to cover farmers having holdings of not more than one hectare and agricultural labourers having a homestead and earning half or more of their income from agricultural wages. Each Project aimed at covering about 20,000 households during the Fourth Five-Year Plan, of which about two-thirds would be from the marginal farmers and the rest from agricultural labourers.

Education and Training for Implementing Developmental Strategy: The acceptance of a progressive democratic approach based upon science and technology implied that the extension workers and the farmers in the whole process of change would work as equal partners. The training of extension workers as well as of farmers was, therefore, emphasized from the very beginning of this programme and a number of special kinds of institutions, including *gram sevaks* and *gram sevikas* training centres, orientation and study training centres, extension wings in the college of agriculture, veterinary and home science, extension educational institutes, national institute or community development, *panchayati raj* training centres, etc. were started in the country. The principles and concepts of selective areas pattern and intensive agriculture pattern threw a new challenge and an integrated training programme for the farmers, farmwomen and young farmers has recently been initiated through the country through farmer's training centres. The agricultural universities and research institutes are also playing a very important role in organizing and conducting training programmes for farmers and extension workers.

Agricultural Technology Management Agency (ATMA)

ATMA is a society of key stakeholders involved in agricultural activities for sustainable agricultural development in the district. It is a focal point for integrating Research and Extension activities and decentralising day to day management of the public Agricultural Technology System (ATS). It is a registered society responsible for technology dissemination at the district level. As a society, it would be able to receive and expend project funds, entering into contracts and agreements and maintaining revolving accounts that can be used to collect fees and thereby recovering operating cost. ATMA covers 28 districts covering seven states.

Andhra Pradesh	:	Kurnool, Prakasam, Adilabad and Chittoor
Bihar	:	Muzaffarnagar, Madhubani, Munger and Rural patna
Himachal Pradesh	:	Simla, Hamirpur, Kangra and Bilaspur
Mahastra	:	Ahmednagar, Amaravati, Aurangabad and Ratnagiri
Punjab	:	Gurdaspur, Jalandhar, Sangrur and Faridkot
Orissa	:	Khurda, Koraput, Sambalpur and Ganjam
Jharkhand	:	Chaibasa, Dumka, Palamau and Jamtara

Why ATMA?

The ATMA at district level would be increasingly responsible for all the technology dissemination activities at the district level. It would have linkage with all the line departments, research organizations, non-governmental organizations and agencies associated with agricultural development in the district. Research and Extension units within the project districts such as KVKs and the key line departments of Agriculture, Animal Husbandry and Fisheries etc. would become constituent members of ATMA. Each Research-Extension unit would retain its institutional identity and affiliation but programmes and procedures concerning district-wise R-E activities would be determined by ATMA Governing Board to be implemented by its Management Committee.

The Aims and Objectives for which the ATMA is Formed are:

1. To identify location specific needs of farming community for farming system based agricultural development

2. To set up priorities for sustainable agricultural development with a Farming Systems Approach

3. To draw plans for production based system activities to be undertaken by farmers/ultimate users

4. To execute plans through line departments, training institutions, NGOs, farmers organizations and allied institutions;

5. To facilitate the empowerment of farmers/producers through assistance for mobilization, organization into associations, cooperatives etc. for their increased participation in planning, marketing and technology dissemination.

6. To facilitate market interventions for value addition to farm produce.

7. Create suitable mechanism to ensure location specific adaptive indigenous knowledge based research.

8. Ensure adequate linkages and frequent interaction between scientists, extension functionaries and technicians and farmers, in order to prepare an integrated plan to effectuate their linkage, support each other, better understanding and appreciation of their problems, means adopted to sort out problems and plans etc., and to develop a mechanism of feed back.

9. Ensure capacity building of the ultimate users- the farmers in terms of physical, financial and skill resources base by way of adequate financial support channelised through credit institutions, private investments and training for skill upgradation.

10. Facilitate farmers' organization to take lead role on mobilizing support services and resources.

11. Facilitate the processing and marketing activities of the agricultural, livestock, dairy, poultry, silk and allied produce of the farmers with the help of private sector institutions.

12. Generate resource in order to bring financial sustainability through charging for selected services rendered to beneficiaries by ATMA.

13. Create administrative, technical, ministerial and other posts in the ATMA and make appointments thereto in accordance with the rules and regulations of the State Government.

14. Make rules and bye-laws for the conduct of the affairs of the ATMA and add to amend vary or rescind them from time to time.

15. Do all other such things as may be considered by the society (ATMA) and may be incidental or conducive to the attainment of its objectives.

Krishi Vigyan Kendras

Genesis of KVK : The KVK has conceived as vocational training institution for promoting farmers. Mohan Singh Mehta Committee (1974), specially appointed by ICAR for formulating KVK scheme, in order to acclerate the process of transfer of technology in more effective manner and then, Indian Council of Agricultural Research, New Delhi granted its sanction for establishment of the Krishi Vigyan Kendra (Farm Science Centre) in May 1992. However the actual implementation of its programmes could be started during March, 1993 only. Before launching its programmes and activities, a Benchmark survey of the selected villages was done to make a socio-economic appraisal and to understand the existing practices of the farmers. This enabled the KVK to identify the technological gaps and critical needs and requirements of the farmers. This formed the basis for framing operational modality like training, demonstration and on-farm-trials by the KVK. Simultaneously the farm development work was also started as an important requirement for the strengthening of training-cum-demonstration infrastructure of the KVK. Initially the off campus training were given emphasis due to unavailability of infrastructure. However, since 1995, the on campus training has become a core activity of the KVK. Comparatively the demonstrations under the lab to land programmes and filed level demonstrations took off on an earlier note during 1993-94.

Apart from conducting these demonstrations various innovative approaches were undertaken for providing the environment friendly packages to meet the farmers problem. An overwhelming response of the farmers to these eco-freindly practices later paved the way for forming various farmers' interest groups and self help groups. The response to these groups organized under the domain of Krishi Vigyan Mandal further motivated the KVK to establish the Innovative Farmers Club in the year 1996. The club is an informal group of self-experimenting farmers that provides an opportunity for sharing their innovations and practices among themselves. Further the KVK started the Innovative Farm Women's Club for involvement of farmwomen in the dissemination of various technological interventions at faster rate. The KVK has excelled in bringing the modern technological packages at the farmers doorstep with the help of various instructional units. The KVK today has sufficient resources to impart training skills for not only the farmers but also the rural youth. The training schedule typically incorporates the existing needs and problems of the farmers for making a positive impact. The trainings are conducted both at the on-campus and off-campus locations. It invariably emphasizes on providing both the short term as well as long durational courses specifically to impart practical orientation to these courses. The KVK has started the instructional units for not only imparting the skills but also for providing the critical inputs as per the demand and need of the farmers. Wherever the response of the farmers to technologies under the Lab to Land Programmes and the on-farm-trials demonstrated by the KVK was multifold the need to take the assistance from the other funding agencies arose. The KVK took bold initiatives in convincing different State and Central Govt. funding agencies to provide the financial support to undertake the innovative schemes and projects for further extending the extension programmes outside the purview of the KVK selected villages.

All the above programmes and activities have earned the KVK Babhaleshwar a place of pride and helped to serve as a role model for others to follow in the state as well as in the country. The KVK takes pride in putting itself among the forerunner of the frontline extension system. Its work was recognized at National level when it was bestowed with the prestigious Best KVK Award on 17 July 2000 at New Delhi among 461 KVKs across the country. 128 new Krishi Vigyan Kendras (KVKs) were established during the year 2004-05. Presently, there are more than 480 Krishi Vigyan Kendras (KVKs) and, 76 Technology Assessment and Refinement through Institution-Village Linkage Programme (TAR-IVLP) centres and 44 Agricultural Technology and Information Centres (ATICs). Through these institutional mechanisms, a number of steps were taken to bridge the gap between the technology developed at the research institutions and its adoption at the field through KVKs, TAR-IVLP centres and ATICs. The KVK scheme enunciate the following basic concepts :

(i) The kendra will impart learning through work exprience

(ii) The kendra will impart training to only three entension workers, who are already employed, to practicing farmers, farmwomen and fisherman.

(iii) There will be no uniform syllabus for a kendra.

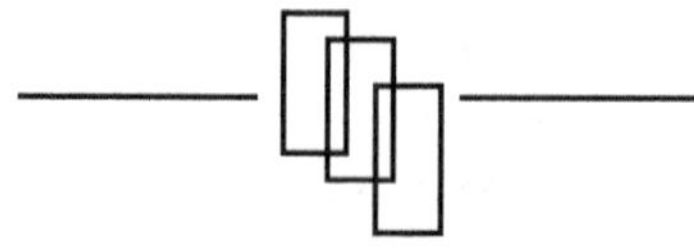

CHAPTER - 13

Cooperatives : Meaning and Objectives

Meaning : Cooperation means working together; 'co' means together and 'operation' means working. Cooperation is a voluntary concept with equitable participation and control among all involved in any business activity or enterprise. According to H. Calvert (1879), it is a form of organisation where persons voluntarily associate together on the basis of equity for the promotion of their own social and economic interests.

History : Cooperation by definition is to be a people's movement. When people began to associate with each other cooperation came into existence. The cooperative movement is a worldwide movement. In 1879 a large number of farmers in Pune and Ahmednagar areas in Maharastra rose; in open hostility against moneylenders, subsequently the Land Improvement Loan Act was passed in 1883. Our farmers have small land holdings. Intensive cropping therefore has been the way of farming and use of production inputs went to increase. Thus, input-output ratio started getting imbalanced. The need of cash increasingly felt to buy inputs in such situations helped moneylenders to exploit farmers and gave new directions to cooperation.

- Cooperative Credit Society Act, 1904 was passed to enact a separate legislation
- Cooperative Societies Act, 1925 was enacted to took care of credit and non credit societies
- Rural Credit Survey Committee was established in 1954, as an integrated scheme of rural credit involves three policies such as: state participation at different levels, coordination of credit with other economic activities and administration through trained and efficient personnel.
- Service cooperatives: are expected to cater to the needs of the farmers, towards banking, storage, supply and marketing.

Social Objectives

(i) To develop democratic leadership and

(ii) To motivate people for voluntary participation and group action

Economic Objectives

(i) To bring welfare of the society through availability of credit and for better prices to farm produce.

(ii) To avoid exploitation by middlemen.

(iii) To provide employment opportunities and distribution of essential commodities.

Educational Objectives

(i) To bring working knowledge among the members

(ii) To develop responsibility and honesty among the members

Principles of Cooperation

The cooperative principle is not something which is only a way of managing credit or marketing. It is a way of life. If you make it a way of life, you not only attempt to solve the country's problems, but also help in the solution of international problems. Cooperation is of course the basic social process. Whenever we work together or aid or facilitate each other, even at the most simple task, we are cooperating.

- *Rochdale Principle :* The modern formal cooperative movement dates to 1844, when 28 poor weavers of Rochdale came together with capital to open a small retail shop. They adopted a set of rules which were later known as Rochdale Principles and which today effectively guide the philosophy and conduct of cooperative societies all over the world.

- *Voluntary Membership :* Membership of a cooperative society should be voluntary to any person possessing the requisite qualification for being a member.

- *Democratic control :* Cooperative societies are democratic organizations, where day-to-day administration is carried out under power derived through general meetings.

- *Limited interest on share capital :* Share capital should receive only a limited rate of interest; hence dividend on share capital is restricted under Cooperative Societies Act.

- *Patronage divided :* It is based on the equity surplus, i.e. the surplus of the society should be distributed amongst its members in proportion.

- *Promotion of education :* Cooperative societies should make provision for imparting education to their members.

- *Mutuality :* Cooperative societies should actively cooperate with other cooperatives at various levels.

- *Association of person :* In cooperative societies emphasis is on man and not on the capital the member contributes.

- *It is an understanding :* Cooperative enterprise is run by members themselves at their own expense and risk.
- *The basis is equality :* Within the members, the relations between them are governed by a rule of equality.
- *It is a socio economic movement :* The cooperative movement aims at bringing about revolutionary changes in social and economic aspects of the community by peaceful means.

Dairy Cooperatives

Dairy Cooperatives account for the major share of processed liquid milk marketed in the country. Milk is processed and marketed by 170 Milk Producers' Cooperative Unions, which federate into 15 State Cooperative Milk Marketing Federations. The Dairy Board's programmes and activities seek to strengthen the functioning of Dairy Cooperatives, as producer-owned and controlled organisations. NDDB supports the development of dairy cooperatives by providing them financial assistance and technical expertise, ensuring a better future for India's farmers. Over the years, brands created by cooperatives have become synonymous with quality and value. Brands like Amul (Gujurat), Vijaya (Andhra Pradesh), Verka (Punjab), Saras (Rajasthan); Nandini (Karnataka), Milma (Kerala) and Gokul (Kolhapur) are among those that have earned customer confidence. Some of the major Dairy Cooperative Federations include:

- Andhra Pradesh Dairy Development C ooperative Federation Ltd (APDDCF)
- Bihar State Cooperative Milk Producers' Federation Ltd (COMPEED).
- Gujarat Cooperative Milk Marketing Federation Ltd (GCMMF).
- Haryana Dairy Development Cooperative Federation Ltd. (HDDCF)
- Himachal Pradesh State Cooperative Milk producers' Federation Ltd (HPSCMPF)
- Karnataka Cooperative Milk Producers' Federation Ltd (KMF)
- Kerala State Cooperative Milk Marketing Federation Ltd (KCMMF)
- Madhya Pradesh State Cooperative Dairy Federation Ltd (MPCDF)
- Maharashtra Rajya Sahakari Maryadit Dugdh Mahasangh (Mahasangh)
- Orissa State Cooperative Milk Producers' Federation Ltd (OMFED)
- Pradeshik Cooperative Dairy Federation Ltd (UP) (PCDF)
- Punjab State Cooperative Milk Producers' Federation Ltd (MILKFED)
- Rajasthan Cooperative Dairy Federation Ltd (RCDF)
- Tamilnadu Cooperative Milk Producers' Federation Ltd (TCMPF)
- West Bengal Cooperative Milk Producers' Federation Ltd. (WBCMPF)

The Anand pattern of dairy cooperatives has three tier system, starting from village to state level, they are :

I. Primary Milk Producer's Cooperative Society

Each producer from the village usually sends the milk to the cooperatives through vehicles hired by cooperatives. The members of the society elect the members of managing committee democratically: such as secretary, milk collector, fat tester, clerk, inseminator and those who will officiate for one year. It is the basic unit of a milk cooperative organisation functioning at a village level which is involved in the dairy business and dairy activities.

Profile : It includes the (i) name of the primary milk cooperative society, (ii) year of establishment, (iii) number members enrolled, (iv) office bearers, (v) business turnover, (vi) number of general body meetings conducted, (vi) profits or dividends given to members and (vii) any other social activities.

Steps Involved in Organisation and Registration of Cooperative Society

- Survey for assessing the potentiality
- Organization of Gram Sabha
- Meeting with the enrolled members
- Adoption of registered bylaws for the society
- Construction of adhoc management committee and election of chairman
- Appointment of society staff
- Opening of bank account in the name of the society
- Running the society at least for 3 months and assessing for the viability for registration
- Affiliating of the society with the district union
- Appointing local auditor

Functions:

A. *Technical/Operational Function*

- Collection of milk both in the morning and the evening throughout the year
- Testing the milk for its quality
- Despatch of milk
- Payment for milk supply either weekly or fortnightly
- Accounting
- Maintaining cleanliness
- Providing artificial insemination and health coverage
- Supply of technical inputs

B. *Managerial Function*

- Maintenance of records of milk payment and dispatch of milk regularly
- Recording total milk collected during flush or lean season
- Profits / dividends distributed to the members
- Preparation of balance sheet

C. *Extension Function* : *Providing technical information to the members*

D. *Social Development Function* : *Providing temporary loans, organizing games, cultural meets etc.*

Distribution of Profits : Out of the net profit, 25% is allocated to its reserve fund and not exceeding 12% is paid as dividend to the shareholders. Out of balance 65% is paid as bonus to the members, 10% to the cattle development fund, 10% to charity and 10% to cooperative propaganda.

II. District Dairy Cooperative Union

Each village cooperative society will be a member of the District cooperative union. The union is controlled by a board of directors of 16 to 17 members of which 12 are democratically elected representatives of the village societies and remaining five comprising the managing director, secretary, two representatives from financial institutions, a representative of the registrar of cooperative societies and a representative of the federation. The board of directors elects the chairman. One-third of the elected board members retire every year by rotation, so that each member functions for three years to ensure continuity of management.

Functions : The major functions of district cooperative union are procurement, processing and marketing of milk, providing technical inputs, strengthening of milk cooperative movement, Organisation of extension activities and rural developmental services.

Distribution of Profits : Out of the net profit, 25% is allocated to its reserve fund and not more than 12% is paid as dividend to the shareholders. Out of the balance 80% is paid as bonus to the members, 10% to charity; 5% to cooperative propaganda, 3% towards research and 2% towards dividend equalisation fund.

III. State Cooperative Federation

At state level, the federation is responsible for evolving and implementing policies on cooperative marketing, price fixing and provision of services and technical inputs to members. The federation board consists mainly of the elected chairman of all the district unions and the federation managing director. Other members are the representative of the registrar of cooperative societies, a representative of financing agency, nominee of NDDB and one nominee of the state government. All those members elect the chairman of the board.

Functions : It meets once in a month and implements policies, plans, inputs and technical services to the members.

Main Objective : Materialising the socio-economic changes with massive dairy development programmes.

General Objective

(i)   To provide assured yearround income and market for surplus milk

(ii)   To promote professional and decentralised management of dairy co-operatives and to function with profitability and trust on industrialisation of rural dairy farming.

(iii) To stimulate milk production and to develop infrastructure

Specific objective

(i)   To carry out activities for promoting production, procurement, processing and marketing of milk and products for the economic development of the farming community

(ii)   To study the problems of mutual interest relating to production, procurement, processing and marketing of dairy products

(iii) To purchase commodities from the members and sell the same

(iv)   To improve animal health care and disease control activities

(v)   To impart training among all the members of the society

(vi)   To assist the members of the union in promoting the organization, planning the developmental strategies, and function as technical, administrative and financial agents and advise the members on price fixation, public relations and allied matters

(vii) To undertake periodic supervision of the members of the union and other societies affiliated to them.

Problems of Dairy Cooperatives

- The societies have not been absolutely loyal to the unions
- Inadequate supply of milk to the plant results in under utilised capacity of plants
- Delay in the payment to the societies by dairy plants
- Adulteration of milk and milk products
- Inaccurate testing of fat by the secretary
- Intervention of middlemen
- Curdling of milk due to delay in transportation

- Cooperatives fail to increase the yield
- Fluctuations in the supply of milk

Suggestions to Overcome the Problems in Dairy Cooperatives
- Insistence of minimum quality of milk by the society
- Introduction of pasteurization facility to avoid spoilage of milk
- The societies and unions must pay attention towards breeding, fodder cultivation and purchase of concentrate feeds
- Societies should be formed within a radius of 30 miles from towns with a population of 30,000 or more.
- Societies should notify to unions at the beginning of each month about the approximate quantity of milk that can be supplied daily, time of disbursement of payment and distributional procedures of subsidy.
- Animals should be milked in the presence of the secretary
- The financial needs of the members should be met by giving loans
- Should suggest suitable milk collection centers in the village

Management of Dairy Co-operatives

The primary societies in a particular milk shed federate and form a dairy co-operative milk producers union. As more district unions were organised in Gujarat State, it was felt necessary to organize a federal body at the state level. This federal body exists to co-ordinate the overall activities of the district unions, to provide a platform for sharing common benefits, to avoid competition between the district unions and to ensure rigid quality control for the production of top quality milk products. The state federation provides the direct link between the district milk co-operatives and the National Milk Grid (NMG). The NMG co-ordinates, at the national level, the supply of milk from the surplus-producing areas to the potential urban consumer markets. It helps to moderate the seasonal and regional imbalance between demand and supply of milk. The National Co-operative Dairy Federation of India (NCDFI), a federal society, was formally established in 1970 as a national level apex organisation. Now it has been restructured through affiliation with its member apex co-operatives at the state levels. The NCDFI, thus, provides the basic institutional framework for better co-ordination, monitoring and guidance, and gives adequate direction to the state federations to ensure a stronger co-operative milk marketing system in the country. The NCDFI is the apex body of all the state dairy federations in the country, which have been entrusted with the management of the NMG activities. To facilitate operations of the NMG, four regional programming committees have been established by the NCDFI, which meet periodically in their respective regions. These committees provide a platform for the participating federations to transact business and share each other's experiences in the management of milk procurement, handling and marketing. The activities of the four programming committees are coordinated by a central programming committee. Even at the profit-sharing level, the distribution is made in proportion to the volume of business contributed by each member; therefore, bonuses etc. are determined from the value of the commodity supplied by the members. This in turn ensures that while the cooperative does business, it also makes its members quality conscious.

Gujarat Cooperative Milk Marketing Federation

The Anand pattern cooperatives have also taken into consideration the capabilities of each tier, vis-à-vis the systems they should own. These systems include the processing, marketing, advertisement and input organisation etc. along with the large capital-based operations that are owned either by the district union (the second tier) or the state federation (third tier). The primary co-operatives, on the other hand, act as procurement units for the individual members and as retail outlets for the union/federation to ensure that inputs reach the individual members at the village level on time. This two-way constant communication, between the primary unit at the village level and the district/ state level bodies, has guarded the Anand pattern co-operatives against the dangers faced by large co-operatives which tend to drift away from individuals, as they grow larger. Thus, the Anand pattern co-operatives ensure that the services required, to market the produce or to improve production, reach their members. Today in Gujarat, under the Anand Pattern system, there are 11 thousand village level co-operatives with a total membership of 2.1 million milk producers affiliated to 12 district level unions . These unions federate into a state level apex marketing organisation known as the Gujarat Co-operative Milk Marketing Federation (GCMMF). The GCMMF was established in 1973 with the objective of providing the milk producers of Gujarat with their own marketing and distribution network. This aimed to give them access to the most important link in the system-the customer. The farmers had realised that marketing was the key to the success of the Anand Pattern and to their success when they had control over the marketing system. The results are evident. Today, GCMMF is India's largest food products marketing organisation with an annual sales turnover exceeding Rs 22 billion (about US$ 483.5 million). The Amul brand is among the most popular brands in the country.

Objectives and Business Philosophy of GCMMF

The main stakeholder of GCMMF is the farmer member for whose welfare GCMMF exists. GCMMF states that its main objective is the 'carrying out of activities for the economic development of agriculturists by efficiently organising marketing of milk and dairy produce, veterinary medicines, vaccines and other animal health products, agricultural produce in raw and/or processed form and other allied produce'. GCMMF aims to market the dairy and agricultural products of co-operatives through:

- Common branding
- Centralised Marketing
- Centralized quality control
- Centralised purchases and
- Efficient pooling of milk.

GCMMF has Declared that its Business Philosophy is as Follows :

- To serve the interests of milk producers and
- To provide quality products that offer the best value to consumers for money spent.

The biggest strength of GCMMF is the trust that it has created in the minds of its consumers regarding the quality of its products. Amul stands for guaranteed purity of whatever products it produces. None of its products are adulterated. In India, where such trust has been hard to come by, this could provide a central anchor for GCMMF's future business plans.

Organisational Structure of GCMMF

GCMMF is a lean organisation, a strategy that is believed to provide it with a cost advantage. At its headquarters in Anand, four general managers (GMs) and four assistant general managers (AGMs) assist the managing director (MD). The four AGMs look after the functions of marketing, systems, co-operative services and technical projects, respectively. The four GMs are in charge of marketing (dairy products), human resources development and marketing, finance and quality assurance, respectively. The whole country is divided into five zones, each headed by a zonal manager responsible for the sales of all products within his zone. These managers report to the MD but functionally each also reports to the various AGMs/GMs at the headquarters. There are 50 sales offices spread across the country (of which only two are in Gujarat); a sales manager heads each office and is assisted by sales officers and field salespersons. The entire country has been represented in this structure. GCMMF has one overseas office in Dubai.

Facts about the Kaira District Cooperative Milk Producers' Union

- Established in 1946 - two societies collected 250 litres of milk.
- Competed with another vendor, Persons to supply milk to Bombay.
- 1952 - Bombay Government terminated Persons contract and signed with AMUL.
- 1955 - dairy and milk powder plant was established with aid from the United Nations Children's Fund (UNICEF).

- 1960 - AMUL pioneered production of milk powder and baby food from buffalo milk.
- AMUL - meets producer demand for critical inputs, veterinary services, artificial insemination and feed.
- Today AMUL members supply more than 1 million lures of milk per day.
- AMUL sells 400 tones of cattle feed every month.

The Amul Story

In the 1940s, in the district of Kaira in the State of Gujarat, India, a unique experiment was conducted that became one of the most celebrated success stories of India. At that time, in Gujarat, milk was obtained from farmers by private milk contractors and by a private company, Polson's Dairy in Anand, the headquarters of the district. The company had a virtual stranglehold on the farmers, deciding the prices both of the procured as well as the sold milk. The company arranged to collect, chill and supply milk to the Bombay Milk Scheme, which supplied milk to the metropolis of Bombay, and to cities in Gujarat. Polson's Dairy also extracted dairy products such as cheese and butter. Polson's Dairy exploited its monopoly fully; the farmers were forced to accept very low prices for their products, and the decisions of the company regarding the quality and even the quantity of the milk supplied by the farmers were final.

In 1946, inspired by Sardar Vallabhbhai Patel, a local farmer, freedom fighter and social worker, named Tribhuvandas Patel, organised the farmers into co-operatives, which would procure milk from the farmers, process the milk and sell it in Bombay to customers including the Bombay Milk Scheme. Purely by chance, in 1949, a mechanical engineer named Verghese Kurien, who had just completed his studies in engineering in the USA, came to India and was posted by the Government of India to a job at the Dairy Research Institute at Anand. Settling down in Anand was hardly a part of his career plans; however, a meeting with Tribhuvandas Patel changed his life and changed India's dairy industry. What Mr Patel requested of Dr Kurien was hardly to bring about such a revolution. All he wanted was help in solving various problems with bringing into working order some of the equipment just purchased by his co-operative, especially the chilling and pasteurising equipment. These items of equipment malfunctioned, leading to the rejection of large quantities of milk by the Bombay Milk Scheme. Dr Kurien's involvement with the Kaira District Co-operative Milk Producers' Union Limited (KDCMPUL; the registered name of the co-operative) grew rapidly. Initially he merely provided technical assistance in repairing, maintaining and ordering new equipment but subsequently he became involved with the larger sociological issues involved in organising the farmers into co-operatives and running these co-operatives effectively. He observed the exploitation of farmers by the private milk contractors and Polson's Dairy, and noted how the co-operatives could transform the lives of the members.

The most important feature of these co-operatives is that they are run purely as farmers' co-operatives, with all the major decisions being taken by the farmers themselves. The co-operatives are not 'run' by a separate bureaucracy with vested interests of its own; the farmers are truly in charge of their own decisions. Any farmer can become a member by committing to supply a certain quantity of milk for a certain number of days in a year and shall continue to be a member only if he keeps up this commitment. Each day, the farmers (or actually, in most cases, their wives and daughters) bring their milk to the village collection centres where the quantity of milk is checked in full view of all and quality (milk fat content) is checked using a simple device, again in full view of all. The farmers are paid in the evening for the milk they supplied in the morning, and the next morning for the evening's milk. This prompt settlement of cash is a great attraction to the farmers who are usually cash starved. Thanks to the above system, there are no disputes regarding quantity or quality of the milk supplied by each farmer. It was soon realised that it was not enough to merely act as the collection and selling agents for the farmers. A variety of support services were also required to enable the farmers continue selling milk of adequate quality and to avoid disasters such as the death of their cattle (for a family owning just one or two cattle and depending on the milk for their income, the death of a cow could indeed be a disaster). The farmers were progressively given new services such as veterinary care for their cattle, supply of good quality cattle feed, education on better feeding of cattle and facilities for artificial insemination of their cattle. All these were strictly on payment basis; none of the services were free. This experiment of organising farmers into co-operatives was one of the most successful interventions in India. A very loyal clientele was built up who experienced prosperity on a scale they could not have dreamt of 10 years earlier. With good prices paid for their milk, raising milch cattle could become a good supplementary source of revenue to many households. The co-operatives were expanded to cover more and more areas of Gujarat and in each area, a network of local village level co-operatives and district level co-operatives were formed on a pattern similar to that at Anand (the so called Anand pattern). In 1955, KDMPUL changed its name to Anand Milk Union Limited, which lent itself to a catchy abbreviation, Amul, which meant; priceless in Sanskrit. The word was also easy to pronounce, easy to remember and carried a highly positive connotation. It became the flagship brand name for the entire dairy products made by this union.

In 1954, Amul built a plant to convert surplus milk produced in the cold seasons into milk powder and butter. In 1958, a plant to manufacture cheese and one to produce baby food were added—for the first time in the world, these products were made from buffalo milk. Subsequent years saw the addition of more plants to produce different products. Starting from a daily procurement of 250 litres in 1946, Amul had become a milk giant with a large procurement base and a product mix that had evolved by challenging the conventional technology. On his visit to Anand in 1965, the then Prime Minister of India, Lal Bahadur Shastri, was impressed by what he saw—a system that procured, processed and delivered high quality milk to distant markets cost efficiently. Shastri could also see the difference that the income from milk had made to the standard of living of farmers in the area. What impressed him the most was that Amul had done all this without government assistance, in marked contrast to a number of government sponsored dairy programmes that were doing poorly in terms of procuring and marketing good quality milk and boosting farmers' incomes. Shastri asked Dr Kurien to replicate Anand's success all over India.

A pattern similar to the Anand pattern was to be built in other states of India. This was carried out under a programme launched by the Government of India, entitled 'Operation Flood'. The operation was co-ordinated by the National Dairy Development Board (NDDB), a body formed by the Government of India with this specific objective.

Structure of the Anand Pattern

The basic unit in the Anand Pattern is the village milk producers' co-operative—a voluntary association of milk producers in a village who wish to market their milk collectively. All of the village milk producers' co-operatives (primaries) in a district are members of their district co-operative milk producers' union. Every milk producer can become a member of the co-operative society. At a general meeting of members, representatives are selected to form a managing committee, which frames the policies of the society to govern the day-to-day affairs relating to milk. Milk collection, the testing for milk fat content, sale of cattle feed etc. is handled by paid employees from the same village. Each society also provides artificial insemination (AI) services and veterinary first aid (VFA). Thus, these primaries also generate local employment in the rural community. Each producer's milk is tested for fat percentage (many also measure solids-not-fat) and is paid for, on the basis of the quality of the milk. Usually, the morning milk is paid for in the evening and the evening milk is paid for the next morning. The primary milk producers' societies are affiliated to a district union, which owns and operates a feeder/balancing dairy cattle feed plant and facilities for production of semen and its distribution. The union also operates a network of veterinary services to provide routine and emergency services for animal health care. (Fig. 13.1).

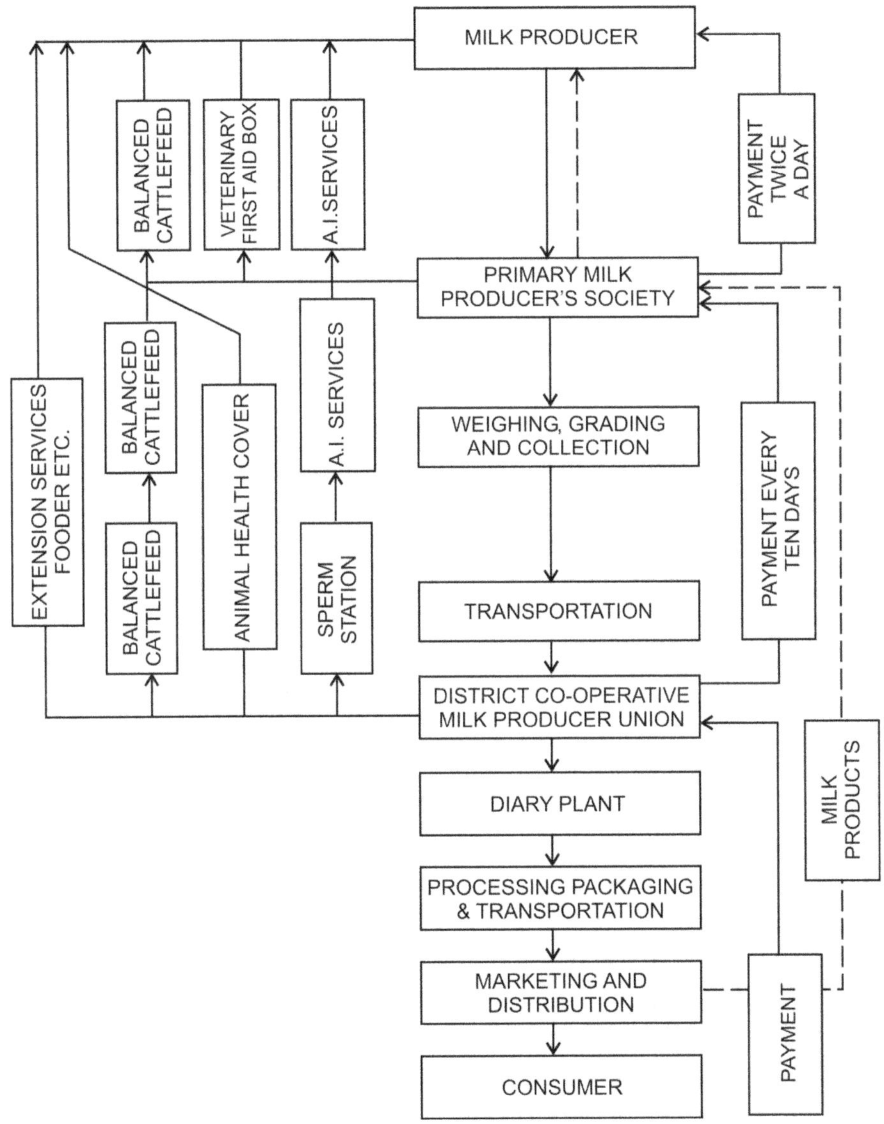

Fig. 13.1 Organisation of Anand pattern.

The chairpersons of village societies elect the board of directors of the union, which frames the policies for the day-to-day management of the union's centralised facilities for milk collection, processing and marketing and also technical inputs. Each union is managed professionally by a managing director, who reports to the elected chairman and a board of directors. The dairy, owned by a union, usually has a milk processing plant to convert seasonal surpluses of liquid milk into milk powder and other conserved products. With the help of the dairy plant, the union is able to ensure that the milk producers get 80–90% of the lean season price even in the flush season. The farmers are, therefore, able to get a good price for the bulk of the milk that is produced in the flush season. This has enabled the farmers to get 20–40% higher prices than they would have if they had not been a part of the co-operative system. Before the co-operatives, the middlemen usually paid only 60–70% of the lean season price in the flush season. Milk producers are able to substantially increase their returns from milk production because of better returns for their milk and lower feeding costs. The milk collected from the village is usually sent to the co-operative dairy using trucks hired by the co-operative union. Each co-operative dairy tries to market the bulk of its milk as liquid milk and converts surplus milk into products with a longer shelf life. Professional managers employed by the co-operative ensure that they get the best returns for their produce. The profits made by the dairy are redistributed to the milk producers as a subsidiary payment. Many societies are able to pay substantial amounts as bonuses to their milk producers, based on the proportion of business contributed to the co-operatives.

National Dairy Development Board

Over the last 25 years or so, the Indian dairy industry has progressed from a situation of scarcity to that of plenty. Dairy farmers today are better informed about technologies of more efficient milk production and their economics. Even the landless and marginal farmers now own highly productive cows and buffaloes in many areas. Along with high increase in the production of both cereals and cash crops, dairying has expanded rapidly with a compound growth rate of over 5 per cent per annum over the last two decades. Application of modern technology and advanced management systems in milk processing and marketing have brought about a marked change in the market place. Consumers now have a wide range of choice of products and packages. The availability of dairy products has become regular. It would only be proper to state that the country has achieved a greater degree of self-reliance in the dairy sector and this has happened essentially because of the sound base that has been created by the dairy co-operatives. How did these changes take place? Many factors contributed, but the key role in bringing about this transformation has been played by NDDB.

Philosophy

- Cooperation is the preferred form of enterprise, giving people control over the resources they create through democratic self-governance.
- Self-reliance is attained when people work together, have a financial stake, and both enjoy the autonomy and accept the accountability for building and managing their own institutions
- Progressive evolution of the society is possible only when development is directed by those whom it seeks to benefit.
- All beneficiaries, particularly women and the less privileged must be involved in cooperative management and decision making
- Technological innovation and the constant search for better ways to achieve objectives is the best way to retain a leading position in a dynamic market
- While our methods change to reflect changing conditions, our purpose and values must remain constant.

NDDB was created in 1965 in response to the then Prime Minister Lal Bahadur Shastri's call to 'transplant the spirit of Anand in many other places.' He wanted the Anand model of dairy development with institutions owned by rural producers, which are sensitive to their needs and responsive to their demands replicated in other parts of the country. He said in a letter addressed to the State Chief Ministers, "We envisage a large programme of co-operative dairies during the Fourth Plan and this will, no doubt, be based on the Anand model. If we can transplant the spirit of Anand in many other places, it will also result in rapidly transforming the socio-economic conditions of the rural areas." He decided that the Government of India would create a body, whose job would be to replicate 'Anand.' It would be headed by Dr Verghese Kurien, the then General Manager of the Kheda District Co-operative Milk Producers' Union Limited (AMUL). In the late sixties, the Board drew up a project called Operation Flood (OF) · meant to create a flood of milk in India's villages with funds mobilised from foreign food donations. Producers' co-operatives were the central plank of the project, which sought to link dairy development with milk marketing. NDDB underwent a structural change in 1988. The Board, registered as a society, and the Indian Dairy Corporation, a Company formed and registered under the Companies Act 1956, were merged by an Act of Parliament, the NDDB Act, 1987. The resulting corporate body bears the old name: National Dairy Development Board. The Act has declared the Board to be an institution of national importance.

Constitution *:* The National Dairy Development Board has been constituted as a body corporate and declared an institution of national importance by an Act of India's Parliament. The National Dairy Development Board initially registered as a society under the Societies Act 1860 was merged with the erstwhile Indian Dairy Corporation, a company formed and registered under the Companies Act 1956, by an Act of India's Parliament the NDDB Act 1987 (37 of 1987), with effect from October 12, 1987. The new body corporate was declared an institution of national importance by the Act.

The greatest contribution of NDDB has been the initiation and fostering of the process of modernisation of India's dairy industry. The co-operative movement NDDB helped to create has become a model for other developing countries and the international agencies that are concerned with dairy development. The Operation Flood programme is a unique approach to dairy development. During the 1970's, dairy commodity surpluses were building up in Europe and Dr Kurien, the founder Chairman of NDDB, saw in those surpluses both a threat and an opportunity. The threat was massive imports of low cost dairy products in India which, had it occurred, would have tolled the death knell for India's struggling dairy industry. The large quantities that India was already importing had eroded domestic markets to the point where dairying was not viable. The opportunity was built into the Operation Flood strategy. Designed basically as a marketing project, Operation Flood recognised the potential of the European surpluses as an investment in building India's dairy industry. With the assistance of the World Food Programme, food aid, in the form of milk powder and butter oil, was obtained from the EEC countries to finance the programme. It was for the first time in the history of economic development that food aid was seen as an important investment resource. Use of food aid in this way is anti-inflationary: it provides a buffer stock to stabilise market, and it can be used to prime the pump of markets that will later be supplied by domestic production.

A number of programmes and policies have each played a role in the process of modernising India's dairy industry. Certainly, the introduction of modern technology, both for the farmer and in the processing of milk and products, has been important. Similarly, establishing an urban market has provided the stability necessary to encourage farmers to invest in increased milk production. The induction of professional managers to serve farmers has reversed the normal pattern where farmers are supplicants and officials are benefactors. Perhaps, the most important factor is the cooperative structure itself. By giving the farmers command over the resources they create, Operation Flood has ensured that they receive the maximum return from each rupee that the consumer spends on milk and milk products. This, in turn, has provided the incentives on which the growth of our dairy industry has been based. Imagine every morning and evening, some 90 lakh farmers carrying potfuls of milk to their co-operatives milk that will travel from remote villages to towns and cities throughout India. Today, these farmers own some of the largest and most successful businesses in India. Their infrastructure has returned a greater share of the consumer's rupee to the farmer. It has built markets, supplied inputs, created value-added products. All this has happened because the farmers' productive capacity has been linked with professional management in co-operatives. While the demand for milk was rising under Operation Flood, the total cattle population remained more or less static. NDDB realised that if milk production had to be increased, the buffalo and our indigenous good milch breeds of cattle had to be upgraded and nondescript cows had to be cross-bred with exotic semen to increase their milk production to make them more efficient converters of feed. With this objective in mind, a thrust was given to intensive research and development in animal husbandry. Today, animal breeding is an integration of three major areas artificial insemination and quantitative genetic techniques, embryo transfer and embryo micro-manipulation techniques, and biotechnology and genetics engineering. The optimal genetic improvement can be achieved by making use of the latest developments in each of these areas.

Realising the significance of embryo transfer in India, NDDB established its main embryo transfer lab at the Sabarmati Ashram Gaushala near Ahmedabad. Embryo transfer in the Indian context is primarily a tool to create a nuclear germ plasm pool to supply future bulls to artificial insemination centres for production improvement in cattle. The techniques have enabled India to rapidly multiply the elite donors among the cross-breds and buffaloes and to produce superior bulls from them in adequate numbers. Production of buffalo calves using non-surgical transfer was achieved successfully in India, for the first time in the whole of Asia. Through the wide network of research laboratories, the transfer of embryos has already moved from the laboratory to the villages where embryos have been transplanted into cows by specially trained staff belonging to the cooperatives. The success of the co-operative movement under the aegis of NDDB has yielded unexpected results. The government has brought other primary commodities like edible oil, fruit and vegetables under the ambit of NDDB. The co-operative umbrella has been extended to tree plantation and even to salt farming. NDDB initiated the 'Restructuring Edible Oil and Oilseeds Production and Marketing' Project in 1979 to increase farmer investment in oilseeds sector through farmer-owned co-operatives. More than 9 lakh farmers have joined nearly 5,500 oilseeds growers' co-operative societies affiliated to 18 unions in Andhra Pradesh, Gujarat, Karnataka, Madhya Pradesh, Maharashtra, Orissa, Rajasthan and Tamil Nadu. Capacities have been created to crush 3,735 TPD oilseeds; solvent-extract 2,180 TPD oil cake; and to refine 778 TPD of edible oil. Storage has been built to handle 1.9 lakh mt. oilseeds and 2.96 lakh mt. oil. The traditional mono-cropping of oilseeds entails risks for both the producer and the co-operative. To reduce these uncertainties, the project has promoted a multi-oilseeds cropping system supported by introduction of non-traditional oilseeds crops.

The NDDB was appointed the Market Intervention Agency for oilseeds and edible oil by the Government of India in 1989 and was entrusted with the responsibility of ensuring remunerative prices to farmers, reasonable prices to consumers and attaining self-sufficiency. The Market Intervention Operation of the NDDB has proved that imported oil cannot be used against the Indian farmers. Over the last few years, the Integrated Policy on Oilseeds and Edible Oil, with its curb on imports, rationalisation of allocations to the Public Distribution System and vanaspati manufacturers and the Market Intervention Operation, have all created a climate of stable, remunerative prices which encouraged the farmer investment in activities that enhance oilseeds productivity. In order to establish a direct link between producers and consumers of oil and thereby reduce the role of oil traders and oil exchanges, NDDB decided to enter the consumer pack market for edible oil through its Dhara refined rapeseed oil and groundnut oil. Within a short span, Dhara has become the market leader in branded edible oil because of its tamper-proof packing and high quality. Dhara received the Grahak Suraksha Award given by the Gujarat State Consumers Protection Centre in 1990. Dhara has been successful in slowly weaning away the consumer from buying oil in bulk. Dhara is teaching them not to store because the consumer is assured of good quality oil at reasonable prices round the year. The Fruit and Vegetables Project of the NDDB became operational in Delhi in January 1988. It seeks to provide a direct link between fruit and vegetable growers and consumers. The project is designed to handle 1,20,000 mt. of fruit and vegetables annually in Delhi. Already nearly 200 specially designed modern retail outlets have been set up, each outlet has a capacity to sell upto 1,600 kg of fruit and vegetables daily and other co-operative products like butter, oil, cheese, chocolates, etc. Substantial quantities of peas, carrot, cauliflower and ocra are frozen and marketed under the brand name 'SAFAL.' Efforts are on to provide a package of practices including improved varieties of seeds, fertilisers, etc., to growers to enable them to produce better quality vegetables.

The Sabarmati Salt Farmers' Society was set up by NDDB as a pilot project at Kharaghoda in 1987 at the request of the Government of Gujarat to improve the socio-economic conditions of the Agarias, the salt producers, in the Little Rann of Kutch in Gujarat. Beginning with only 27 Agarias, it has today 82 Agarias who are no longer dependent on the traders. It has also acted as a catalyst in initiating a few development programmes like medical and health care, education, etc., to improve the quality of life of the salt producers who are a highly exploited group. In sum, NDDB is a unique example of an organisational innovation with a focus on human resource and cooperative development in India. NDDB has helped to lay the foundation of democratic enterprises at the grassroots and shown the way to ensure economic growth with social justice and people's participation. By placing technology and professional management in the hands of the village societies it has helped to raise the standard of living of millions of poor people. These processes prove that true development is the development of the people and this could be achieved only through putting the instruments of development in the hands of the people.

As on March 2001, India's 96,000 dairy cooperatives integrated through a three tier cooperative structure "The Anand Pattern", owned by more than ten million farmers, procure an average of 16.5 million litres of milk every day. The milk is processed and marketed by 170 milk producers' cooperative unions, which, in turn, own 15 state cooperative milk-marketing federations. Since its inception, the Dairy Board has planned and spearheaded India's dairy programmes by placing dairy development in the hands of milk producers and the professionals they employ to manage their cooperatives.

In addition, NDDB also promotes other commodity-based cooperatives, allied industries and veterinary biologicals on an intensive and nation-wide basis. Presently Dr. (Ms) Amrita Patel serving as the Chairman of NDDB; Dr. Verghese Kurien was the founder Chairman.

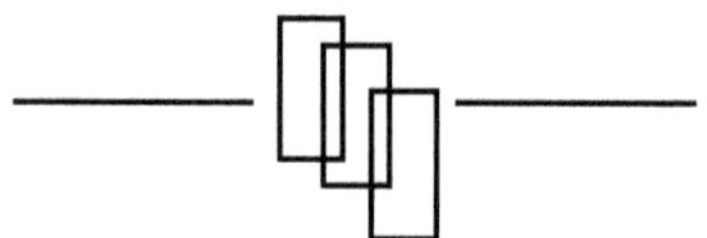

CHAPTER - 14

Livestock Economics and Accountancy

Economic : Meaning, Definitions and Scope

Economics is a study of human wants and their satisfaction. Human wants are the driving force of all economic activity. Man has several wants. He wants food, clothing and shelter for the preservation of life. In addition to these, he requires several other things for leading a healthy and happy life. He wants clothing, shoes, radio, pens, books, furniture and so on. As a matter of fact, there is no limit to human wants. But the goods that he can get are always less than those required to satisfy all his wants. To procure the goods he requires Men is engaged in one form of activity or the other which also form part of economics.

Meaning : Economics does not study religious, social and political activities of man. It does not also study human activities undertaken for the sake of pleasure or out of love. Ex. A Professor doing gardening work and a mother looking after her children. According to Marshall "Economics is a study of man's actions in the ordinary business of life." It enquires how he gets his income and how he spends it. Thus "Wants – efforts – satisfaction" are the subject matter of economics. Wants give rise to efforts. Efforts help us to satisfy our wants. After that new wants spring up, new efforts are put to satisfy them.

Definitions : Some of the eminent economists have defined economics basing on the scope in different ways.

1. *Adam Smith* defined economics as the science of wealth. Wealth, according to classical economics, means material and tangible goods only. In their opinion economics is concerned with the earning and spending of wealth. This definition was seriously criticised by some literary writers like Carlyle and Ruskin. They have condemned it as a disimal science teaching about selfishness. The main defect of the above definition lies in the fact that it places too much emphasis on wealth. It completely ignores man for whom wealth is intended.

2. *Robbins* defined economics as "the science which studies human behaviour as a relationship between ends and scarce means.

Robins definition is based on the following four important facts.

    1. Human wants are unlimited.

    2. The means to satisfy human wants are limited

    3. The means are not only limited but they have alternative uses.

    4. Ends differ in importance.

3. *Prof. Marshal* views economics as a science of man rather than of wealth. He defines economics as a "study of man in the ordinary business of life". Man is primary and wealth is secondary. We produce wealth not for its own sake. It is not an end in itself. It is a means to an end – the satisfaction of human wants. We are concerned with the economic activity of people. Wants are the starting point of this activity. To satisfy these wants man needs goods and services. Many of these goods and services are limited in quantity. They are scarce. These scarce goods and services are known as wealth in economics. The whole of human activity is centered around wealth. People are engaged in production, exchange, distribution and consumption of wealth. So if we want to satisfy the economic activity of people we must also study wealth. Hence Prof. Marshall defines Economics as " on one side a study of wealth and on the other and more important side, a part of the study of man". It enquires how he gets his income and how he uses it. So it is a study of man's actions in the ordinary business of life. It excludes the study of political, social or religious activities of mankind. It studies man not in isolation but as a member of a social group. It examines that part of individual and social action which is most closely connected with the attainment and with the use of material requisites of well – being. Marshal recognizes the supreme important of man and human welfare. Thus economics is a study of causes of material welfare and is concerned with the promotion of economic welfare of the people. This definition was generally accepted as welfare definition till recently.

Subject Matter : Economics studies human beings. It does not study man in isolation, It studies man as a member of society, thus economics is called a social science. Human efforts produce goods, which satisfy their wants. Wants are fundamental basis of our economic life. Economics studies human activities undertaken to satisfy wants.

The Subject Matter of Economics is Divided into Four Parts

- *Consumption :* we study why we work to earn money and how we satisfy our wants.

- *Production :* we know how goods are produced.
- *Exchange :* we study the way in which goods produced in our country and other countries are supplied to us and study the principles that determine the prices of goods in the market.
- *Distribution :* we study how wealth produced is distributed among the agents of production.

Scope of Economics

- Economics is called a social science as it studies human beings.
- It is also called as arts as it studies how people behave in the earning and spending money.
- It is called a positive science as it studies existing conditions and discoveries of truths.
- It is also called a normative science as it studies the objects to be achieved.
- Economic knowledge serves us in managing our personal lives, Understanding society, analysis of national economic issues, explain & predict economic behavior and improving the world around us.

Basic Concepts of Economics

A graduate of economics will have to understand the meaning of some fundamental terms that are frequently used in the subject. All these terms are derived from ordinary language and they are given a special meaning. In this chapter the meanings of these terms are clearly explained.

1. *Wants* : We studied that human wants are the starting point of all economic activity. How do wants arise? In the first place, we have some elementary wants to satisfy. We must have a minimum amount of food, clothing and a certain type of shelter. These wants are physiological in nature. They make our life possible. Secondly as members of society, we are expected to wear a certain type f dress and live in a certain manner. Many of our wants can be traced to this clause. So some wants arise from our economic and social position. Thirdly some wants arise as a result of habit and custom. Many of us are accustomed to smoking, though it is not essential for our existence. Coffee, tea, smoking and drinking come under this head. Finally advertisement and propaganda of producers create some wants in our minds. We often see advertisements in journals and newspapers regarding some goods. When we realise the advantage of such goods a want for them will arise in our mind. We are told for instance, that Colgate toothpaste, Lifebuoy soap, are the best products. Man has wants. To satisfy them he is doing some work or the other. If there had been no wants we would not have worked at all. Thus wants are the mainspring of all economic activities.

Characteristics of Wants

- *Wants are unlimited* : Man is a bundle of desires. There is no limit to wants, if one gets satisfied another shoots up. So man never says that he is completely satisfied.

- *Particular want can be Completely Satisfied* : As our income is limited we cannot satisfy all our wants. But we can satisfy any one want completely. The law of diminishing marginal utility is based on this characteristic.

- *Wants are Competitive* : Man has innumerable wants. But he does not have sufficient money to satisfy all of them. So he has to decide which wants are to be satisfied first and which are to be left off. In others words wants compete with one another for his choice. The law of equimarginal returns is based on this characteristic.

- *Wants are complementary* : If we want tea we need milk, tea powder, sugar i.e. If we want to satisfy one want, we need many things. One of several things is of no use.

- *Wants are recurring* : Some wants occur again and again. Ex: Hunger

- *Wants are alternately* : A want can be satisfied by several ways Ex: A person wants to go to Hyderabad he can go by bus, train or aeroplane.

- *All wants are not equally important* : Some wants are more urgent and some are less urgent. Want for food is more urgent than want for a silk shirt. With limited money we satisfy wants according to their importance.

- *Wants become habits* : If a person goes for a want number of times then it becomes a habit Ex: taking coffee, cigarettes, liquor, opium etc.

- *Wants are the results of customs* : In our country people spend money on ceremonies even though they do not like it or afford it, that is because of custom. Ex- crackers on Diwali.

Classification of Wants

We all known that we cannot satisfy all our wants, as our income is limited. So every person has to draw a distinction between more urgent and less urgent wants. Every person will have his own classification. Our wants or the goods that satisfy these wants are generally classified into necessaries, comforts and luxuries.

Necessaries : There are certain things, which are absolutely essential for our very existence. It means that without consuming them we cannot keep our body and soul together. Necessaries are elementary requirements to make life possible. For example: Food, clothing, and shelter. Some of the necessaries increase the efficiency of the people. They are called necessaries of efficiency. For example, a car for a doctor. Some goods are consumed either as a habit or custom, which are called conventional necessaries. Ex: opium, liquor, cigarettes and marriage expenditure or festival needs.

Comforts : Goods, which satisfy wants of less urgent character, are known as comforts. Their consumption increases once efficiency but not in proportion to their prices. These are not essential, but they increase the comfort of the people. They stand mid way between necessaries of efficiency and luxuries. Thus if a middle class man purchases an electric fan or wears a woolen suit these may be treated as comforts to him.

Luxuries : Luxuries satisfy superfluous wants of a man. These things do not add to the efficiency of a person and he can lead a healthy life even without them. Some luxuries are harmful also. For example, wine. Thus if a farmer uses hair oil or puts on a silk dress we treat them as luxuries to him. At this stage goods can not be divided into these three water tight compartments. The classification is not rigid and they are relative terms. One good may be necessary for one man and comfort for the other and luxury to the third. Ex. a car is a necessary for a doctor and luxury for a clerk. Much depends on the income and social position of the person concerned. They are relative with regard to place also. Woolen dress is a necessary in a cold country and a luxury in hot country like India. The important point to bear in mind is that there is no ethical or moral significance for this distinction.

Standard of Living : The term standard of living refers to the amount of necessaries, comforts and luxuries to which a person or a class of people has been accustomed. The food that a man consumes, the way in which he dresses and the type of house in which he lives indicates his standard of living. The standard of living is not a fixed one. It varies from nation to nation, class to class and from individual to individual. It also varies from one period to another, with the same individual or class or nation. The financial position, habit, education and love of imitation influence the standard of living of a person. The standard of living affects the health and efficiency of people, people: with high standard of living are better fed and more efficient than those with a low standard. The standard of living people of a country indicate its economic prosperity.

2. **GOODS** : Any thing that satisfies a human want is called a good. *Ex :* water, air, books, pens, cycle etc. The services of a doctor or a teacher are also goods. Goods are of two kinds:

   Free Goods : Air, water and sunshine are free goods, which are available in abundance in nature and for which we need not pay.

Economic Goods : Books, cycle and doctors services are economic goods, which are limited in quantity and for which we pay.

- A good may be a free good at one place and economic good at another place Ex: water is free near a river and economic good in city.

3. ***Utility*** : The want satisfying power of a commodity is called utility or the capacity of a thing to satisfy a human want. It is also called **"Value in use"**.

- Utility does not mean usefulness.
- Utility of a commodity varies from person to person.

4. ***Wealth*** : The popular meaning of wealth associates with being rich, but in economics wealth means "Any thing that satisfies a human want and is limited in supplying is called wealth". Rice, cloth, pens, jewelry, land, buildings, factories are all wealth. These are also called economic goods. Thus wealth means economic goods.

Characteristics of Wealth

It must possess utility

It must be limited in supply

It must be transferable

It must be external to man – health for example is not transferable

- Anything which satisfies the above four qualities is called wealth. Wealth is different from capital. Wealth consists of all economic goods. Only those goods, which are used for further production of wealth, are called capital.
- Thus all capital is wealth but all wealth is not capital.

Classification of Wealth

(i)   Individual wealth

(ii)   Social wealth

(iii) National wealth – Individual wealth + social wealth

5. ***Value*** : The word value is used in two senses:

- Value in use: The good possesses the power to satisfy human wants.
- Value in exchange: The purchasing power of a commodity. i.e. capacity to secure other goods in exchange.

6. ***Price*** : Now-a-days, the value of a commodity is expressed in terms of money. The money value of a commodity is called its price. Prices help us to compare the value of two commodities.

7. ***Income*** : It means the amount earned by an individual during a certain period. Wealth is called stock or fund, while income is called flow as it is received regularly. It is of two types:

- If income is measured in terms of money we call it "money income".

- If it is measured in terms of goods it is called "Real income".

8. ***Costs*** : When a man wants to produce a commodity he has to incur certain expenses. The total amount of money spent in making a commodity is its cost of production. Such cost consists of cost of raw material, wages, rent of factory, interest on capital, depreciation of machinery and normal profit of organisation. We require the four factors of production to make any commodity – Land, labour, capital and organisation. The total amount of money spent on these factors is the cost of production of a commodity.

Approaches of Economics

There are two approaches for the study of economics: the Traditional approach and Modern approach.

1. ***Traditional Approach*** : According to the old view the study of economics is divided into four main divisions viz. consumption, production, exchange and distribution. In consumption we study the nature of human wants as well as the principles governing their satisfaction. The law of diminishing marginal utility, law of substitution, the law of family expenditure and consumers surplus is of special importance. We also study the nature of demand, weather it is elastic or inelastic as well as the law of demand. In production we study how man makes efforts to satisfy his wants by producing wealth. In particular, we see how the four agents of production; land, labour, capital and organisation, co-operate and combine in the work of production. We study each of these agents, their importance and the conditions of their efficiency. In the third division of exchange, we discuss how in the various market forms buying and selling are done and how prices are determined. This happens through the interaction of the forces of demand and supply. The fourth division of economics is distribution. It is devoted to the study of the respective shares that go to the four agents – land, labour, capital and organization. These shares take the form respectively of rent, wages, interest and profit. In distribution we study how the share of each agent of production is determined. In addition to the above four divisions we have to study the problems of public finance also. Here we discuss how governments get money and how they spend it. Thus it involves the consideration of taxation and allied questions.

2. ***Modern Approach*** : In latest view, the economics could be explained in the following two ways:

   A. ***Micro Economics*** : Micro economics consists of looking at the economy through a microscope, as it were, to see how the millions of cells in the body economic – the individuals or house holds as consumers, and the individuals or firms as producers – play their part in the working of the whole economic organism. The economy can be compared to a living organism. In this organism hundreds and thousands of individuals acting as consumers and producers are like the blood cells in our body. The blood cells can be studied only with the help of microscope. In the same way the economic behaviour of individual consumers and producers as small groups of persons engaged in an economic activity can be studied through micro economics only. For ex. an individual consumer faces a problem of distributing his income or various commodity which he can buy with in the limited range of his income. He wants to do this with a view to maximising his personal satisfaction. This is a micro level problem and the equilibrium of an individual consumer forms part of Micro Economics. So also an individual producer faces the problem of maximizing his profits. Microeconomics does not go beyond this and cannot study the equilibrium of the entire economy.

   B. ***Macro Economics*** : In macro economics also we are concerned with demand supply and equilibrium but levels at which these terms are taken up are different. In macro economics we speak of aggregate supply i.e., supply of all goods and services taken together. We also speak of aggregate demand which is the sum total of individual demand. For ex. in macroeconomics we want to know the total demand for eggs, milk or vegetables etc. taken together. There is no way of adding all these quantities because they are measured in different units. Therefore, the only way of achieving this function is to express the values of all these quantities in terms of money and then adding these values together. Similarly aggregate supply can also be expresses in terms of money. If demand and supply expressed in terms of money are not equal, we face the problem of disequilibrium. If aggregate demand is less the prices will fall and in turn income will also fall. If supply is less, the prices will rise and producers will be encouraged to produce more and this leads to start more business units and the level of employment increases. This is also known as "Aggregative economics".

Consumption

Meaning :The existence of human wants is the starting point of all economic activity. Every want gives rise to a desire, consequently it leads to an effort on the part of man to obtain the good that satisfies his desire. So wants give rise to efforts and efforts secure satisfaction. Consumption or satisfaction of human wants is thus the end of all economic activity. Consumption involves using up of commodities and services in satisfaction of human wants. When we eat ice cream we are said to consume it. In the same manner when we read a book or smoke a cigarette we are said to consume them. Goods that satisfy human wants possess utility and when we use these goods we simply use their utilities. Consumption may, therefore, be defined as using up of utilities. Consumption does not mean destruction of matter. Man cannot create matter nor can he destroy it. He creates utilities and he uses those utilities. All consumption, therefore, involves not the destruction of matter but the destruction of utilities. This destruction may take place in a very short period as in the case of consumption of food or it may spread over a much longer time as in the case of a house.

Definition :Wants-satisfying quality and utility of a commodity is called as consumption. For example: drinking milk, sitting on a chair, getting veterinary service for an animal

Consumption may be useful or wasteful, such as, useful: consuming a mango (destruction of utility) and wasteful: destruction of good due to fire (destruction of matter).

Types of Consumption

(i) *Direct or final Consumption :* Goods, which satisfies human wants directly and   immediately. For example, consuming a mango

(ii) **Indirect or Productive Consumption :** Goods, which are not meant for final consumption but for producing other goods, which will satisfy human wants directly. For example, using a gas stove and cooker to boil an egg

Engle's Law of Consumption : Earnest Engle was a German statistician. He made a systematic study of a number of family budgets. The budgets indicates the proportion in which the different classes of society distributed their expenditure among necessaries, comforts and luxuries. From his study, Engle has laid down the following propositions. As the income of a family increases

1. The proportion spent on food decreases
2. The proportion spent on clothing remains approximately the same

3. The proportion spent on fuel, lighting and rent remains fairly constant and

4. The proportion spent on comforts and luxuries such as education, health and recreation tends to increase

Thus if the family income is small, a greater part of it will be spent on necessaries of life. We must carefully note that the law refers to the percentage variation in expenditure and not to the variations in the amount of expenditure on different items.

Demand

*Meaning :*In the ordinary speech we use the term 'demand' to express our wish or desire for a thing. But in economics, a mere desire is not demand. Thus, a poor man may desire to have a motor car, but he cannot afford to purchase it. Such a desire is a mere wish and not effective desire. Such a desire becomes effective only when he has adequate purchasing power we call that effective desire as demand. Demand implies more than mere desire. It means that the person is willing and able to pay for the object he desires. Demand often confused with desires. Demand means, person is willing and able to pay (eg. a businessman's desire to travel by air), while desires means person is willing but unable to pay (eg. a beggar's desire to travel by air).

Definition : Demand for any thing, at a given price, is the amount of it, which will be bought per unit of time at that price.

Functions : The quantity demanded for a commodity (D_a) is the function of demand i.e. depends upon the

- Price of good(s) in question (P_a)
- Price of all other goods $(P_b, P_c, ...)$
- Household income (I)
- Household taste (T)

This statement may be expressed in symbols what is called demand function.

$$D_a = f\,(\,P_a, P_b, P_c, ...\,;\,I\,;\,T\,)$$

Kinds of Demand

(i) *Price Demand* : Price demand refers to the various quantities of a commodity or service that a consumer would purchase at a given time in a market at various hypothetical prices.

Assumption: $P_b, P_c,$ I and T remain unchanged

(a) *Individual demand* : The demand of individual customer.

(b) *Industry demand* : Aggregate demand of all customers for the commodity or service.

(c) *Individual seller's demand* : The total demand for the product of an individual firm at various prices.

(ii) *Income Demand* : Income demand refers to the various quantities of a commodity and service, which would be purchased by the consumer at various levels of income.

Assumption : P_a, P_b, P_c and T remain unchanged

Superior goods : command brisk sale when income increases.

For example, *Gold*

Inferior goods : command large sales when income decreases.

For example, *Silver*

(iii) *Cross Demand* : Cross demand means the quantities of a commodity or service which will be purchased with reference to changes in the price not of this good but of other related goods.

Assumption : P_a, I and T remain unchanged

For example, a change in price of tea will affect demand for coffee

Demand Schedule

The demand schedule represents the conditions of demand only at a particular time. There will be various influences working to change the state of demand. Such influences or changes in the tasters or habits or income or the prices of competing goods. When there is a change in these conditions the old demand schedule becomes invalid. A new one must replace it.

If we look at the demand schedule we find that more units are bought with every fall in the price. Why is it so ? The answer to this question will be furnished by the law of diminishing marginal utility. The law states that the utility of additional units of good decreases with every increase in the stock of it. It means that the successive units of a good give us less and less satisfaction. So we are prepared to pay a lower and lower price for the additional units. Thus more units will be bought with every fall in the price of a good. More over with every fall in the price of a good it will come within the reach of people with lower incomes. So new customers will buy the good with a fall in the price. So the demand for a commodity will be more at a lower price than at a higher price. Just as we have drawn the demand schedule of an individual we can draw a demand schedule of a market for any good. It too shows that more units will be bought with every fall in the price of the good concerned.

Individual Demand Schedule: It is a list of various quantities of commodity, which an individual household purchases at different alternative prices. For example, The following is the demand schedule of an imaginary consumer of milk. (Table 14.1).

Market Demand Schedule: It can well be obtained by adding up all the demand schedules of the individual households in the market. For example,

We can the divide the consumers into two main categories: 'A' with monthly income above Rs. 2000 (above poverty line) and 'B' - below Rs. 2000 (Below Poverty Line). (Table 14.2).

Table 14.1 Individual Demand Schedule

Price of milk per litre (Rs.)	Quantity Demanded per day (litre)
14.00	0.5
12.00	1.0
10.99	1.5
8.00	2.0
6.00	2.5
4.00	3.0
2.00	3.5

Table 14.2 Market Demand Schedule

Price of Milk per Litre (Rs.)	Quantity demanded per day (litre)		
	Category A	Category B	Total
14.00	0.5	Nil	0.50
12.00	1.0	0.25	1.25
10.00	1.5	0.50	2.00
8.00	2.0	1.00	3.00
6.00	2.5	1.50	4.00
4.00	3.0	2.0	5.00
2.00	3.5	2.5	6.00

Difficulties in the Construction of a Demand: An individual cannot positively say how much he would purchase since the prices in the schedule are unreal. Market schedule is even more hypothetical.

Practical Utility of Demand Schedule

- Helps to find out the effect of different rates of taxes on sales of a commodity; based on the effect our government imposes the tax.
- Helps the businessman to forecast the profits and to arrange production

Law of Demand : At this stage we can frame the law of demand. This law says that demand varies inversely with price. If price rises demand contracts and if the price falls demand expands. "Other things remain the same, with a fall in price demand expands and with a rise in price demand contracts" the condition that other things remaining the same is very important. If there is a change in tastes, incomes, population or season the demand may change even though there is no change in price. The producer regulates the supply of a good according to demand at a given price. When a tax is imposed on a good the price of it will rise and it may lead to a very great or very small fall in its demand. If the demand falls very much the tax will bring less revenue to the state. So the government will study the demand condition of a good before tax is imposed. At any given time the demand for a commodity or service at the prevailing price is greater than it would be at a higher price and less than it would be at a lower price. The law of demand states that more will be demanded at a lower price and less when price is higher.

Exceptions to Law of Demand : The law of demand seems to be applicable to all situations concerning a consumer's demand. But there are certain exceptions:

(i) ***Conspicuous Consumption*** : Where a consumer regards the consumption of a commodity as a mark of distinction, he will go in for a high-price commodity, a diamond.

(ii) ***Giffen Paradox*** : Demand for a commodity is increased, when its price goes up. eg. In the mid-19th century , when price of bread increased, the low-paid workers in Britain spent more on it (since it was their staple food) and they cut on meat. That is, they substituted bread for meat.

(iii) ***Changes in Expectations*** : When prices are expected to continue rising, people, in order to avoid paying more in future, buy more even though the price has risen.

(iv) ***Trade Cycle*** : People buy more even when the price goes up, since peoples' incomes have gone up.

(v) ***Different Brands*** : Some persons conscious of their higher status buy more of the higher-priced brand than the lower-priced brand, the former are regarded as status symbols.

(vi) ***Other Miscellaneous Situations :***
   - The law of demand does not apply to articles of display like diamonds.
   - It will not be true if people expect further rise or fall in price.
   - The law of demand holds good only when other things are constant. The law assumes that things like incomes, tastes and fashions of the consumers, population, prices of other articles etc. do not change. If there is any change in these things the law will not be true.
   - If a commodity goes out of fashion, its demand may fall despite its falling prices.
   - When a shortage of commodity is feared, more of it may be purchased even if its price in the market is rising.
   - People may not buy more of a commodity, notwithstanding a fall in its price, simply because they may not be aware of such a fall.

All the above types of exceptional demands arise under special circumstances are represented by exceptional demand curves, i.e., curves rising upward instead of sloping downwards.

Elasticity of Demand

The law of demand states that a change in price of a commodity causes a change in the quantity demanded. A fall in price causes an increase in demand and a rise in price causes a contraction of demand. But the rate of response of demand will be different in the case of different goods. For some goods, like rice and salt, the rate of response will be small. While for some other goods like soaps and silk cloths, the rate of response will be great. The degree of changes in demand in response to changes in price, thus differs with different goods. The relationship between changes in demand and changes in price is known as elasticity of demand. Elasticity is a measure of relative change in the amount demanded due to a change in price of a good.

Elastic and Inelastic Demands : Demand is said to be elastic if a little change in price causes a great change in the amount demanded. On the other hand, if the demand changes very little (or no change) due to a change in price, the demand is said to be inelastic. The terms very great and very little are vague. There must be a definite way of measuring elasticity of demand. There are two well-known methods of measuring it.

First Method :

$$\text{Elasticity (E)} = \frac{\%\ \text{Change in demand}}{\%\ \text{change in price}}$$

$$E = \frac{\text{Change in demand}}{\text{Amount demanded}} \bigg/ \frac{\text{Change in price}}{\text{price}}$$

If the change in demand is proportionate to a change in price elasticity is equal to unity or one. Example if a change of 10 % in price causes a change of 10 % in amount demanded elasticity is unity. Unit elasticity is the dividing line between elastic and inelastic demand. If the elasticity is one or more than one the elasticity is said to be elastic. If the elasticity is less than unity or one, it is said to be inelastic.

$D_1\ D_1$ -   Inelastic demand

$D_2\ D_2$ -   Moderately elastic

$D_3\ D_3$ -   Highly elastic.

Second Method : Elasticity may also be measured in terms of the total amount of money spent on a good.

(a) When the total amount of money spent increases with a fall in price (or decreases with a rise in price) elasticity is greater than unity or the demand is elastic.

(b) When the total amount of the money remains the same elasticity is said to be unity.

(c) When the total amount of the money decreases with a fall in price (or increases with a rise in price) elasticity is said to be less than unity or the demand is inelastic. This may be graphically represented as under

Example :

Price	Demand	Total Amount Spent	
Rs. 5	2000 units	Rs. 10.000	(a)
Rs. 4	3,000 units	Rs. 12,0000 ⎤	
Rs. 3	4,0002 units	Rs. 12,0000 ⎦	(b)
Rs. 2	5,500 units	Rs. 11,0000	(c)

Factors Determining Elasticity

1. For necessaries the demand is inelastic Ex: wheat , rice, salt etc.

2. For luxury the demand is elastic. A little fall in their price leads to a great increase in demand and little rise in price leads to a great fall in demand Ex: silk cloth, hair oils etc.

3. The demand for a commodity is elastic if it has substitutes ex: tea, coffee, bus and train travel etc.

4. If a commodity can be put to several uses demand tends to be elastic. With a fall in price such a commodity will be put less and less urgent uses. Ex: Electricity, LP Gas etc.

5. The demand for a commodity will be elastic if its use can be postponed. Thus if the price of shoes is very high we postpone purchasing a new pair.

6. Elasticity of demand also depends on the level of prices. If the price is too high (ex: cars) only rich people will purchase the good. A fall or rise in price causes no great change in demand. So the demand for them is inelastic.

Interconnected Demand : Often the demand for one commodity may be connected with the demand for another there is two such cases to be considered.

A. *Joint Demand* : In this case, several commodities are required to fulfill one demand, such as: bricks, cement, wood and iron are required for construction of house.

B. *Composite Demand* : In this case, one commodity will meet several demands, such as: electricity will meet the demands like lighting, heating, transport, industrial purpose etc.

Degrees of Elasticity

1. *Perfectly Elastic Demand* : When a small rise in price causes to a big fall in demand or zero, and conversely a small fall in price causes a large increase in demand it is said to be perfectly elastic. Thus the demand hypersensitive. Ex : 5 % increase in demand and with no fall in price.

   $E_p = 5/0 = ¥$; this is a rare practice but has great theoretical importance.

2. *Perfectly Inelastic Demand* : When demand is unaffected even after a substantial charge in price, the elasticity of demand is said to be zero or inelastic. Ex. If price is changed by 50 %, change in demand is zero.

   $E_p = 0/50 = 0$

3. *Highly Elastic Demand* : When a small change in price is followed by large change in demand, the demand is said to be highly elastic. Ex: A 40 % increase in demand is the result of 20 % fall in price.

   $E_p = 40/20 = 2$

4. *Inelastic Demand* : When a large change in price is followed by a small change in demand we can say that the demand is inelastic. Ex:- If the demand increased by 5 % following a 10 % rise in price

   $E_p = 5/10 = 0.5$ (Less than unity or 1)

5. *Unit Elasticity of Demand* : When a change in price leads to a equal change in demand the elasticity of demand is unit. Ex: - 20 % rise in price caused 20 % rise in demand.

   $E_p = 20/20 = 1$

Determinants of Price Elasticity of Demand

1. Degree of elasticity
2. Proportion of consumers income spent
3. Habits
4. Substitutes

5. Number of uses of commodities
6. Durable goods
7. Time
8. Range of prices

Practical Importance of Elasticity of Demand : The concept of elasticity of demand place a very important role in theoretical discussions in economics. A more remarkable point about this concept is that it serves as a very important guide for policy formulations in a number of areas.

1. *Pricing under imperfect competition* : To fix prices under monopoly
2. *Policy formulations by the government* : To impose taxes.
3. *Resources Prices* : Interest rates, transport rates etc.
4. *Terms of trade* : Foreign trade- quantity and quality and value of exports and imports
5. *Rate of exchange* : Changes in the exchange rates of other currencies
6. *Public utilities* : For fixing rates of services

Supply

Meaning : Supply means the quantities that a seller is willing and able to sell at different prices. It is obvious that if the price goes up, the seller will offer more for sale. But, if price down, he will be reluctant to sell and will offer to sell less. Just as demand implies willingness and ability to pay, the supply implies both willingness and ability to deliver the goods. Like demand, supply is also relative to a person, place and time.

Definition : The supply of any good may be defined as a schedule of respective quantities of the good which people are ready to offer for sale at all possible prices.

Distinction between Supply and Stock

Supply	Stock
Quantity actually offered for sale at a certain price	Total quantity which can be offered for sale if the conditions are favourable

The stock will change into supply and vice-versa according as the market price rise or falls. In case of perishable articles, like fresh milk and vegetables, there is no difference between stock and supply. The entire stock is supply and has to be sold off, for unless it is disposed of quickly, it will perish.

Supply Schedule

It is a list of various quantities of commodity, which a seller offered for sale at different alternative prices. For example, a supply schedule of the milkman in. a village. We notice that as the price falls, less milk is being offered for sale, and as price rises, the milkman is prepared to sell more.

Table 14.3 Supply schedule (Quantity of milk offered (in litres) by a milkman)

Price of milk per litre (Rs.)	Quantity offered per day (litre)
15.00	40
12.00	30
9.00	20
6.00	10
3.00	NIL

Supply Curve : The supply schedule can be represented in the form of a curve, as given below.

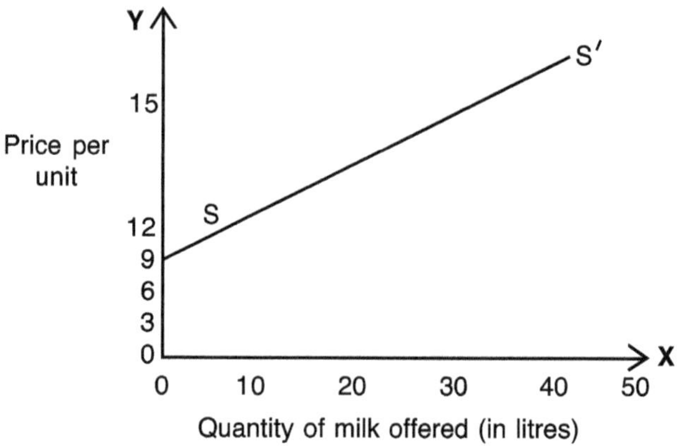

Quantities of milk offered for sale are measured along OX axis and prices along OY axis. The supply curve SS^1 slopes upwards as we go from the left to the right. This means that as the price rises, more is being offered for sale and *vice-versa*.

Supply Curve : The supply curve of an industry depicts the various quantities of the product offered for sale by the industry a various prices at a given time. Following are the supply curves of an industry.

- Short-run supply curve
- Long-run supply curve
- Supply curve of an increasing cost industry
- Supply curve of a decreasing cost industry

Law of Supply : In a given market, at any time, the quantity of any goods which people are to offer for sale generally varies directly with the price. Varies directly means that as price rises the quantity offered increases, and as it falls, the quantity offered decreases. It should be noted that, in case of demand, the quantity demanded varies inversely with the price, i.e., as price rises, demand decreases and vice-versa.

Causes of Changes in Supply : We cannot attribute changes in supply to changes in price, because when supply changes in consequence of a change in price, it is called extension and contraction, and not increase or decrease. In order to account for increase or decrease in supply, we have to discover the factors, which bring about a change in the very conditions of supply.

(i)   *Natural Calamities* : If rainfall is plentiful, timely, and well distributed, there will be bumper crops. On the contrary, floods, droughts, or earthquakes and other natural calamities are bound to affect production adversely. This is one set of conditions which brings about a change in the supply.

(ii)   *Technical Progress* : In manufacturing industries, this is a very important factor. A new machine may have been invented a new process discovered or a new material found, or perhaps a new use may have been found for a by-product. The discoveries of synthetic dyes, artificial rubber and wool are some such discoveries or improvements in technique.

(iii)   *Changes in Factor Prices* : If the factors of production become cheap, the supply will increase and *vice-versa.*

(iv)   *Transport Improvements* : It reduces the cost and increases the supply of the product.

(v)   *Calamities* : Calamities like war or famine must also affect the supply of goods. Even at higher prices adequate supplies are not forthcoming.

(vi)   *Monopolies* : The monopolists may deliberately increase or decrease the supply as it suits them. Thus exercise of monopolistic power brings about a change in supply.

(vii)   *Fiscal Policy* : The fiscal policy of the government also may affect the supply. For instance, a higher import duty will restrict the supply and a lower duty will stimulate it.

Interrelated Supply

Joint supply : Joint supply refers to goods supplied or produced jointly. Some commodities have a common origin and are produced in the same process. For example, sheep provides mutton, wool and manure.

Composite Supply : When there are different sources of supply of a commodity or service, we say that its supply is composed of all these sources. You can get light from electricity, gas, kerosene oil, candles, *etc.* All these sources go to make up the supply of light. It is a case of composite supply. Whenever there are substitutes or rival sources of supply, the supply is composite. For example, gas, electricity and oil all are useful or supplied for lighting.

Interaction of Demand and Supply

Price, under conditions of perfect competition is determined by the interaction of demand and supply. In this context, Marshall (1985) gave a famous analogy of a pair of scissors which is worth quoting: "We might as reasonably dispute whether it is the upper or the under blade of scissors that cuts a piece of paper, as whether value is governed by utility or cost of production. Neither the upper blade nor the lower one taken individually can do the work of cutting; both have their importance in the process of cutting. The lower blade may be kept stationary and only the upper one may be moved, yet both are indispensable for the process of cutting. The only really accurate to the question whether it is supply or demand which determines price is that it is both. The demand of all consumers and the supply of all firms together determine the prices which are then taken as given by each of them.

The price at which demand and supply are equal is known as an **equilibrium price**, since at this price the forces of demand and supply are balanced, or are in equilibrium. The quantity bought and sold (or the amount supplied or demanded) at this equilibrium price is known as **equilibrium amount**. An example in terms of both schedule and curve will make the whole thing clear. A glance at the following table for the demand and supply schedules relating to a variety of cotton cloth will show how the price is determined by demand and supply.

Table 14.4 Schedule for interaction of demand and supply

Price per Metre (Rs)	Quantity demanded (millions of metres / month)	Quantity supplied millions of metres / month	Pressure on price	Market equili-brium
25	9	18	Failing	Above market Equilibrium
20	10	16	Falling	
15	12	12	Neutral	Market Equib
10	15	7	Rising	Below market] equilibrium
5	20	0	Rising	

It will be seen that, when price is Rs. 15 per metre, 12 million metres are supplied and 12 million metres are demanded; the quantity demanded is equal to the quantity supplied. Rs. 15 per metre, therefore, is the equilibrium price. Price is in equilibrium at Rs. 15 or price of Rs. 15 will persist in the market, because at this level there is no tendency for it to rise or fall. Only at this price, there will be no unsatisfied buyers or sellers, who are prepared to let price alter to satisfy their desires. Of course this equilibrium price may not be reached at once. There may have to be an initial period of trial and error or oscillations around this equilibrium level before price finally settles down and supply balances demand. The equilibrium between demand and supply, often called market equilibrium determines the price in the market. Price comes to settle in the market at the level where demand and supply curves intersect each other.

Cost

Meaning : It plays an important role in the decisions of the farmers. Explicitly or implicitly most of the producers keep in mind the cost of producing additional units of output. In a competitive market, prices are not in the control of an individual farmer because there are large number of farmers who are individually producing very small proportions of total production of a commodity, this additional production must therefore sell at the same or even lower prices, even though additional production might involve higher costs. The second alternative is to reduce the cost of production through rationalisation of resource use with low cost production factors. The cost of production of different commodities relative to their prices is an important topic for discussion. Cost of production means the expenses incurred per unit of output. And these expenses are calculated with relation to a particular amount of product and in a particular time period.

Concept : The cost of production exists because the supply of productive resources is scarce relative to their demand. The costs have relevance to a specific time period and the cost in any production period include the value of the resource services transformed into product in this single period rather than the value of the resources itself. The resources used in the production may be (i) Poly-period resources, which represents stock services and only a part of these stock of services is transformed into product in each distinct production period and (ii) Mono-period resource, which represent a stock of services but here the entire stock of services is transformed into a product in a single period. In order to make rational choice from amongst alternatives, costs should therefore, be considered with relation with relation to a specified time period.

Categories of Cost : Several kinds of costs are involved in even the most simple production process. There are two major categories of costs namely:

Fixed Costs *:* are those which do not change in magnitude as the amount of output of the production process changes and are incurred even when production is not undertaken. These costs can be cash or non-cash costs. Fixed cash cost includes land taxes, interests, and insurance premiums. While non-cash fixed costs include depreciation on building, machinery equipment, interest on capital investment, cost of family labour and cost of management. It should be emphasised that costs do not become fixed until they have been incurred or committed. In the long run, however, all costs become variable.

Variable Costs *:* are the costs of using the variable inputs. These costs vary with the level of production. Higher the production, more will be the variable costs; lower the production, lower will be the variable costs. These costs include items such as feed, medicines etc.

Total Costs *:* Fixed costs plus variable costs equal to total costs. Total costs are required for computing net revenue. During long-run planning period, all inputs are variable. Thus in the long-run, there are no fixed costs.

Opportunity Costs *:* The farm resources normally have a number of alternatives uses. If they are not paid as much as they can earn in other farms, a particular farmer will find that the productive factors tend to shift to the lines of production in which they are better paid. We refer to the price that will be required to prevent the transference of factors to other uses as opportunity cost or alternative cost.

Total Fixed Cost *:* The TFC are the same for all output levels. Thus, once computed from the farm or enterprise budget, TFC is unchanging.

Total Variable Cost *:* TVC are computed by multiplying the amount of variable input used by the price per unit of input. TVC is zero, when output consequently the input is zero. It increases as output increases.

Total Costs *:* TC are the sum of total of total fixed cost and total variable cost (TC = TFC + TVC).

Average Fixed Cost *:* AFC is worked out by dividing total fixed costs by the amount of output. It is fixed cost per unit of output. AFC will vary for each level of output. As output increases, AFC decreases. Thus, when economists refer to increasing output as a method of spreading fixed costs, they mean increasing production to divide total fixed costs among more and more units of output thereby reducing cost per unit.

Average Variable Costs *:* AVC is worked out by dividing total variable costs by the amount of output. AVC varies with the level of production; When AVC decreases the efficiency of the variable input increases and it is at a maximum when AVC is at a minimum.

Average Total Cost : ATC can be computed in two ways: (a) By dividing total costs by output; (b) by adding AFC and AVC. ATC deceases as output increases, attains a minimum and increases thereafter. ATC is often referred to as the unit cost of production-the cost of production of unit of output. (ATC = AFC + AVC).

Marginal Cost : MC is the change in total cost in response to a unit increase in output at the margin. It is the change in cost associated with an increase or decrease of an additional unit of output. It is in other words, the cost of producing an additional unit of output and is computer by dividing the change in total cost by the corresponding change in output i.e., MC = ? TC / ?Y. Marginal product ratio = ?Y / ?X.

Cost Function : It can be presented in three ways; i.e. (i) Arithmetically (in tabular form), (ii) Geometrically (in graphic form) and (iii) Algebraically (in equation form).

Price Determination

Price determination is a process of balancing demand with supply. The demand of all consumers and the supply of all firms together determine the prices by their interaction.

Approaches in Price Determination

1. Tabular approach
2. Graphical approach
3. Simultaneous equations approach: The simplified market model may also be experienced in three equations.

Demand:	Qd	=	f x P
Supply:	Qs	=	f x P
Equilibrium condition:	Qd	=	Qs

Where,

Qd	-	Quantity demanded
Qs	-	Quantity supplied
P	-	Price
f	-	Frequency

Other Factors Involved In Price Determination

1. *Tax Incidence :* If tax is levied, seller will receive on each unit sold the market price of the commodity minus the amount of tax. If producers are to receive the same amount per unit as they were receiving prior to tax, the market price will have to raise the full amount of tax.

2. ***Equilibrium Price in Agriculture*** : The production of agricultural goods is subject to large variations due to factors beyond human control i.e., natural calamities. These factors reduce the agricultural output to a level well below the planned level. On the other hand, favourable conditions can cause production to be well above the planned level. Unplanned fluctuations in output will cause price variations in the opposite direction to the output changes.

3. ***Cobweb Behaviour in Production*** : Variations in agricultural production mostly due to the factors beyond human control. In general, there are fairly definite cycles of production. If we analyse a definite period, say 30 or 50 year period, we will find a peak and trough about every 10-15 years. This cyclical behaviour is demonstrated by what is called the "cobweb" theorem of production. According to this theorem, a commodity, which has a short production period, is usually subject to more violent fluctuations in production and prices than a commodity, which has a long period of production. The cobweb theorem is most appropriate in the case of non-storable goods.

Classification of Price Determination

Price determination can be classified in two ways i.e. under perfect competition, there are market price as well as normal price for perishable goods and non-perishable or reproducible goods. While under imperfect competition, there are monopoly, monopolistic competition and oligopoly.

Under Perfect Competition : There is a large number of producers. The supply of each producer constitutes only a small proportion of the total supply. Hence under perfect competition, no one seller can influence the price by changing his own supply (Table 14.5).

Table 14.5 Market price versus normal price.

S.No	Market Price	Normal Price
1.	Market price is determined by the equilibrium between demand and supply in a market period or very short run.	Normal or natural value of a commodity is that which economic forces would tend to bring about in the long run. Normal prices are those prices to which one expects prices to tend.
2.	Determined by temporary equilibrium between the forces of demand and supply at a time.	Result of the long-run equilibrium between demand and supply when the supply conditions have fully ad justed themselves to the changed demand conditions.

Table 14.5 Contd....

S.No	Market Price	Normal Price
3.	Reached at a give moment of day as a result of the temporary equilibrium.	Long-run normal price, in practice, may never reach.
4.	Governed by temporary causes and passing events.	Influenced by permanent and persistent causes.
5.	Fluctuates from day to day due to temporary changes either on the side of demand or on that of supply.	Under given permanent conditions demand and supply, remains the same.
6.	Market price may be above or below the marginal and average cost production depending upon the demand conditions.	The long-run normal price must be equal to both the marginal cost and the minimum long-run average cost.
7.	All commodities have a market price.	Only reproducible commodities can have normal price.

N.B.:Average prices will be an arithmetical average of all actual prices.

If commodities cannot be reproduced at all, their supply is fixed for all time. If the output of a commodity cannot be varied in response to changes in demand, there is no sense in speaking of their normal price. Such commodities may be pictures of old masters, unique diamonds, old manuscripts, etc. These commodities have market price but no normal price.

Under Imperfect Competition

(i) *Monopoly*

Monopoly may be defined as that market form in which a single producer controls the whole supply of a single commodity which has no close substitutes. Power to influence price is the essence of monopoly. But monopoly can either fix the price or fix the supply.

(ii) *Monopolistic Competition*

There is a large number of firms – but not too large. The commodity produced by these firms is not identical but is a bit differentiated. For example, household fan: different brands, namely, Orient, Usha, Polar, Compton Creaves, Ganga, *etc.*

Under monopolistic competition, different firms produce different varieties of products, therefore, different prices for them will be determined in the market depending upon their respective demand and cost conditions. Each firm will set price and output of its own product.

(iii) *Oligopoly*

The number of sellers is small. Since price-output decisions by one firm affect the decisions of other firms, nobody can be sure of their reaction.

Production

Meaning : Production in economics means production of wealth or value, and not merely utility.

Definition : Production is best defined as the creation or addition of values or of wealth. It may consist not only of goods but also of the services as of doctors, teachers etc. Production, in short, does not mean creation of all utilities, but only such utilities as have value-in-exchange.

From the above, it is clear that the act of production is not complete till the commodity reaches the hands of the consumers.

Factors of Production / Traditional Classification

Factors of production are classified as land, labour, capital and organisation (or enterprise). These factors called 'inputs' and the production is called 'output'.

(i) *Land* : Land means not merely soil, but all the natural resources on land, in water and air available to man.

(ii) *Labour* : It includes both mental and manual work undertaken for getting an income. Thus, any type of work undertaken for earning an income is called 'labour' in economics.

(iii) *Capital* : Capital means not only cash used in business but it also includes tools, machinery and appliances used in production.

(iv) *Organization* : Organisation or enterprise is the work of bringing the above three factors together and making the work harmoniously.

The factors are supplied by landlords, labourers, capitalists and organizsers (or entrepreneurs). Their respective incomes are called rent, wages, interest and profit.

Are these four factors reducible to two?

Some economists put forward the view that capital is the result of saving. Labourers work on land for the purpose of earning income and from that income only they are saving. Hence capital is the product of land and labour. As for organisation or enterprise, it is pointed out that it is only a form of labour. We are thus left only with two Actors, viz., land (nature) and labour (man). The other two factors, capital and enterprise, are included in them.

The above position is quite logical, but not realistic enough. Capital in modern times is of tremendous importance. Organisation is a type of mental work which includes taking risk. Taking risk is not mere labour. We thus come back to the view that there are not two but four factors of production: land, labour, capital and organisation. In modern view, economic thinkers of today contend that the factors of production are land, labour, capital and organisation. Anything, which goes into the production is a factor. Every acre of land is a factor by itself, and so is every worker, each rupee, and each individual entrepreneur.

I. Land

The word, land not only stands for surface of the earth but it also includes all nature, living and lifeless. The term 'land' thus embraces all that nature has created on the earth, above the earth and below the earth.

Definition : According to Marshall, land means not merely land in the strict sense of the word, but the whole of the materials and forces which nature gives freely for man's aid in land, water; in air and light and heat.

In India, agricultural land area is 141.23 million hectares. Of which, 40 per cent is irrigated and 60 per cent is rain fed (As on March'2005).

Characteristics of Land

(i) ***Land is a Free Good of Nature*** *:* It is not a 'produced' or man made agent. We have to accept it as it is. No doubt man tries to improve and modify nature. But he cannot completely master it.

(ii) ***Land is Limited in Area*** : Efforts have been made to reclaim land from the sea, and thus add to the total land surface. Some land in Holland has been reclaimed from the sea, but it is a small percentage of the total land surface of the world.

(iii) ***Land is Permanent*** : All other factors of production are destructible, but land cannot be completely destroyed.

(iv) ***Land Lacks Mobility*** : Land lacks geographical mobility. But it can put to many alternative uses and is thus mobile from a different point of view.

(vi) ***Land is Infinite Variety*** : Different pieces of land present infinite variation. None can say where the sandy soil ends and the clay begins.

II. Labour

Meaning :In economics, the term labour has a wider meaning. It does not merely mean manual labour. It includes mental work too. Doing one's own work or labour of love, however hard it may be, is not labour in the economic sense. Unless work is done for some consideration, i.e., payment in cash or kind, it cannot be called labour.

Definition : Labour may thus be defined as "any exertion of mind or body undergone partly or wholly with a view to earning some good other than the pleasure derived directly from the work.

According to 2001 census, total population in India is 102.53 crores. Of which, number of total workers is 40.25 crores (rural: 31.07 crores and urban 9.18 crores). According to latest World Development Report, in India, about 64 per cent of total labour force is dependent on agriculture, 16 per cent on industries and the rest about 20 per cent on trade, transport and other services. Livestock sector provides regular employment to 11 million people in principle status and 9 million in subsidiary status. About 6 million people are employed in fisheries sector (As in March' 2005).

Productive and Unproductive Labour

All the labour in the economic sense is productive. The term unproductive labour is now applied to wasted labour or misdirected labour or labour which has failed to achieve its purpose. Even misdirected labour is productive from the point of view of the labourer, as he gets payment for it; it is unproductive from the point of view of society only.

Characteristics of Labour

(i) *Labour is Inseparable from the Labourer* : A labourer's work has to be delivered in person. A labourer cannot stay at home and let his 'labour' work in the fields. A doctor has to attend his patients in person.

(ii) *The Labourer Sells his Services and not Himself* : Whatever has been spent on a labourer's training cannot be recovered by selling his labour.

(iii) *Labour is more Perishable than any other Commodity* : If time time flies labour flies with it. A day lost without work means a day's work gone for ever.

(iv) *Labourers have not the same power of bargaining as their employer* : This is because labour cannot be stored up, and labourers are poor and ignorant. Collective bargaining removes the weakness of labour.

(v) *Man, not a Machine* : A labourer differs from a machine in that he has feelings and likings. He works best when he is happy and puts his heart in the work.

(vi) *Less Mobile* : Labourer does not want to leave his home and his kith and kin.

(vii) *Supply Independent of its Demand* : The supply of labour is almost independent of its demand and cannot be easily and quickly increased or decreased. Labourers cannot be 'made to order' as other goods.

(viii) *Difficulty of Calculating Cost of Production.* It is not easy to calculate the cost of production of labourers. Here again labour is a peculiar commodity. How can the cost of bringing up children be calculated?

   (ix)   *Labourers Differ in Efficiency* : Unlike tools and machines, they are not exactly interchangeable. Workers are of varying degree of efficiency.

III. Capital

Meaning : Capital has been defined as that part of a person's wealth, other than land, which yields an income or which aids in the production of further wealth. Obviously, if wealth is left unused or is hoarded, it cannot be considered capital. Capital serves as an instrument of production. In ordinary language, capital is used in the sense of money. But when we talk of capital as a factor of production, capital includes money, machinery, tools and raw materials, transport equipment, factories and dams *etc.*

Are Securities and Shares Capital?

Securities and shares possessed by a man yield income to him. But they cannot be called capital, because they represent only titles of ownership rather than factors of production. Capital is not a primary or original factor of production since it has also been defined as "produced means of production" and as "man-made instrument of production". But land and labour are considered as primary or original factors of production.

Characteristics of Capital

   (i)   *Capital is Man-Made* : It is, therefore possible to increase its supply when the situation requires.

   (ii)   *Capital Involves the Element of Time* : That's why payment for capital is calculated in terms of so much percent per annum.

   (iii)   *The use of Capital Increases the Efficiency and Productive Power* : Frequent use of capital increases the efficiency and productive power of all the factors with which it is combined and used.

Classification of Capital : Capital can be divided into fixed capital and working capital.

   (i)   *Fixed Capital :* Fixed capital is the durable-use producer goods which are used in production again and again till they wear out. Machinery, tools, railways, tractors, factories etc., are all fixed capital. Fixed capital does not mean fixed in location.

   (ii)   *Working Capital :* Working capital includes the single-use producer goods like raw materials, goods in process, and fuel. They are used up in a single act of production. Moreover, money spent on them is fully recovered when goods made with them are sold in the market.

Wealth and Capital

From the definition of capital, it is clear that capital consists of valuable economic goods, which are scarce. Such goods are called wealth in Economics. All capital, therefore, is wealth. But all wealth is not capital. Only that part of wealth, which is used productively, is called capital. The car which is used for personal enjoyment is wealth but not capital. Wealth and capital, therefore, not synonymous.

Income and Capital

Capital is a fund (or stock) and income is a flow. Income flows in at regular intervals. Ex. a factory that a man owns is his capital, but the profit that he gets out of it every year is income. Income is calculated at regular intervals. It is calculated per week, per month or even per year.

Table 14.6 Is land equal to Capital?

S.No.	Land	Capital
1.	Nature's free gift to man	Man-made
2.	Lacks mobility	Fairly mobile
3.	Has no supply price	Supply of capital varies with its Price

For all these reasons, land can be distinguished from capital and is not regarded as capital.

Capitalist and Entrepreneur

The Capitalist is the owner of capital. He invests capital and receives interest irrespective of profit or loss from it. Thus he runs no risk.

The entrepreneur alone shoulders the risk. The whole loss, if any, falls on him just as the whole profit, if any, goes into his pocket.

Actually, the capitalist and entrepreneur may be one and the same person. In real life, the entrepreneur usually invests in the business some of his own money. He is thus partly a capitalist too, and earns interest, besides his profit, if any.

Theory of Production

Production of goods depends on the cost of production, which in turn depends on the prices of inputs or the factors of production. Cost of production is determined by the physical relationship between inputs and outputs.

Production in economics is generally understood as the transformation of inputs into outputs. Apart from physical changes of matter, production also includes services like buying and selling, transporting and financing. But the economic analysis we restrict the use of the term production to the production of goods only, because in the production of goods we can precisely specify the inputs and also identify the quantity and quality of outputs. Theory of production includes factors of production (land, labour and capital) and their organisation.

Importance of Theory of Production

- Helps the analysis of relation between costs and volume of output; it tells us how a manufacturer combines various outputs in order to produce a given output in an economically efficient manner, i.e., at the minimum unit cost.
- Provides a base for the theory of the demand of firms for production.

Production Economics : It is a broad division or field of specialisation within the subject of economics. It is concerned with the choice of production patterns and resource use in order to maximise the objective function of the society or the nation, within a framework of limited resources. It is concerned with choosing of available alternatives or their combinations with a view to maximising the returns and/or minimising the cost. Production economics is concerned with two broad categories of decisions in the production process:

1. How to organise resources in order to maximise the production of a single commodity i,e., to make choices from among various alternative ways of using resources.
2. What combination of different commodities to produce.

Goals of Production Economics

1. To provide guidelines in using the resources most efficiently, and
2. To facilitate the most effective use of resources from the stand point of the economy.

Objectives of Production Economics

- To determine and define the conditions which provide for optimum use of resources.
- To determine the extent to which the existing use of resources deviates from the optimum use.
- To analyse the factors or forces which are responsible for the existing production patterns and resources use and
- To delineate means and methods for changing the existing use of resources to the optimum level.

The principles or laws that help attain these objectives are the same on a micro as well as on a regional or national level. On a micro level where intra-farm resource allocation and production pattern involved; for example, if poultry is recommended to the farmers being a paying enterprise, this may increase poultry production so high that through lower prices the income might slowly decline.

Production Function : In the production function, output is dependent upon inputs or use of resources. The production function is of two types:

- ***Continuous Function*** : It can be explained by the response of production to the inputs. Labour is also quite divisible because it can be used in terms of man hours. Indivisible inputs can also be used as divisible units. For example, a tractor is not a divisible unit as an input. If it is to be used in production, no less than one can be employed. However, the tractor input can be varied by using a tractor more or leser number of hours per day.

- ***Discontinuous or Discrete Function*** : It is obtained for input factors or work units which are used or done in whole numbers such as one feeding or number of feedings. One can shift from one point to another. In discrete function, the alternative decision points get limited, otherwise same method of analysis is followed in both types of functions.

Merchandising and Product Planning

Merchandising : Merchandising is product planning i.e the internal company planning to get the right product or service to the market at the right time, at the right place and in the right colors and sizes. Merchandising is one of the two most loosely used terms in the marketing vocabulary. (1. Merchandising 2. Sales production). Merchandising means many things to many people. To some it is synonymous with marketing. To others it is an all inclusive term and marketing is only part of merchandising. It is used in several senses as a verb, noun or adjective. Ex. Marketing of new products, marketing success, marketing plans, marketing managers etc. However merchandising is synonymous with product planning i.e it includes all planning activities manufactures and middlemen designed to adjust their products to the market demand.

Merchandising by Manufactures and by Middle Men : Manufacturers are concerned with what to make and middlemen are concerned with what to buy and both of them are concerned with what to sell and want their merchandise to be market oriented – to please a consumer by satisfying his wants. Retailers wholesellers can provide ideals to manufacturers by knowing the tastes of the consumers while taking orders from customers.

Product Planning : Product planning is the process of determining that line of products which can secure maximum net realization from the intended markets. It is an " Act of marking out and supervising the search, screening, development & commercialization of new products; the modification of existing lines and discontinuation of marginal or unprofitable items". The managerial decision making in this area centers around deciding the type of products the company should develop and sell so that product serves as an instrument of achieving marketing objectives. It also involves surveillance of product behaviour and deciding whether it deserves to be on the product line, abandoned or repositioned. By this the management aims at insuring itself to maintain a strong competitive position. For this the company has to conduct market and product research and contribute to the fulfillment of consumer and business needs. The scope of product planning and product development includes decision making in the following areas.

1. Which products should the firm make and which should it buy ?
2. Should the company expand or simplify its line?
3. What new uses are there for each item ?
4. What brand, package and label should be used for each product ?
5. How should the product be styled and designed and in what sizes, colours and materials should it be produced ?
6. In what quantities should each item be produced ?
7. How should the product be priced

Process of Product Planning

Any company, before introducing a new product into the market, should conduct a customer oriented product research. The following figure reveals that the process of product planning starts with the conception of product ideas. The ideas so generated are screened for their commercial and technical merits by the marketing department. If the ideas appear to have both types of merits, marketing research is activated so as to probe their market potentialities with regard to consumer acceptance to scan market segments and to identify possible price ranges. The favourable marketing research finds lead management to reappraise these ideas form the view point of the company; facilities and constraints in the areas of finance, production, raw materials and state regulations. Reappraised product ideas are then turned over to marketing executives for approval. Marketing executive pass them on to the business analysis group for their appraisal in terms of economic viability. Then the ideas are transmitted along with consumer specifications to engineering people for developing product prototypes. The prototypes are then evaluated and approved for test product runs on a small scale so as to test market it and gather market reactions and response. The results of the test market are analyzed and necessary changes, if necessary, are made in the prototype before full skill production runs are under taken and product delivered to consumers.

The above described customer oriented product planning process reveals that it is composed of a sequence of six activities – namely

1. Idea generation

2. Screening
3. Business analysis
4. Product development
5. Test marketing
6. Commercialisation

Standardisation and Grading of Products

Standardisation makes sale by description possible. It assumes quality. It promotes uniformity of products. It widen the market for commodities. Standardisation means setting standards of quality.

Grading means separating products according to established standards. Grading enhances marketing efficiency. In agricultural marketing standardisation and grading plays a very important role.

Standardisation and grading are very important in product planning and development. Standardisation assures the quantity, shape, design, taste and colour of the products. Grading is a part of standardisation. Grading means division of products into lots as per their standards and quality. The following are the advantages of standardisation and grading.

1. Consumers can be confident to purchase standardised goods. So the sales may increase
2. It is easy to inform the consumers about the standardised goods through advertisements.
3. The distribution of standardised & graded goods is very easy.
4. It is easy to value the standardised goods and make insurance claims in case of loss of stocks.
5. The marketing expenses will be reduced when the products are graded.

Branding : The definitive identity which is used to identify an article to the consumers is called Brand. Brand may include a name, symbol, picture, logo, design or a combination of the above. For example, Vimal, Tata. If brands is registered under law, it is called a trademark.

Agmark : Agmark is the symbol given by the government to marketable agricultural products, certifying that they are upto marked standards of quality in respect of quantity, quality, taste, flavour, colour, processing etc. The producers have to apply to the concerned authorities sending the samples of their products periodically. If the authorities are satisfied with the standards of their products, the certificate will be given which is called an AGMARK certificate. The producers will print the symbol on their products to attract the consumers. Consumers also will be confident about the standards of the product if it has an AGMARK certificate. The producers and traders will be benefited in advertising and selling such products.

ISI : ISI means Indian Standards Institution. It is also the symbol given by the Government to the marketable manufactured products, certifying that they are up to marked standards of quality, in respect of quantity, quality, taste, flavour, colour, texture, processing etc. The manufacturers have to apply to the concerned authorities sending the samples of their products periodically. If the authorities are satisfied with the standards of the products in all respects the certificate will be given which is called an ISI certificate. The manufactures will print the symbol of ISI on their products to denote that their products have reliable standards. Consumers also will be of the opinion that the ISI products are of consistent quality. The manufactures and traders will be facilitated in the their propaganda and sales. The consumers need not be motivated nor educated to buy the AGMARK and ISI goods as the symbol itself does every thing.

Strengths & Weaknesses of Livestock sectors

India has hugeovine population of 209 millions accounting for about 51 % of Asia and 19 % of the world. The growth rate of buffaloes is more pronounced than cows. While, buffaloes constitutes less than 40 % of bovine population they account for more than half of the total milk production. The development of the dairy industry has startled the entire world and has been acknowledged as one of the most successful processes in the world. As per 1997 livestock census, the country has 56.8 million sheep and 115.52 million goats, ranks sixth and first in the world respectively, while Andhra Pradesh ranks second in sheep population after Rajasthan in the country and in case of goats, north-western region comprising Punjab, Hariyana, Rajasthan and Gujurat has dominated. According to National Commission on Agriculture (1997), income from sheep rearing in India is estimated to be Rs. 140 crores per annum, which is based on the annual production of 132 million kgs of mutton, 44 milion kgs of wool, 37 milion pieces of skins and 20 million tons of manure. Whereas the income from goat rearing in the country estimated to be Rs. 2443 crores, which is based on the annual production of 458 thousand metric tons of meat, 85 thousand metric tons of manure, 1500 thousand metric tons of milk and various others like skin and mohair.

The Dairying as a sub sector occupies an important position in the agricultural commodity contributing to the G.N.P. next only to rice. Dairy sector in India provides regular employment – 9.8 million people in principal status and 8.6 m people in subsidiary status which together constitute 5 % of the total work force. Why the share of agricultural output to the total G.D.P. has been declining the share of livestock out put to agriculture has been increasing and now it accounts to 26 % of agricultural output (around 6 % of total G.D.P.). Milk alone contributes Rs. 450 Billion to the G.N.P. of the country. 88.6 m. tons production in 2001-2002 projected india as the largest milk producer in the world, but the plan investment in agricultural and dairying is only a meager of 5 % of agriculture. Milk production between 1950-51 to 1970-71 is only 2 m. tons and began raising after Operation Flood Programme and reached 88.6 m. tons by 2001-02.

Strengths

1. 30 breeds of cattle – rich in genetic diversity of cattle
2. Economic symbiosis of crop and livestock production.
3. For small farmers cattle is perhaps the only tangible asset and mainstay for their socio-economic security.
4. Provides draught power, manure and cash income.
5. Dairy animals produce milk by converting the crop residues and by-products from the crops which otherwise be wasted.
6. Largest number of buffaloes in the world.
7. India has abundant labour force, which includes rural women forceengaged in rearing of dairy animals.
8. Well established cooperative network in the dairy sector has its routs in more than 270 Districts – serves to improve the marketing efficiency.

Weaknesses

1. Low productivity of cattle – average productivity of a cow is only 987 K.G. / as against the world average of 2038 K.G. / Lactation.
2. Raise in the non descript cows 80% and buffaloes 50 %
3. Chronic storage of feed and fodder.
4. Poor nutritive value
5. Low fertility
6. Processing inefficiencies.
7. India has a bad reputation of being simultaneously the largest producer as well as the largest waster of several agricultural and livestock products. It is devoid of a backup of world class processing technology to avoid waste or add value to many of its resources.

Accountancy

Accountancy : It is an art of recording, classifying and summarising in a significant manner and in terms of money, transactions and events which are of financial character and interpreting the results thereof.

Account : It is a statement of various dealings, which occur between a trader and other head of account of personal, real and nominal nature. It is a clear and concise record of the transactions relating to a person or firm, assets or liabilities, and expenses or incomes.

Book Keeping *:* is the art of recording transactions systematically in the books of accounts maintained by the trader for his purpose.

Ledger : Ledger is a book consisting of different accounts of personal and impersonal nature.

Journal : A journal may be defined as recording of business transactions with the application of rules of debit and credit chronologically after their analysis and classification.

Proforma of Journal				
Date	L.F.	Particulars	Debit Amount	Credit Amount

Debit				Credit			
Date	JF	Particulars	Amount	Date	JF	Particulars	Amount

Transaction : Transactions are the activities of business, which involve transfer of money or money's worth (goods or services) between two persons or two accounts.

Income : The gross inflow of cash receivables or other income arising in the course of the ordinary activities of an enterprise from the sale of goods, from the rendering of services and from the use of enterprises resources by others. It can be classified into capital income and revenue income. For example, capital, bank loans, debentures, interest, royalties, dividends, received.

Expenditure : An amount spent on any item (Account) by the trader to get benefit out   of it. It can be classified into capital and revenue expenditure.

Ex : purchase of Machinery, furniture, and payment of salaries, wags, rent & purchase of goods, etc.

Purchases : Act of buying goods purely meant for sale and not for personal use or office use. In other words buying of goods by the trader for selling them to the customer. Cash purchases will be recorded in cashbook. Credit purchases will be entered in purchases book.

Sales : When goods purchased are sold out it is called as sales. Here the title or the ownership will be transferred to the buyer. It is also known as "Business turnover". Cash sales are recorded in cashbook. Credit sales are entered in sales book.

Proforma of sales book				
Date	L.F.	Particulars	Outward Invoice No.	Amount

Purchases Returns : When the goods purchased on credit are found to be spoiled or of less quality such goods will be sent back to the supplier. They are called purchases returns. (Returns outwards). They are entered in purchases returns book.

Proforma of purchases returns book				
Date	L.F.	Particulars	Debit Note No.	Amount

Sales Returns : When goods sold to dealers or customers are found to be damaged they may be returned back to the trader. They are called sales returns (returns inwards). They are entered in sales returns book.

Proforma of sales returns book				
Date	L.F.	Particulars	Debit Note No.	Amount

Invoice : It is a business document prepared up the seller and sent to the buyer. It gives the particulars of the goods such as quantity, unit price and total amount, any deductions such as trade discount. It will be numbered and dated. Invoice on purchases is called as inward invoice and on sales is called as outward invoice.

No.		Proforma of Invoice		Date:
S.No.	Particulars of goods	Quantity	Price per unit	Amount

Debit Note : It is a document prepared by buyer sent to the seller. It consists of the particulars of goods returned out of purchase.

No.	Proforma of Debit Note		Date :	
Your account is debited with Rs. 50/- being the value of goods returned to you with the following particulars				
S. No.	Particulars of goods returned	Quantity	Price per unit	Amount
Reason: Being goods are of low quality 500				
Ref No. : Invoice No.				

Credit Note :   This is a document sent by vendor to the bayer with the particulars of good sold but returned by the customers.

No.	Proforma of Credit Note		Date :	
Your account is debited with Rs. 80/- being the value of goods returned by you				
S. No.	Particulars of sales returns	Quantity	Price per unit	Amount
Reason: Being goods damaged in transit 80				
Ref No. : Invoice No.				

Voucher : A serially numbered document being the written authorisation for each expenditure and serves as an evidence. It consists of the head of account the amount and date of payment.

Stock :   The value of unsold goods purchased or manufactured at the end of a period, which is called closing stock and the same becomes opening stock at the commencement of the next year.

Capital : The amount, which the proprietor invests in the business. It may be in the form of cash or other assets. It can also be calculated as excess of assets over liabilities. Capital is the liability of the business to its proprietor.

Liabilities : Liability means the amount that a business concern has to pay legally. It is the amount, which is owed to others on different accounts. They may be categorized as 1. Current liabilities   2. Long-term liabilities and   3. Owner's equity (capital)

Assets : Assets are the material things or possessions or properties of business including the amounts due from others. They may be categorized as 1. Current assets   2. Fixed assets and   3. Intangible assets.

Equity : It denotes the claims of various parties against the assets. Equity can be divided into two types.

  (a) *Owners Equity* : It is the claim of the owners against the assets of the business (Ex. Capital)

  (b) *Outsiders Equity* : It is the claim of outside parties such as creditors, debenture holders against the assets of the business.

Debtors : Debtors are the personal accounts which show debit balances and who receives a benefit without giving money or moneys worth. They owe us.

Creditors : Creditors are the personal accounts which show credit balances and who give a benefit without receiving anything immediately we owe them.

Bills of Exchange : A bill of exchange may be defined as an unconditional order drawn by a person over another to pay a certain sum of money only to or to the order of a certain person or to the bearer of the instrument. Creditors draw bills on their debtors to recover the debt. They can be divided as bills receivable and bills payable. Bills receivable are received by us from our debtors and bills payable are accepted by us to our creditors. Bills receivable are assets and bills payable one liability. The record of bills receivable is made in Bill Receivable Book and bills payable is made Bills Payable book

Investments : If the surplus funds of the business are retained in the business they are idle and reap no income. So they are invested, otherwise in private or public sectors. Ex. shares, debentures, government security, N.S.S. etc. They yield some income and can be encashed whenever need arises. Investments are assets.

Cash Book : All the cash transactions of the business are recorded in the cashbook. It is two in one. i.e. cash book is a journal and a ledger too. It means original entries and ledger postings also are made in it. It belongs to real account category and the rule of debit what comes in and credit what goes out applies to the cashbook. It can be divided into 3 kinds.

1. Simple cash book ; 2. Two columnar cash book and 3. Three columnar cashbook

1. Simple Cash Book

Debit				Credit			
Date	LF	Particulars Amount	Amount	Date	LF	Particulars	Amount

2. Two Columnar Cash Book

Debit					Credit				
Date	LF	Particulars	Dico. Given	Cash /bank	Date	LF	Particulars	Discount received	Cash /bank

3. Three Columnar Cash Book

Debit						Credit					
Date	LF	Particulars	Dico. Given	Cash	Bank	Date	LF	Particulars	Dis recor ded	Cash	/Bank

Classification of Accounts

Before dealing with the types of accounts we must know how to identify the transaction. In double entry system of book keeping, the transactions relating to various business activities can be classified into three categories.

1. Transactions relating to individuals and organisations
2. Transactions relating to properties, goods and cash
3. Transactions relating to expenses or losses and incomes or gains.

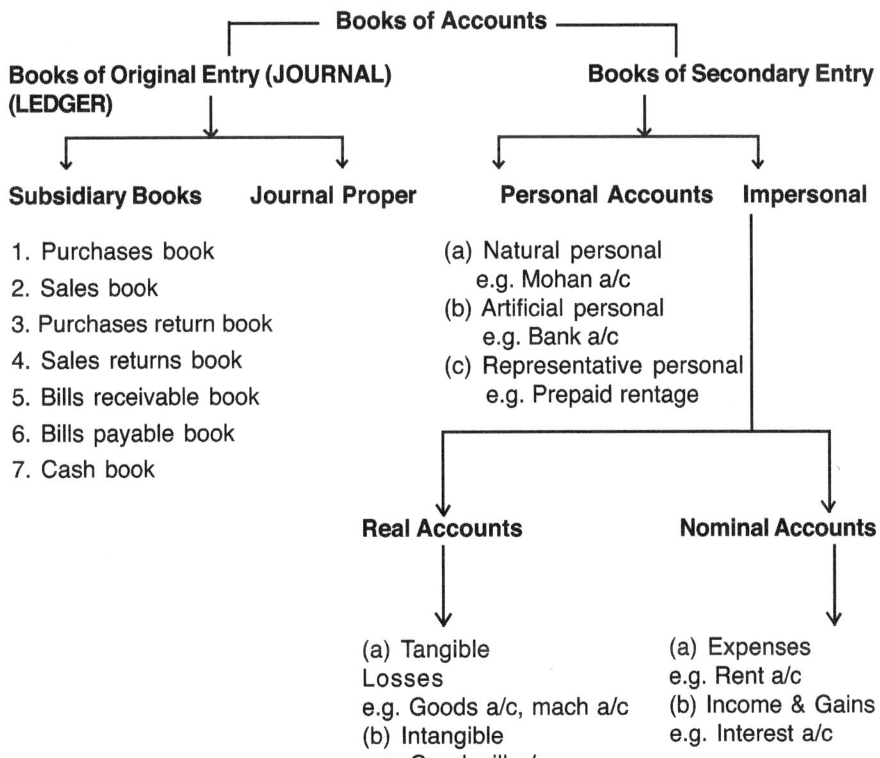

Rules of Debit and Credit

Personal Account : Debit the receiver, while credit the given.

Real accounts : Debit what comes in, while credit what goes out.

Nominal Accounts : Debit all expenses and losses, while credit all incomes and gains

Table 14.6 Examples of Different Types of Accounts.

Name of a/c	Type of a/c	Name of a/c	Type of a/c
Rent a/c	Nominal	Mohan a/c	Personal
Cash a/c	Real	Salaries a/c	Nominal
Buildings a/c	Real	Cattle a/c	Real
Furniture a/c	Real	Feed a/c	Real
Equipment a/c	Real	Medicine Exp. A/c	Nominal
Interest a/c	Nominal	Machinery a/c	Real
Capital a/c	Personal	Bank a/c	Personal
Advertisement a/c.	Nominal	Wages a/c	Nominal
Electricity Exp A/c	Nominal	Stationary a/c	Nominal
Discount a/c	Nominal	Sales returns a/c	Real
Commision a/c	Nominal	Prepaid Exp. a/c	Personal
Repairs a/c	Nominal	Dividend a/c	Nominal
Carriage a/c	Nominal	Dividend a/c	Nominal
Office expenses a/c	Nominal	Travelling Exp. A/c.	Nominal
Insurance a/c	Nominal	Postage a/c	Nominal
O.s commission a/c	Personal	Prepaid Interest	Personal
Interest recd in	Personal	Customs duty a/c	Nominal
advance a/c	personal	Customs duty a/c	Nominal
Bills receivable a/c	Real	Bills payable a/c	Real

Table 14.7 Examples of Various Nature of Accounts based on Transaction.

Transaction	A/c involved	Nature of account	Debit / Credit
Cash paid for rent Rent a/c	Cash a/c Nominal	Real Nominal	Credit Debit
Farm building purchased	Building a/c Cash a/c	Real Real	Debit Credit
Paid Salaries in cash	Salaries a/c Cash a/c	Nominal Real	Debit Credit
Goods sold to Suresh	Sales a/c Suresh a/c	Real Personal	Credit Debit
Capital introduced	Capital a/c Cash a/c	Personal Real	Credit Debit

Table 14.6 *Contd....*

Transaction	A/c involved	Nature of account	Debit/Credit
Interest received	Interest a/c Cash a/c	Nominal Real	Credit Debit
Telephone charges paid	Tel. Charges a/c Cash a/c	Nominal Real	Debit Credit
Commission received	Commission a/c Cash a/c	Nominal Real	Credit Debit
Machinery sold	Machinery a/c Cash a/c	Real Real	Credit Debit
Feed purchased	Feed a/c Cash a/c	Real Real	Debit Credit
Wages paid	Wages a/c cash a/c	Nominal Real	Debit Credit

Final Accounts

At the end of financial year the ledger accounts are balanced. Some of the accounts show debit balances and some other credit balances. Trial balance is prepared with the balances of ledger accounts. Debit balances will be listed in debit column and credit balances in credit column. Trial balance may be defined as a statement of ledger balances. The totals of debit and credit columns will be equal as per the rule of double entry system of book – keeping. The agreement of the two columns of the trial balance is the proof of accuracy of the accounts prepared. If there is any disagreement in the trial balance, it means that there are some errors. Even if the trial balance agrees then may be some errors. The errors and their rectification will be dealt with in a later chapter. The final accounts will be prepared after the agreement of trial balance. The final accounts constitute three parts.

1. *Trading Account :* To find out the result of trade (Purchases, manufacturing and sales) trading account is prepared with the incomes and expenses concerning the trade. The result of trading account is called gross profit or gross loss. The proforma of trading account is as under.

2. *Profit and Loss Account :* To find out the result of whole business the Profit & loss a/c is prepared with general incomes and expenses of Administration, selling and distribution and other indirect expenses. The gross profit / loss will be brought down from trading a/c. the balance shown by profit & loss a/c will be net profit or net loss. For example, proforma of profit & loss account is as under.

3. *Balance Sheet :* Balance sheet is a statement prepared with the values of assets and liabilities. It explains us about the financial position of the trader i.e. solvency and liquidity. It also explains the resources of the business through which funds are received and the method of their investment. For example, proforma of balance sheet is give here under.

Table 14.8 Balance Sheet.

	Capital & Liabilities			Properties & Assets	
1.	**Current liabilities**		1.	**Current Assets**	
	Outstanding expenses	xxx		Cash	xxx
	Bills payable	xxx		Bank	xxx
	Bank O.D.	xxx		Bill receivable	xxx
	Creditors	xxx		Investments	xxx
				Debtors	xxx
				Out standing incomes	xxx
				Closing stock	xxx
2.	**Loans**		2.	**Fixes assets**	
	Long term loans	xxx		Tools & Equipment	xxx
	short term loans	xxx		Vehicles	xxx
				Furnitures	xxx
				Plant	xxx
				Land & Buildings	xxx
3.	**Reserves & Provisions**		3.	**Nominal assets**	
	General reserves	xxx		Good will	xxx
	Specific reserves	xxx		Patents, copyrights	xxx
4.	**Capital**				
	Op. capital -	xxx			
	Add Addl. Capital -	xxx			
	Add Interest on capital -	xxx			
	Add Net profit -	xxx			
		xxx			
	Less drawings + interest -				
	xxx	xxx			xxxx
	Less net loss -				
	xxx				

As per the rule in double entry system of book keeping, the totals of Assets side and liabilities side will be equal. Otherwise, some errors might have taken place in the accounts.

Adjustments : The final accounts will be prepared as per the balances given in the trial balance. Some of the transactions relating to incomes & Expenses pertaining to a particular accounting year might not be taken into account. Unless they are also taken into consideration, the results shown by the final accounts will not be correct. So such unaccounted incomes and expenses may be given as adjustments annexed to the trial balance. Such adjustments are to be passed as original entries, which are called adjustment entries. The usual adjustments are listed as under.

1. Closing stock
2. Outstanding expenses
3. Prepaid expenses
4. Outstanding income
5. Income received in advance
6. Depreciation on assets
7. Bad debts
8. Reserve for bad debts
9. Interest on capital
10. Interest on drawings
11. Loss of fire & Insurance coverage

The transactions given as adjustments must be passed through debit and credit i.e. they will be adjusted in any two places in trading a/c; profit & loss a/c and balance sheet.

1. *Closing Stock* : It is to be posted on the credit side of trading a/c and shown on the assets side of B/s. The adjustment entry is

2. *Outstanding Expenses* : Some of the expenses like salaries, rent etc. maybe due relating to this year. Such expenses are to be added to the respective items in trading & profit and loss account and shown on the liabilities side of the Balance sheet.

3. *Prepaid Expenses* : Some of the expenses may be paid in advance, e.g. Insurance premium. A portion of the insurance paid may belong to next year which is to be deducted from it and shown as an asset in Balance Sheet.

4. *Outstanding Incomes* : Some of the incomes may be occurred but not received during this year, e.g. Commission, rent etc., It is to be added to the respective item in the profit and loss account and shown as an asset in Balance Sheet.

5. *Income received in Advance* : Some of the incomes may be received in advance which belongs to next year. It is to be deducted from the respective income in profit and loss account shown as a liability in Balance Sheet, ex. apprentice premium or commission received in advance

6. *Depreciation on Assets* : The value of the assets may be reduced due to several reason like wear and tear, outdated, climate, transport etc.

which is called depreciation. It is to be treated as a business expense. It is to be shown on the debit side of Profit and Loss account and debited from the value of the asset in Balance Sheet. If there is an increase in the value of the asset and is realized such increase should be taken as a profit.

7. *Bad Debts* : When goods are sold on credit debtors will arise who owe us. Some of the debtors may not turn up and may become insolvents. Where by the amounts due from them become bad. It is a business loss. It is to be posted on the debit side of profit and loss account and deducted from debtors in the balance sheet.

8. *Reserve for Bad Debts* : In anticipation of loss through bad debts the trader keeps a part of the present profits aside, which is called reserve for bad debits. It is to be shown on the debit side of Profit and Loss account and deducted from debtors in the Balance Sheet.

9. *Interest on Capital* : If the trader starts business with borrowed capital, he has to pay interest on it. Even it he invests his own funds he may take interests on his capital. Such interest would be an expense in the business. It is to be shown on the debit side of profit and loss account and added to the capital.

10. *Interest on Drawings* : When interest on capital is charged, it is logical to calculate interest on drawings also. This is income to the business. It is to be shown on the credit side of profit and loss account and added to drawings before deducting from capital.

11. *Loss by Fire / Accident and Insurance Claim* : A trader suffers loss if any fire or accident takes place. If there is no insurance coverage the trader has to bear all the loss. It is to be shown on the credit side of trading account and on the debit side of a/c. If the loss is covered by insurance, insurance company makes good the loss completely or partially. Thus the insurance claim is to e deducted from loss on the debit side of Profit and Loss account and shown on the assets side of the balance sheet.

Errors and their Rectification : Trial Balance is prepared to verify the accuracy of the accounts already prepared. If the debit and credit totals are not equal, it is evident on the face of it that there were some errors. Then they will be rectified and the trial balance will be set right. The agreement of the trial balance cannot be taken as evidence of the accuracy of the books. That means there may be errors even if the trial balance agrees. Hence the errors may be classified into two categories.

1. Errors disclosed by the disagreement of the trial balance.

2. Errors not disclosed by the trial balance.

1. **Errors Disclosed by the Disagreement of Trial Balance** : The errors which lead to a difference in the trial balance are as under.

   (a) **Errors of Omission** : The errors, which are omitted to be posted to a ledger account lead to a difference in trial balance. For example, purchase of goods was entered in purchases a/c but not in personal a/c

   (b) **Errors of commission** : Posting a wrong amount in a ledger a/c leads to a difference in trial balance. For example, purchase of goods for Rs. 1000/- from x was wrongly posted as Rs.100/- in his a/c

   (c) **Posting on the Wrong Side of the Ledger A/c** : this type of errors leads to a difference of double the amount in trial balance. For example, purchase of goods from Y was wrongly debited in his a/c

   (d) **Errors in Casting Subsidiary Books** : If there are errors in totaling the subsidiary books, it leads to a difference in trial balance. For example, total of sales book is under cast by 1000/-

   (e) **Errors in Totaling the Ledger Accounts** : This also leads to a difference in trial balance. For example, cashbook was over cost on the debit side.

   (f) **Errors in Balancing the Ledger Accounts** : This leads to a difference in trial balance. For example, the balance of interest a/c was shown as 400/- instead of 410/-

   (g) **Omission of an Account to be Brought into Trial Balance:** This leads to a difference in the trial balance. For example, purchases return a/c showing a credit balance of Rs. 1500/- not brought to Trial Balance.

2. **Errors not Disclosed by Trial Balance** : As was already stated, there may be errors even if the trial balance agrees. The effect of these errors will be dual i.e. debit & credit. So these errors do not lead to a difference in trial balance. They are as under.

   (a) **Errors of Omission** : Ommission of a transaction to be recorded in journal leads to this type of errors. For example, purchase of goods from 'X' for Rs. 300/- was not entered in purchases book

   (b) **Errors of Commission** : A transaction will be entered in journal but posted to a wrong account leads to this type of errors. For example, sold goods to 'X' for Rs. 800/- and was posted to 'Y' account.

   (c) **Errors in Original entry:** these errors take place if a wrong amount is recorded in journal. For example, purchased goods for Rs. 750/- was recorded in purchases book as Rs. 570/-

(d)   *Errors in Changing the Nature of Transactions:* If the nature of the transaction is changed this type of errors takes place. For example, purchase of goods for Rs. 500/- from z was recorded in sales book.

(e)   *Errors of Principle :* If the transactions are recorded in original books without observing the rules of accounts, these errors take place. For example, purchased furniture from B for Rs. 5000/- and recorded in purchases book.

(f)   *Compensating Errors :* If an error takes place on the debit of one account and another error on the credit of another account for an equal amount, it will be called as compensating error, which does not result in a difference in trial balance. For example, purchase a/c is over cast by Rs. 1000 and sales a/c is over cost by Rs. 1000/-

If the final accounts are prepared without rectifying the errors, the profit or loss would not be reliable. If the errors are found before the preparation of final accounts, they will be rectified and then the final accounts will be prepared. Some of the errors may not be located or found before the preparation of final accounts. The management proceeds to prepare the final accounts. As and when the errors are located, they will be rectified and the profit /loss also is to be adjusted by preparing the adjusted Profit & Loss account. Some errors will have no effect on profit /loss while some errors inflate the profit /loss and some errors reduce the profit /loss. The profit /loss will be adjusted accordingly to arrive at the correct profit /loss through a statement or adjusted Profit and Loss account.

Analysis of Financial Accounts and Different Financial Test Ratios

Preparation of financial accounts helps the trader to find out the results of his business and the financial position of the business. But the trader needs more than that for the efficient management of his business. He needs different rates and ratios to compare the present position with the past results and set standards if any and assess the situation. If there are any adverse variances he has to find out the causes and set them right. For this he needs different ratios of different periods for the purpose of analysis and makes proper decisions or changes in the decisions already made. This is called ratio analysis. The purpose of ratio analysis is to compare between different years of the business, with other businesses, with the industry, with standards set previously and among different departments of the business. By ratio analysis, the trend of the business can be understood.

The ratios can be Divided into Three Categories:

I.   Balance sheet ratios

II. Profit and loss ratios

III. Combined ratios

I. Balance Sheet Ratios

$$\text{(a) Current Ratio} = \frac{\text{Current assets}}{\text{Current liabilities}}$$

The standard of this ratio is 2 : 1, that means the value of current assets must be double the current liabilities. For example, current assets are Cash, Bank, B/R, Debtors, Investments, Advances and Current liabilities are Bank O.D., B/P, Creditors, provision for I.T., dividends declared, dividends unclaimed, advance income, O/S EXP., short term liabilities.

$$\text{(b)}\quad \frac{\text{Quick Ratio}}{\text{(Acid test Ratio}} = \frac{\text{Liquid assets}}{\text{Current liabilities}}$$

The standard of this ratio is 1 :1

$$\text{(c) Liquidity Ratio} = \frac{\text{Cash\&bank Balance + Realisable Securities}}{\text{Current liabilities}}$$

The standard of this ratio is 1 :1

$$\text{(d) Ratio of Inventory to Working Capital} = \frac{\text{Inventory}}{\text{working capital}}$$

The standard of this ratio is 1 :1

$$\text{(e) Ratio of Current Assets to Fixed Assets} = \frac{\text{Current assets}}{\text{fixed assets}}$$

This ratio has no standard

$$\text{(f) Debt equity ratio} = \frac{\text{Outsider's funds}}{\text{Share holders funds}}$$

The standard of this ratio is 2 :1

$$\text{(g) Proprietary ratio} = \frac{\text{Share holders funds}}{\text{Total assets}}$$

The standard of this Ratio is 1 : 3

$$\text{(h) Capital Gearing Ratio} = \frac{\text{Equity share capital}}{\text{Liabilities having fixed interest}}$$

II. Profit and Loss Ratios

(a) Gross profit Ratio = $\dfrac{\text{G.P.}}{\text{Net sales}} \times 100$

(b) Operating Ratio = $\dfrac{\text{Cost of goods sold} + \text{OP. Exp.}}{\text{Net sales}} \times 100$

If the operation ratio is less, the business is better

(c) Operating profit Ratio = $\dfrac{\text{Operating profit}}{\text{Sales}} \times 100$

More is Operating profit ratio, more profitable is the business

(d) Net profit Ratio = $\dfrac{\text{Net profit after taxes}}{\text{Sales}} \times 100$

More is Net profit ratio, more is the profitability.

III. Combined Ratios

(a) Stock turnover Ratio = $\dfrac{\text{Cost of goods sold}}{\text{average stock held}}$

(b) Fixed assets turnover Ratio = $\dfrac{\text{Total sales}}{\text{Net assets}}$

More is Fixed assets turnover ratio means more is the efficiency of the business.

(c) Capital turnover Ratio = $\dfrac{\text{Total sales}}{\text{Total capital}}$

Total Capital = Share capital + L.T. loans

(d) Returns on share Holders investment Ratio

$= \dfrac{\text{net profit}}{\text{share holders funds}} \times 100$

Share holders funds = E.Share capital + Pref. Share capital + Capital reserve + Accumulated Profits.

(e) Returns on capital employed Ratio = $\dfrac{\text{Net profit}}{\text{Capital employed}} \times 100$

Capital employed = fixed assets + current assets – current liabilities

Returns on capital employed ratio explains the efficient use of capital and is used to take budget decisions.

Preparation of Project of Dairy Farm

Establishment of a Dairy farm with 10 Jersey cows by educated unemployed youth of Rural Background under partnership (alleast five partners) type of organization.

Total Capital Requirements

Fixed Capital :

1. 10 cows x 25,000	2,50,000
2. Sheds for 20 animals with pillars and asbestos sheets	2,50,000
3. Tools and implement (buckets, Feed tubs, Chopper etc)	20,000
TOTAL	**5,20,000**

Working Capital (Estimated per Annum)   5,20,000

1. Feed 10 cows + 10 Calves × 12 months × Rs. 500	60,000
2. Electricity 12 months × Rs. 500	6,000
3. Salary of workers 3 × 1000 × 12 months	36,000
4. Green fodder 5 acres base @ Rs.10,000/- P.A	50,000
5. Hey (wheat & Rice) 10 Cows + 10 Calves × 12 Months × 300	36,000
6. Other incidentals including medicines 12 months × 1000/-	12,000
7. Veterinary consultation 12 X 500/-	6,000
TOTAL	**7,26,000**

Total capital approximately	7,50,000
Own capital (atleast two partnets	1,00,000
Loan from the bank	6,50,000
TOTAL	**7,50,000**

Estimated Annual Income

1. Sale of milk

   10 cows @ 10 litres per day @ Rs. 12 per litre for

   300 days i.e., 10 × 10 × 12 × 300 — 3,60,000

2. Sale of manure

   100 Trucks @ Rs. 300 each i.e. 100 x 300 — 30,000

3. Sale of calves (2 years old) @ Rs.5000

   i.e. 10 × 5000 — 50,000

Total Income   4,40,000

Estimated Annual Expenditure

1. Feed 10 Cows × 12 months × Rs. 500   60,000
2. Electricity 12 Months × Rs. 500   6,000
3. Salary of Workers 3 workers × Rs. 1000 × 12 months   36,000
4. Green fodder (per year) 5 acres lease @ Rs. 10,000   50,000
5. Hay 10 Cows + 10 Calves × 12 months × Rs. 300   36,000
6. Other incidentals including medicines 12 months × 1000   12,000
7. Veterinary Consultation 12 Months @ Rs. 500   6,000
8. Depreciation @ 10%
    Sheds   25,000
    Cows   25,000
    Tools   2,000
9. Bank interest @ 8 % per annum   64,000

Total Expenditure   **3,22,000**

Approximate Expenditure   Rs. 3,20,000

Estimated net Profit = Total Income – Total expenditure

$$= 4,40,000 - 3,20,000$$

$$= \mathbf{1,20,000}$$

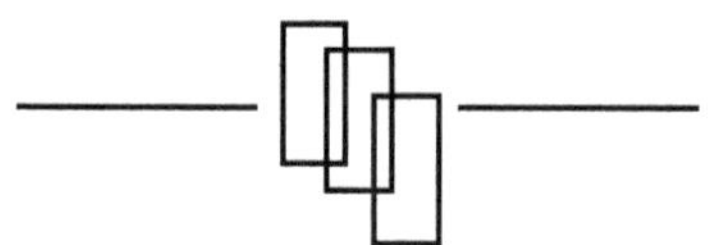

Livestock Farming : Types, Systems and Marketing

On the basis of land area, farmers can be categorised into viz. marginal farmers (less than 1 ha), Ssmall farmers (between 1 to 4 ha) medium farmers (4 to 10 ha) and large farmers (above 10 ha). The categorisation of land into different sizes depends on the purpose for which we classify the farm holdings.

Large Scale Farming

Farm size is a topic of extreme interest in agriculture, because size of farm is the vital element in determining the earning capacity of the farmer as well as efficiency of a farming unit. There has been debate over what should be the appropriate size of a farm; economists and farmers advocated large scale farming for efficient operations, satisfactory income and afford ability to the consumer in terms of reasonable prices. But, on the other hand, some persons strongly advocate small scale farming on the ground of social justice. A great majority of farm families with low incomes live on under-sized and inadequate units. Since the amount of income is dependent on the size of the units, small and tiny holdings are mainly responsible for the under-development of poor in the developing countries.

Agricultural Holding: Agricultural holding normally implies, the management unit i.e. the area of the land for cultivation as a single unit by an individual, joint family or more than one farmer on joint basis. The land may be owned, taken on lease or may be partly owned and partly rented. The term agricultural holding has been used to mean the total area of land owned by an individual or joint family whether cultivated by the family or rented out to tenant farmers.

Economic Holding : The Agrarhian Reforms committee (1949) defined an economic holding as one which could provide a reasonable standard of living to the cultivator and give full employment to a family of normal size.

Costs and Profits Related to Size : Under most circumstances a small farm has greater costs for each unit of crop or livestock produced than a larger one. Ordinarily costs per unit reduce, as the size of the farm or single enterprise increases.

Advantages of Large-Scale Farming

1. *Increased efficiency and full utilisation of labour (labour efficiency)*: Labour can be fully utilised and also efficiently because of the large size of the farm or enterprise. Family labour can be fully employed on the farm.

2. *Lower Machine Cost as a Result of Greater Annual Use* : On large farms, fixed costs can be spread out over more acres, making machinery services more economical. Total fixed costs could be spread out over more acres and a greater output.

3. *Building Economics (Economical use of Building)* : Livestock housing and grain storage on large scale. The livestock building per acre declines with acreage. The investment in farm buildings may not increase as rapidly as size of the farm increases. For example, doubling the size of the farm does not mean, that building costs will be doubled.

4. *Buying and Selling (Economy in Buying and Selling)*: Large farms enjoy a further advantage in buying and selling. In buying some types of inputs (supplies) a reduction in price is possible though large volume of purchases. On the selling side, the principal economics are likely to come in transportation.

5. *Management* : On most of the farms the farmer is both labourer and manager. When the farm is small the labour function is generally as important as the management function. As the size of the farm increases, management tends to become more important then labour. On a large farm these is always the possibility of reducing the many of the tasks to a routine. The manager, being relieved from physical effort, may pay his attention more largely to planning and solving the problems connected with the business. But the small farm is without its advantages even in management.

6. *Economics in Financing (Easy to Get Finance)* : Large farmers can borrow money and obtain credit with greater case and at a lesser expense. Further, due to bigger turnover and greater total profit, large farmer is better placed as far as self–financing is increased.

7. *Economic in the use of By-products* : On every farm there are certain products or materials which are produced incidentally along with main products. Straw and cattle manure are examples of by products. These are non marketable, as a rule, unless transformed. The economic utilisation of these by products depends upon the quantity available for transformation. In this respect a large farm has greater advantages over a small farm.

Disadvantages of Large Scale Farming

1. There are greater losses during depression period, because the total costs are relatively large. There is diversity of business and hence loss from every direction would add to the total loss.

2. When the yields per lactage or per animal are low due to abnormal weather conditions or attack of pests or diseases, the farmers would suffer a greater loss.

3. There is an increasing difficulty of supervision over wider area leading to some slackness on the part of the labour force. The inefficient use of labour reduces the advantages of large farmers.

4. The farm suffers a great loss if the manager cannot make full use of available equipment and labour on the farm.

Apart from the above disadvantages; If the size of farm is extended beyond a limit, some of the resulting economics discussed earlier tend to be offset by inefficiencies. The pre-occupation of farmer into living things, the dependence on weather and season, dispersal of labour over large areas and the absence of continuous routine activities present a set of conditions in agriculture, which come in the way of reaping the economics of large scale production.

Small Scale Farming

On the economic side, small size of holdings have came as a major hurdle in the way of introduction of new and improved technology. Smallness of holdings on the one hand deters the use of mechanisation and on the other hand do not allow the use of new varieties of seeds fertilisers and other modern inputs due to lack of purchasing at power the levels of small farmers. In comparison with small holders, the new agriculture technology is well accepted and adopted by the owners of large holdings leading to increase in their incomes. This has created wide gulf between the income levels of large farmers and small farmers growing rise to serious social implications. These small land holders who cannot make their subsistence from their respective farm holdings either sell or lease out their holdings and become agricultural labours or migrate to urban areas in search of some work. Farms with an area up to 10 hectares can be considered as small-scale farming. Small size holdings predominate the Indian agriculture.

Advantages Small Scale Farming

1. *Close Attention and Supervision :* As the size of the firm is small, close attention and supervision by the farmer is possible. Close attention and supervision of the farm help increasing the productivity of the resources.

2. ***Efficient use of Family Labour :*** The family labour can be efficiently used on the farm. The family labour participation increases the productivity of the farm. The dependence on the hired labour decreases.

3. ***Better use of Farm By-Products :*** small farms usually keep certain number of cattle on the farm. Farm by-products like straw can be feed to the cattle.

4. ***Higher Productivity :*** The productivity on small farms is more when compared to large farms. The additional inputs used on the farm results in higher productivity.

5. ***Recycling or Organic Matter :*** almost all the organic matter available on the farm is converted into manure, which is used for crop production.

6. ***Low Market Dependence :*** The productive resources required for farm production are produced on the farm itself. for example, seeds & manures. Hence the dependency on the market for inputs is reduced.

7. ***Better Family Satisfaction :*** All the family requirements are produced on the farm and consumed by the family members, which leads to the satisfaction of the family members.

8. ***Intensive Cultivation is Possible :*** The land will be put to more intensive use as the family depends on the farm. More number of crops are grown on the farm during a year.

9. ***Social Justice :*** Disparities in income and wealth in rural area are brought down through distribution of landed property among the rural people.

10. ***Higher Family Labour Employment :*** Cooperation pattern and type of family adopted on the small farms invariably include raising of livestock. Use of labour is intensive methods besides exchange of family labour between farms provides more or less continuous employment of family labour.

Disadvantages of Small Scale Farming

1. Inadequacy of productive resources
2. Unremunerative farming.
3. Limited scope for development.
4. Shortage of farm implements.
5. Under employment
6. Market diseconomies
7. Shortage of finance
8. Relatively more overheads
9. Untanable for high investment

Classification of Farming

Broadly speaking farming may be classified on the basis of livestock raising and mode of economic and social functioning. Based on these two factors farming is classified into 2 groups.

1. Type of farming
2. Systems of farming

Types of Farming

The types of farming refer to the nature on degree of product or combination of products being produced at the farm and the methods and practices used for the same.

1. *Specialised Farming :* When major resources are devoted to or income is derived from a single enterprises; it is known as specialised farming. Under this types of farming the major enterprise generally contributes more than 50 % to total farm income. For example, dairy farming, poultry farming, sugar cane farming etc. wherein a particular enterprise is predominant.

   Advantages

   1. Better use of land
   2. Better marketing
   3. Better management
   4. Less equipment and labour are needed
   5. Cost efficient machinery can be kept
   6. Efficiency skills are increased

   Disadvantages

   1. Greater risks
   2. Productive resources are not fully utilised
   3. Fertility of soil can not be maintained for lack of crop rotation
   4. By-products cannot be properly utilized for lack of sufficient livestock on the farm.

2. *Diversified Farming :* In the case of diversified farming a number of enterprises are taken up on a farm and no single enterprise is relatively more important. In other words no single product or source of income equals as much as 50 % of the total receipts.

   Advantages

   1. Better use of productive resources.
   2. Risk is reduced
   3. Regular & quicker returns are obtained from various sources.

Disadvantages

1. Marketing is insufficient
2. Supervision becomes ineffective
3. Better equipping of the farm is not possible

3. ***Mixed Farming :*** It is such type of farming under which crop production is combined with livestock raising. The livestock enterprise is complementary to crop production. If a farm is to be categorised as mixed type of at least 10 % of the livestock must contribute to its gross income. This contribution is no case should exceed 49 %.

Advantages

1. Milch cattle provide draught animals for crop production and rural transport.
2. Crops provide feed to livestock by the later give manure to the crops. Thus both add to the productivity of each other.
3. It gives balanced labour load throughout the year for the farmers by his family.
4. If permits proper use of byproducts.
5. If offers higher returns on farm business.

4. ***Dry Farming :*** It refers to an area which receive 20 or less of annual rainfall. The major farm management problem in these tracks where crops are entirely dependent on rainfall is the conservation of soil moisture. Dry farming involves the adoption of the following practices.

Advantages

1. Timely preparation of the land to a condition in which it is best able to receive and conserve the available moisture.
2. Timely and proper inter-culture during the growth of the crop.
3. Application of organic manure to provide nutrients as well as to improve the soil texture.

5. ***Irrigated Farming :*** Irrigated farming refers to the harnessing of artificial resources of water for crop raising in areas where rainfall is insufficient or not well distributed over a period of time.

6. ***Ranching :*** It means practice of grazing animals on public lands. Ranch land is not utilized by filling or raising crops. Ranching is very common in Australia & Tibet. Some parts of Rajasthan, ravines of river chambal, and some portions of Agra division are the examples of this type of farming in our country.

Systems of Farming : The systems of farming refer to the organisation set up under which a farm is run.

1. *Co-operative Farming :* Means a system under which all agricultural operations (or) part of them are carried on jointly by the farmers on a voluntary basis. The farmers would pool their land, labour and capital, management is under a democratic constitution.

   The four types of co-operative farming societies prevalent in India are as follows :

   (a) *Co-operative better farming :* The land is not pooled and each farmer carries on cultivation separately. The co-operative better farming society promotes the interests of the members through the adoption of better farming practices. It arrange for the purchase of seeds, fertilizers, joint use of machinery. A member is free to follow his own way of farming except in respect of the purpose for which he joins the society.

   (b) *Co-operative Joint Farming :* The land and other resources of the members are pooled for joint cultivation. The ownership of each member over his land is recognised. The members work under the direction of a management committee. Management committee formulates the schemes and does the duties of administration. A member receives daily wages for his daily work and the profit in the end is distributed according to his share in land.

   (c) *Co-operative Tenant Farming:* In this system land belongs to the society. The land is divided into plots, which are leased out for cultivation to individual members. The society arranges the purchases of agriculture inputs and marketing of the produce. Each member in responsible to the society for payment of the rent of his plot.

   (d) *Co-operative Dependent Farming :* Members do not have any right on land and they can't take farming decisions independently but are guided by general body. Profit is distributed according to labour and capital invested by the members.

2. *Collective Farming :* It implies the collective management of land where in large number of families or villages residing in the same village pool their resources. The members work together under a management committee elected by themselves. The resources are pooled do not belong to any farmer but to the society or collective. If any farmer wants to dissociate from it, he can do so but he can't go with his share (resources). Money in lieu of his share will be given to him. The main drawback is

that the individual has no voice. Farming is done on a large scale and therefore is mostly mechanised. This system is not prevalent in India but is common in communist countries. In Russia collectives are called " Kolkhozes".

3. *Capitalistic Farming (Estate Farming)* : The ownership of such farms is under rich persons or capitalists. The size of such farmers is sufficiently large and the management is also quite efficient. Individuals or groups of individuals or shareholders own these farms. Resources are plenty. Latest technical know-how is used. Sugar factory farm, rubber, coffee, tea, plantations are common examples of such system. The advantages of such farming are good supervision, strong organizational set up, sufficient resources etc., its' weakness are that it creates socio-economic imbalances in the actual cultivation is not the owner of the farm.

4. *State Farming* : As the name indicates, it is managed by government. Here the operation and management is done by the government officials. All the labourers are hired on daily or monthly basis in they have no right in deciding the farm policy. Such farms are not very paying because of lack of incentives.

5. *Peasant Farming* : This system of farming refers to the type of organization in which an individual cultivates is the owner, manager by organization of the farm. He makes decisions and plans for his farm depending upon resources. The biggest advantage of this system in that the farmer himself is the owner and therefore free to take all sorts of decisions. A general weakness of this system in that the resources with the individuals are less in comparison to those of other systems.

System of Farming	Type of ownership	Type of operationship
1. (a) Co-operative better farming	Individual	Individual
(b) Co-operative joint farming	Individual	Collective
(c) Co-operative tenant farming	Collective	Individual
(d) Co-operative Collective Farming	Collective	Collective
2. Collective Farming	Society	Society
3. Capitalistic farming	Individual	Individual
4. State farming management	State	Paid
5. Peasant farming	Individual	Individual

Table 15.1 Differences between Types and Systems of Farming

Type of Farming	System of Farming
1. It refers to the nature and degree of products and their combination being produced on the farm and the practices and methods followed in production.	1. If refers to organisational set up under which the farm is being run. It involves questions like who owns the land, resources are pooled or used independently and who makes the managerial decisions etc.
2. Concerned with production process.	2. Decision making process.
3. Classified on the basis of similarities in crop and livestock raising.	3. Classified on the basis of mode of economic and social functioning.
4. Ex: Specialized, diversified, mixed, ranching and dry farming.	4. Ex: co-operative farming collective, state, peasant and capitalistic farming.

Functions of Livestock

Milk production and marketing provide the household with a regular daily source of cash throughout the season/year, which can be used for small expenditures, while crop production results in a lump sum only after the harvest. Animals are not kept exclusively for milk production, however. Other possible functions are their use for draught, stock, manure, meat, hides, hair and wool. They can also be kept for investment purposes, which is an increasingly important consideration for livestock ownership, especially in countries with unstable economies. Livestock are not only useful for their owners, but they may have benefits for non-livestock-keeping households as well. For instance, grazing on harvested plots of non-livestock owners may maintain or improve the fertility of the soil. In such cases, it is important to identify the relations between these households. Knowledge of the organization of these relationships is necessary: what are the rights and duties and the costs and benefits for each person or household?

With the proceeds from economically good years farmers tend to invest in animals for conversion into cash in adverse times. This attitude may also be the reason why the introduction of improved breeds meets with resistance on occasion. Local animals are better adapted to adverse conditions than high-bred ones and may survive longer when feed is scarce. As improved breeds are also considerably more expensive, the same amount of money would permit a farmer to buy more local animals and under conditions of stress the sale of a local animal would not reduce the capital to a large extent. The importance of each function differs per situation and is related, among other things, to the farming system, production purposes and strategies and rights and duties of family members. Not surprisingly, considerable differences in interest may be found between men and women, which may lead to unexpected or at least unforeseen results.

Marketing Methods of Urban and Rural Societies

Introduction

Marketing, as a concept, is based on two fundamental beliefs all activities of a firm (or producer), including planning, operations and policies, should be oriented towards the consumers (or customers); and profitable sales volume should be the goal of every firm. Consequently, all the firm's activities should be devoted to determining what the consumers' wants are and to satisfying these wants while still making a reasonable level of profit. In the case of livestock producers, especially if they are smallholders, the public sector has a role to play in advising the farmers on what products are in demand and in assisting them to develop and promote consumption of new livestock-based products whenever feasible. Because of its strategic role in economic development, marketing development has come to be accepted as a complementary activity to production development. Hence marketing may be viewed as a social and managerial process through which individuals and groups obtain what they need and want by creating and exchanging products of value with each other. Marketing management can then be viewed as the process of planning and executing the conception, pricing, promotion and distribution of ideas, goods and services to create exchanges that satisfy individual and organisational objectives. The process thus involves analysis, planning, implementation and control, covering not only physical goods and services but also ideas.

Marketing is the handling of business transactions involved in the movement of products from producers to consumers. Market is the place where a product is offered a fair price. In any manned economic development programme, exchange of goods assumes a very important role in maintaining equilibrium between production and consumption. From an analysis of different definitions marketing, it can be reasonably considered that it is the process of discovering consumer requirements, and other service through which the ownership of goods and services are transferred by their physical distribution from producer to consumer. Marketing channels are routes through which agricultural products move from producers to consumers. The length of the channel varies from commodity to commodity, depending on the quantity to be moved, the form of consumer demand and degree of regional specialisation in production.

The four elements of the marketing mix are *product, place, price and promotion* otherwise referred to as the 4 Ps. They provide a marketing executive with a set of effective tools with which to capture the desired market shares. The four elements are normally referred to as the "controllable variables" of a marketing executive's job. This is so because the management of any business is a very challenging task of satisfying consumers, and of also deciding on any product-related issues; setting prices and changing them whenever desired; and the distribution of products and services through various channels to reach whatever markets are wanted. They can also decide whether to undertake promotion of their products or not and there are almost limitless alternatives of promotional mixes that they can employ if they wish. Marketing executives do not work in a vacuum, they work in an environment in which their ability to manipulate the 4 Ps is invariably constrained by a number of factors which include:

- Socio-cultural environment
- Political and legal environment
- Economic environment and
- Business structures, resources and objectives of the firm.

These constraints constitute a set of 'uncontrollable variables' that minimise a marketing executive's ability to make free decisions with regard to one or more of the 4 Ps. From a managerial perspective, livestock production should be viewed as a business organisation. Ideally, the critical job of a marketing executive is to assist his business organisation in ensuring that it comes up with (or develops) the right product which is then appropriately priced and promoted in the right way and is available where needed at the right time from the consumers' point of view. Hence, the author is of the opinion that a consideration of the marketing options for livestock products must examine the appropriateness of the various permutations and combinations of the elements of a marketing mix. This would be an endeavour to develop practicable marketing strategies for livestock products, given the environment within which these strategies have to be implemented.

A review of available literature suggests that most of the previous marketing studies on livestock development in sub- Saharan Africa have focused on marketing systems. They have tended to give little consideration to marketing options in terms of:

- Production and marketing of alternative products.
- Use of alternative distribution methods and systems.
- Application of alternative pricing strategies.
- Application of alternative promotional methods.

A marketing channel may be defined in different ways, According Moore etal., the chain of intermediaries through whom the various food grains pass from producers to consumers constitutes their marketing channels. Kophs and Uhl have defined marketing channel as alternative routes of product flows from producer to consumers.

Marketing Options and their Efficacy

Issues

The type of product to be considered in this case is the live animal itself. Large stock (cattle and buffalo) and small stock (sheep and goats) are the most important animal species in the market in the case of livestock development. From a marketing management perspective, the body condition and the live weight of the animal are the key factors that influence the marketing of live animals. The two factors will be related to the age and the sex of the animal and such other secondary body parameters like fleshing. These factors will thus directly influence the price at which the live animals can be sold. The channel through which the animal reaches the market place will influence both the body condition and the live weight of the animal that is being marketed. Primarily the domestic marketing infrastructure and the type of the ultimate market being considered will influence the types of channels used. Traditional or informal marketing systems refer to systems in which governments do not substantially intervene, either directly through trading or indirectly through regulation.

Trekking of live animals is a common method employed to move live animals from grazing or production areas to the market places, especially under traditional livestock marketing systems. Road and/or railway transportation systems are sometimes used, either exclusively or in supplementation to the trekking during the process of livestock marketing, depending on local circumstances. Advanced transportation facilities (trucks and/or trains, and even shipping to export markets) are more commonly employed in the formal livestock marketing systems. The choice and the use of different types of channels and transportation modes in the marketing of live animals are likely to influence the net returns to livestock producers through their effects on the cost and net prices realized in the marketing process.

Experiences

Producers have the option in many of the countries in eastern and southern Africa to market their live animals either through the formal or the informal livestock marketing systems. Earlier inferences on the economic performance of informal marketing systems, as measured through the marketing efficiency criterion, tended to suggest that such systems performed relatively poorly. It was therefore argued that the only way to improve performance was through increased government intervention. However, such inferences were largely based on casual impressions of activities in traditional market places. This study also gives a less favorable impression of the efficiency of informal marketing systems for live animals. Marketing options for live animals appear to be relatively limited with respect to flexibility and ability to manipulate elements in the marketing mix. The channels of distribution are relatively limited and so is the ability to ask for different prices in the same market. One is able to differentiate live animals primarily on the basis of sex, age, body condition and live weight. This factor limits the scope for product promotion, even though prices may differ on the basis of the factor itself. However, flexibility in pricing is more likely to be possible in the informal marketing system, but one expects that the pricing structure in the formal marketing system will somehow influence the geographical pattern of prices in the informal marketing system.

The scope for the promotion of the marketing of live animals would appear to exist at the institutional level. A certain region (or large farm) may become known nationally for producing good quality animals, as could a certain country become known internationally for producing good quality animals. Such a factor can be exploited when promoting the product (live animals) in both domestic and export markets, through the creation of individual, regional and national brand images.

Dairy products (fresh milk and milk products) and meat (especially beef) are the major types of livestock products, which have been accorded development priority in most countries in sub-Saharan Africa. This is especially true of donor-aided projects. As in the case of live animals, there are both formal and informal marketing systems for livestock products in most of the countries in the region However the channels through which livestock products are marketed are relatively more diverse and complex than in the case of live animals.

Dairy Marketing Options

The main types of dairy products consumed in most of the countries are fresh milk, butter, cheese and yogurt. The relative importance of the consumption of the different types of products (or their local equivalents[1]) in different countries may be expected to vary, but fresh milk consumption is common in all countries. Dairy producers have the option either to sell fresh milk or to process it into various types of dairy products and then sell such products, depending on local market circumstances.

The marketing options for the producers and merchants of the milk and milk products that were sold through these marketing systems were:

- Direct sales to consumers by producers
- Direct sales to consumers by the officials (i.e. government-designated) sales outlets (i.e. Outlets other than grocery stores, super markets and small private shops or kiosks)
- Direct sales to consumers by itinerant traders who sell milk produced by other people
- Direct sales to consumers by small private shops and kiosks
- Direct sales to consumers by grocery stores and supermarkets.

Meat Marketing Options

Both formal and informal channels for meat marketing exist in India. Experiences indicate that the formal channel often faces stiff competition from private butchers and thus experiences major difficulties in meat marketing. The major cause of operational problems for the formal meat marketing channels is the fact that they often have to follow official prices, so that their marketing margins tend to be rigid and often low. This, of course, may be the case for most commodities that are traded through formal channels, but the problem appears to be more acute in the case of meat marketing. The scope of adopting different marketing strategies appears to be greater for meat than for milk and milk products. Meat can be stored and marketed over longer time periods than fresh milk, especially if kept under refrigerated conditions. Promotion in meat marketing can be undertaken in many and varied forms:

- Meat shop (butchery) brand name
- Company and/or private seller's prestige
- Customer prestige, where emphasis is on meat cuts for different classes of consumers (based on their incomes).

The possibility of processing and packaging meat greatly facilitates the ability to promote meat marketing. The ability to promote meat marketing almost certainly depends on the ability to charge or set different prices, according to the market being aimed at.

Egg Marketing Options

The choice of egg marketing method will depend upon several factors like

1. Number of eggs to be marketed
2. Distance to market
3. Availability of labour

Marketing Institutions

These are big business organizations, which have come up to operate the marketing machinery. In addition to individuals, corporate, cooperative and government institutions are operating in the field of agricultural marketing. They perform one or more of the marketing functions. They assume the role of one or more marketing agencies, described earlier in this section. Some important institutions in the field of agricultural marketing are

- State trading corporation
- Food Corporation of India
- National Agricultural cooperative marketing federation
- Agricultural processed products and exports development agency
- Directorate of marketing
- Directorate of inspection
- State agricultural marketing departments
- Agricultural marketing boards
- Cooperative marketing societies

Marketing Channels

Marketing channels are routes through which live animals and livestock products move from producers to consumers. The length of channel varies from commodity to commodity, depending on the quantity to be moved, the form of consumer demand and degree of regional specialization in production.

A marketing channel may be defined in different ways. According to Moore, it is the chain of intermediaries through whom the various products pass from producers to consumers constitute their marketing channels. According to Kohls Uhl, marketing channels are alternative routes of products flows from producers to consumers. (Fig. 15.1) and (Fig. 15.2).

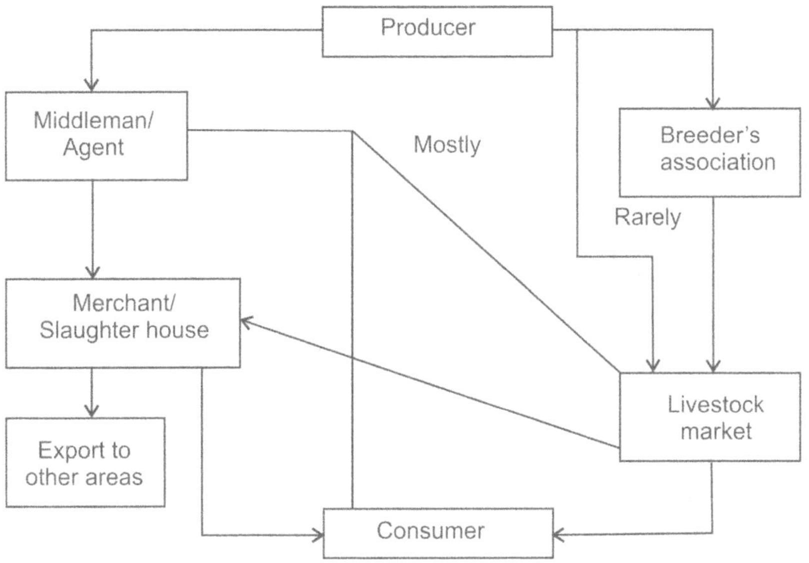

Fig. 15.1 Marketing Channel for Live Animals.

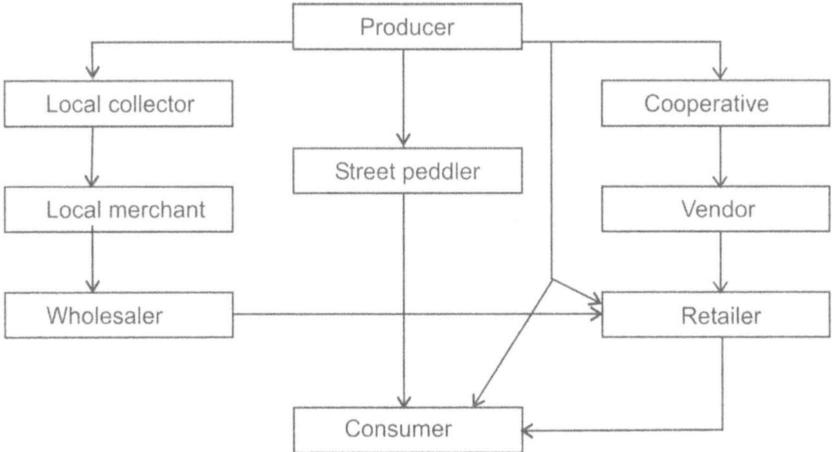

Fig. 15.2 Marketing Channel for livestock Products

Factors affecting length of marketing channels

Marketing channels for live animals or livestock products vary from product to product, country to country, lot to lot and time to time. For example, the marketing channels for eggs are different from those for milk. The level of the development of a society or country determines the final from in, which consumes demands the product. For example, consumers in developed countries demand more processed foods , while in developing countries, consumers demand for raw form of food. With the expansion in transportation and communication network, changes in the structure of demand and the development of markets, marketing channels for farm products in India have undergone a considerable change, both in terms of length and quality.

Marketing Channel for Live Poultry

Most of the studies on the identification of marketing channels for poultry commodities have concentrated on a concept of marketing channel, which defines the flow of the produce from the producer to the consumers. The marketing channels that are emerging go across state or even national boundaries, this apart unless quantities flowing into various channels are estimated, the relative importance of the alternative channels cannot be assessed. There are 28 marketing channels, village traders appear in 8 channels, wholesalers appear in 18 channels, processors appear in 15 channels (Fig. 15.3).

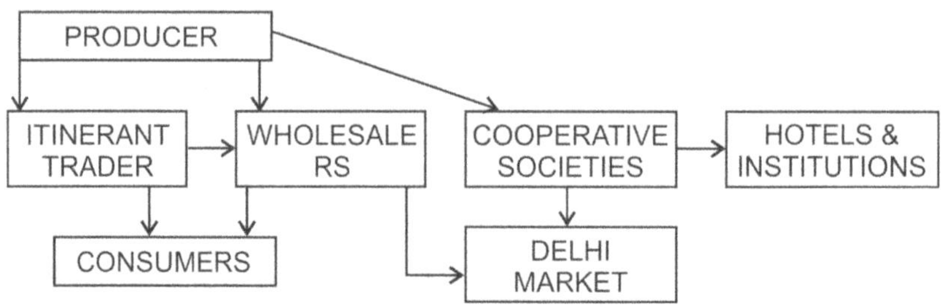

Fig. 15.3 Marketing Channel for Live Poultry.

Marketing Channels for Eggs : The prevalent marketing channels for eggs are:

- Producer to consumer
- Producer to retailer to consumer

- Producer to wholesalers to retailers to consumers
- Producer to cooperative marketing society to wholesalers to retailers to consumers
- Producers to egg powder factory.

Rural Marketing

The concept of marketing and its influential role in the transferring of market, consumer and Indian economy is increasingly felt. It is a pervassive element in contemporary life of every one. Marketing provides an opportunity to contribute to society as well as to an individual company. In the context of present competitive environment, marketing has become the key in deciding the success and health of a corporate. Corporates in India have recognized this fact and thus are laying greater emphasis on marketing. Marketing has both micro aspects. In its micro aspect, an efficient system of marketing enables the producer of the commodity to get higher price for the product and thereby enables him to earn larger income and helps him maintain higher standard of living. In its macro aspect, enables society to get different products, which its members need, at reasonable prices and thereby enables them to improve their consumption standards and levels of living.

Concept of marketing : The term 'Marketing' connotes different meaning to different people; to some it is shopping; to others it is selling and still others understand, it as purchasing as well as selling. Marketing is the activity undertaken by the companies to make exchange, transaction, comsummate and adding to bring out greater output at a minimum cost. Marketing in its most general definition "is the directing of flow of goods and services from the producers to consumers of users". Marketing is human activity directed at satisfying needs and wants through exchange prices or "Marketing is a total system of business designed to plan, promote and distribute want satisfying products, services and ideas to target markets in order to achieve organizational objectives."

The definition of rural as given by the census of India is not being urban. There could several approaches to understand the rural market. It will not be an exaggeration to state that if the whole India excluding the metropolitan cities, various district headquarters and the large industrial townships are considered as the rural market. Rural market is not an independent entity by itself. In fact, there are sociological and behavioural factors, which affect the economy of rural market. By and By and large, markets may be described as an environment of the country side the habitants thereof. The indian rural market, with its vast size and demand, offer great opportunities to marketers. More than three-fourth of country's consumers live in rural areas and more than half of the national income generated by them. The rural market is made two broad componints 1. The market for consumer goods, which includes both durable and non-durable goods and 2. The market for agricultural and livestock inputs and other investment goods.

Attractiveness of rural market

The rural markets have become the new targets to corporate enterprises, mainly because of following reasons:

1. An urban market has become congested with too many copmpetitors.
2. An urban market has reached the saturation point.
3. Urban markets are existing based on Darwin's principle: Survival of the fittest.
4. A rural market has become the main street with potential for consumption of variety of products and services.
5. Rural market makes market entry easy.
6. Large population, raising prosperity, change in life-style, growth in consumption, easy approach because of more transport facilities and more growth rates have made rural markets viable.

Rural and Urban markets - Differences

It may be of relevance to present the salient features of rural market and the difference between rural and urban marketing.

Characteristics of Urban marketing

- Large congruous settlement units of towns or urban agglomerations.
- High infrastructure levels.
- High density of population and are heterogeneous in nature.
- Good physical connectivity
- High mobility
- Large number of interactions with people, but less frequent with the same people.
- Individuals are less known and identified between members in the social system.
- Social norms are less visible.
- Status is acieved.
- Caste influence is indirect and of less strength, generally subjected to economic influence.
- High exposure to branded products, advertisements, marketing researchers and information.
- More convenient buying, high rate of retail outlets, high market reach and availability of wide range of products.
- Most resources to be purchased.
- Access is a function of purchasing power.
- Low dependence on employment and incomes on natural factors
- Occupations mostly include employment.
- Frequency of income receipts prodictable and at regular intervals.

Characteristics of Rural Marketing

- Small contiguous settlement units of villagers far from cities, widely dispersed.
- Low infrastructure levels.
- Low density of population and are homogeneous in nature.
- Poor physical connectivity with other villages and towns.
- Low mobility
- Less number of inter-personal interactions, more frequent interactions between the samer people.
- Individuals are better known and identified.
- Social norms influencing individuals are more visible.
- Status is ascribed.
- Caste influence is direct and strong.
- Low exposure to branded products, advertisements, marketing researchers and information.
- Less convenient buying, low rate of retail outlets, low market reach and availability of limited range of products along with imitation products.
- Abundance of natural resources and high dependency on them for household needs.
- Differential access to resources based on caste, political and money power etc.
- High dependence on employment and incomes on natural factors
- Mostly agricultural occupation (small land holdings per household).
- Acute seasonality in income receipts; high chance element in income receipt (because of the dependence on agriculture and natural factors).

CHAPTER - 16

Survey Techniques and Sampling Methods

Social Survey

The word survey was derived from the Latin words 'sur/sor' means 'over' and 'veeir/veoir' means 'see' i.e. means to 'look over or to see over'

As stated by Mark Abrams, social survey is a process by which quantitative facts are collected about the social aspects of a community's composition and activities. In other words, it is a fact-finding study dealing with the nature and problems of the community.

Characteristics of Social Survey

1. Social survey is confined to the immediate problems of the community.
2. It focuses upon a given locality/geographical area which helps in preparation of programmes from the data obtained through the survey which will help in suggesting improvements.
3. It deals with the social life of people.
4. It collects and describes the social facts about situation of people.
5. It must be collected with an objective spirit without any subjective likes or dislikes.

Purpose of Social Survey

1. To obtain the facts regarding the social aspects of the community.
2. To study social problems and economic conditions of farmers and the factors responsible for these conditions.
3. To evaluate the programme and assess the benefits desired from it. Altogether a social survey helps in bringing about social welfare and the betterment of the people.

Steps in Social Survey

- Selecting the problem
- Definition of aim
- Determination of scope
- Definition of time limit

- Examination of means of sources of information
- Determination of unit of survey
- Determination of the amount of refinement
- Preparedness of the respondents
- Construction of tools for data collection
- Fieldwork and data collection
- Processing and analysis of collected data
- Interpretation and report writing

Types of Social Survey

Social survey is classified into various types according to subject matter, technique of data collection, area covered and regularity etc.

1. General and Specific survey
2. Regular and Adhoc survey
3. Preliminary and Final survey
4. Census and Sample survey
5. Direct or Indirect survey
6. Primary or Secondary survey
7. Initial or Repetitive survey
8. Public or Confidential survey
9. Postal or Personal survey
10. Benchmark or Baseline survey

General Survey : When a survey is conducted for collecting general information about any population, institution or any phenomenon without any objectives is known as general survey. The Government institutions for getting regular data on many socio-economic problems under take such survey. For example, Census survey of population conducted for every 10 years.

Specific Survey : These are conducted for specific purposes or problems for testing the validity of some theory. These survey are naturally more pinpointed for a specific purpose. For example, survey conducted to assess the rate of mortality among livestock affected with a particular disease, hyginic milk production, calf management etc.

Regular Survey : Some surveys are regular in nature and must be repeated at regular intervals. Such survey is undertaken when continuous data are required to study the trend and effect of time upon the phenomenon. For example, collecting data regularly to assess the impact of artificial insemination. in improvement of local cattle or census survey of population conducted for every 10 years.

Adhoc Survey : These survey are undertaken once for all. They are also conducted in phases when the area of investigation is large. The whole process of survey can be completed in one or two installments. This could not be termed as a regular or repetitive survey unless the same information is collected over and over again. Adhoc surveys are undertaken mostly for supplementing some information with regard to any research problem. For example, survey to assess the causes of calf mortality in a particular area.

Preliminary Survey : Also known as pilot study and is forerunner of the final survey. The purpose of this survey is to get firsthand knowledge of the universe and population to be surveyed. It assists the researcher to get acquainted with problems and nature of respondents from whom the information is to be collected. It is very useful in preparing a schedule or questionnaire and organising the survey on proper lines.

Final Survey : Final survey is made only after the pilot study is completed.

Census Survey : In census survey every single unit of the universe is contacted and information is collected.

Sample Survey : In sample survey, a portion of the universe is contacted and the information is collected. This sample represents the entire population and is called as sample survey. These surveys are becoming popular because of their convenience, time saving and low cost etc.

Direct or Indirect Survey : In case of direct survey quantification is possible whereas in an indirect survey quantitative description is not possible. For example, Direct survey - Demographic survey and Indirect survey - Health and Nutritional status survey.

Primary Survey : In the primary survey, the task of survey is taken up afresh and the surveyor himself sets the goals and collects relevant facts.

Secondary Survey : If the facts are already available and there is no need to examine them afresh by a new survey then the survey is called secondary survey. Primary survey is far more reliable than secondary survey.

Initial and Repetitive Survey : If the survey is being made for the first time it is called as initial survey and if it is being made second or third time it is called an repetitive survey.

Public Survey : Survey, which is not highly personal in nature and accordingly no secrecy is maintained in the collection of data or publication of results; is known as public survey.

Confidential Survey : If the nature of survey is such that the information collected is not to be revealed to public, then it is known as a confidential survey.

Postal and Personal Survey : When the data is collected through dispatch of questionnaires by post it is known as postal survey, whereas if the information is collected through direct interview of the respondents usually through schedule, it is known as personal survey.

Benchmark or Baseline Survey : The survey conducted to know the resources and traditional methods of farming of the people. The usual method for studying the existing situation of the clientele is the Benchmark or baseline survey. The survey conducted in extension programmes based upon the needs and interests of the people is known as benchmark or baseline survey.

Subject Matter of Social Survey : There are four broad areas of subject matters to conduct social survey, they are:

(a) *Demographic Characteristics* : Type of family, size of family, marital status, Age, Gender etc.

(b) *Social Environment :* Occupation, income, housing conditions, social amenities etc.

(c) *Social Activities :* Expenditure pattern, savings, habits, use of leisure time etc.

(d) *Psychological parameters :* Opinion, attitude, motives, and expectations towards social and economic factors.

Utility of Social Survey

- Survey is the only practical way to collect any type of information.
- Survey facilitates drawing generalisation about a large population.
- Survey helps the researcher to find out the new problems.
- Survey helps to construct a plan for the development of the society.
- A social survey helps the extension worker to come in contact with the rural people.
- This method permits greater objectivity, as the data collected is not influenced by one-man views and beliefs.
- It helps the extension worker to find out the extent to which the rural people are affected by a particular problem.
- Based on the problem that emerges from a social survey it helps the extension worker to come out with suitable solutions.

The cover letter is an essential part of the survey. To a large degree, the cover letter will affect whether or not the respondent completes the questionnaire. It is important to maintain a friendly tone and keep it as short as possible. The importance of the cover letter should not be underestimated. It provides an opportunity to persuade the respondent to complete the survey. If the questionnaire can be completed in less than five minutes, the response rate can be increased by mentioning this in the cover letter. Flattering the respondent in the cover letter does not seem to affect response. Altruism or an appeal to the social utility of a study has occasionally been found to increase response, but more often, it is not an effective motivator. There are no definitive answers whether or not to personalise cover letters (i.e., the respondents name appears on the cover letter). Some researchers have found that personalised cover letters can be detrimental to response when anonymity or confidentiality are important to the respondent. The literature regarding personalisation is mixed. Some researchers have found that personalised cover letters with hand-written signatures helped response rates. Other investigators, however, have reported that personalisation has no effect on response. The signature of the person signing the cover letter has been investigated by several researchers. Ethnic sounding names and the status of the researcher (professor or graduate student) do not affect response. One investigator found that a cover letter signed by the owner of a Marina produced better response than one signed by the sales manager. The literature is mixed regarding whether a hand-written signature works better than one that is mimeographed. Two researchers reported that mimeographed signatures worked as well as a hand-written one, while another reported that hand-written signatures produced better response. Another investigator found that cover letters signed with green ink increased response by over 10 percent.

It is commonly believed that a handwritten postscript (P.S.) in the cover letter might increase response. One older study did find an increase in response, however, more recent studies found no significant difference.

1. Describe why the study is being done (briefly) and identify the sponsors.
2. Mention the incentive. (A good incentive is a copy of the results).
3. Mention inclusion of a stamped, self-addressed return envelope.
4. Encourage prompt response without using deadlines.
5. Describe your "confidentiality/anonymity" policy.
6. Give the name and phone number of someone they can call with questions

Social Sampling and Sampling Methods

Scientists collecting data to analyse a problem have a choice between making a complete enumeration (census) and taking a sample. Taking entire population is impracticable and uneconomical, thus sampling technique is adopted and the objective is to determine a model that best describes the experimental results. For example to assess the attitude of livestock farmers towards rearing of cross bred cows in a particular region of the state, it is impractical and uneconomical to obtain answers by asking each and every farmer for an opinion. So we can obtain answers to our questions by sampling only a small percentage of farmers and using their findings to generalise about the entire population. Since we use a small fraction of the population, to generalise about the entire population care must be taken in determining the type of sample we draw from the population. So it is necessary to study about different sampling techniques, and understand their properties for making valid generalisations.

There are some important *sampling concepts* that need to be clarified before sampling techniques can be understood.

Population : A population or Universe is defined as an entire group of persons / things / events having at least one characteristic in common. For example, Livestock farmers, Rural people, Veterinary students etc.

Sample : A sample is defined as a small part of the population or universe selected by some rule /plan. For example, among 1000 livestock farmers of a region taking few farmers as a   sample of study. Thus a sample is a part of population. We select the sample, measure the sample and from these, we can make inferences about the population.

Sampling : The process of choosing the units of the target from the universe which are to be included in the study.

Reasons for Sampling

- In many cases, a complete coverage of the population is not possible.
- Complete coverage may not offer substantial advantage over a sample survey.
- Sampling provides a better option, since it addresses the survey population in a short period of time and produces equally valid results.
- Studies based on samples require less time and produce quicker results.
- Samples provide more detailed information and a higher degree of accuracy because they deal a with relatively small number of units.

Principles of Sampling

Samples are to be chosen by means of sound methodological principles.

- Sample units must be chosen in a systematic and objective manner.
- Sample units must be easily identifiable and clearly defined.
- Sample units must be independent of each other, uniform and of the same size and should appear only once in the population.
- Sample units are not interchangeable; the same units should be used throughout the study.
- Once selected, units cannot be discarded and
- Selection process of units should avoid errors and bias

It is incumbent up on the researcher to clearly define the target population. There are no strict rules to follow, and the researcher must rely on logic and judgment. The population is defined in keeping with the objectives of the study. Sometimes, the entire population will be sufficiently small, and the researcher can include the entire population in the study. This type of research is called a census study because data is gathered on every member of the population. Usually, the population is too large for the researcher to attempt to survey all of its members. A small, but carefully chosen sample can be used to represent the population. The sample reflects the characteristics of the population from which it is drawn.

Sampling methods are classified as either *probability* or *nonprobability*. In probability samples, each member of the population has a known non-zero probability of being selected. Probability methods include simple random sampling, systematic sampling, stratified sampling, multi-stage sampling and cluster sampling. In non-probability sampling, members are selected from the population in some non random manner. These include convenience sampling, purposive sampling, judgment sampling, quota sampling, and snowball sampling. The advantage of probability sampling is that sampling error can be calculated. Sampling error is the degree to which a sample might differ from the population. When inferring to the population, results are reported plus or minus the sampling error. In non-probability sampling, the degree to which the sample differs from the population remains unknown.

Probability Sampling method is any method of sampling that utilizses some form of *random selection*. In order to have a random selection method, you must set up some process or procedure that assures that the different units in your population have equal probabilities of being chosen. Humans have long practiced various forms of random selection, such as picking a name out of a hat, or choosing the short straw. These days, we tend to use computers as the mechanism for generating random numbers as the basis for random selection. Before I can explain the various probability methods we have to define some basic terms. These are:

- N = the number of units in the population i.e. population size

- n = the number of units in the sample i.e. sample size

- $_NC_n$ = the number of combinations (subsets) of n from N

- f = n/N = the sampling fraction

Simple Random Sampling

The simplest form of random sampling is called simple random sampling. Simple random sampling is the purest form of probability sampling. Each member of the population has an equal and known chance of being selected. When there are very large populations, it is often difficult or impossible to identify every member of the population, so the pool of available subjects becomes biased. Sample units are selected by means of a number of methods like lottery method, random number methods (tippet) and computer method.

For example, let's say you want to select 100 clients to survey and that there were 1000 clients over the past 12 months. Then, the sampling fraction is $f = n/N = 100/1000 = .10$ or 10%. Now, to actually draw the sample, you have several options. You could print off the list of 1000 clients, tear then into separate strips, put the strips in a hat, mix them up real good, close your eyes and pull out the first 100. But this mechanical procedure would be tedious and the quality of the sample would depend on how thoroughly you mixed them up and how randomly you reached in. Perhaps a better procedure would be to use the kind of ball machine that is popular with many of the state lotteries. You would need three sets of balls numbered 0 to 9, one set for each of the digits from 000 to 999 (if we select 000 we'll call that 1000). Number the list of names from 1 to 1000 and then use the ball machine to select the three digits that selects each person. The obvious disadvantage here is that you need to get the ball machines. Neither of these mechanical procedures is very feasible and, with the development of inexpensive computers there is a much easier way. Here's a simple procedure that's especially useful if you have the names of the clients already on the computer. Many computer programs can generate a series of random numbers. Let's assume you can copy and paste the list of client names into a column in an EXCEL spreadsheet, then sort both columns i.e. the list of names and random number. This rearranges the list in random order from the lowest to the highest random number. Then, all you have to do is take the first hundred names in this sorted list; pretty simple. You could probably accomplish the whole thing in under a minute.

Simple random sampling is simple to accomplish and is easy to explain to others. Because simple random sampling is a fair way to select a sample, it is reasonable to generalise the results from the sample back to the population. Simple random sampling is not the most statistically efficient method of sampling and you may, just because of the luck of the draw, not get good representation of subgroups in a population. To deal with these issues, we have to turn to other sampling methods.

Systematic Sampling

It is often used instead of random sampling. It is also called an kth name selection technique. After the required sample size has been calculated, every k^{th} record is selected from a list of population members. As long as the list does not contain any hidden order, this sampling method is as good as the random sampling method. Its only advantage over the random sampling technique is simplicity. Systematic sampling is frequently used to select a specified number of records from a computer file.

Procedure :

- Arrange the units in a systematic order (Ascending / descending / Alphabetical)
- To get the number 'k', divide the population size (N) with the required sample size (n) i.e., k = N/n.
- Then select any one item between 0 and k; it is the first sampling unit.
- Then go on selecting the items, by adding the number 'k' to the first selected sampling unit, until the required sample size is reached.

All of this will be much clearer with an example. Let's assume that we have a population that only has N = 100 people in it and that you want to take a sample of n = 20. To use systematic sampling, the population must be listed in a random order. The sampling fraction would be f = 20/100 = 20%. In this case, the interval size, k, is equal to N/n = 100/20 = 5. Now, select a random integer from 1 to 5. In our example, imagine that you chose 4. Now, to select the sample, start with the 4th unit in the list and take every k-th unit (every 5th, because k = 5). You would be sampling units 4, 9, 14, 19, and so on to 100 and you would end up with 20 units in your sample.

For this to work, it is essential that the units in the population are randomly ordered, at least with respect to the characteristics you are measuring. Why would you ever want to use systematic random sampling? For one thing, it is fairly easy to do. You only have to select a single random number to start things off. It may also be more precise than simple random sampling. Finally, in some situations there is simply no easier way to do random sampling. For instance, I once had to do a study that involved sampling from all the books in a library. Once selected, I would have to go to the shelf, locate the book, and record when it last circulated. I knew that I had a fairly good sampling frame in the form of the shelf list (which is a card catalog where the entries are arranged in the order they occur on the shelf). To do a simple random sample, I could have estimated the total number of books and generated random numbers to draw the sample; but how would I find book #74,329 easily if that is the number I selected? I couldn't very well count the cards until I came to 74,329! Stratifying wouldn't solve that problem either. For instance, I could have stratified by card catalog drawer and drawn a simple random sample within each drawer. But I'd still be stuck counting cards. Instead, I did a systematic random sample. I estimated the number of books in the entire collection. Let's imagine it was 100,000. I decided that I wanted to take a sample of 1000 for a sampling fraction of 1000/100,000 = 1%. To get the sampling interval k, I divided N/n = 100,000/1000 = 100. Then I selected a random integer between 1 and 100. Let's say I got 57. Next I did a little side study to determine how thick a thousand cards are in the card catalog (taking into account the varying ages of the cards). Let's say that on average I found that two cards that were separated by 100 cards were about .75 inches apart in the catalog drawer. That information gave me everything I needed to draw the sample. I counted to the 57th by hand and recorded the book information. Then, I took a compass. (Remember those from your high-school math class? They're the funny little metal instruments with a sharp pin on one end and a pencil on the other that you used to draw circles in geometry class.) Then I set the compass at .75", stuck the pin end in at the 57th card and pointed with the pencil end to the next card (approximately 100 books away). In this way, I approximated selecting the 157th, 257th, 357th, and so on. I was able to accomplish the entire selection procedure in very little time using this systematic random sampling approach. I'd probably still be there counting cards if I'd tried another random sampling method.

Stratified Sampling

It is a commonly used probability method that is superior to random sampling because it reduces sampling error. A stratum is a subset of the population that shares at least one common characteristic. Examples of stratums might be males and females, or managers and non-managers. The researcher first identifies the relevant stratums and their actual representation in the population. Random sampling is then used to select a *sufficient* number of subjects from each stratum. "*Sufficient*" refers to a sample size large enough for us to be reasonably confident that the stratum represents the population. Stratified sampling is often used when one or more of the stratums in the population have a low incidence relative to the other stratums. For example when the researcher wants to study the profile of livestock farmers, to ensure that there should be representation from different livestock farmers (i.e., land less, marginal, small, medium, and large sized) in the sample, the livestock farmers are divided into groups based on land size and then from each group of farmers draw sample proportionately to get the required size. Drawing a sample from each stratum in proportion to the stratum share in the total population is known as proportionate stratified sampling. This is suitable when the population is large, heterogeneous and overlapping in nature.

Procedure :
- Population is divided into a number of strata according to groups in the population.
- A Random sample is selected from within each stratum.
- Pool all the samples, which are selected from various strata to get required sample size.

For example, let's say that the population can be divided into three groups: Large, Medium, and Small farmers. Furthermore, let's assume that both the medium and small farmers are relatively small minorities of the population (10% and 5% respectively). If we just did a simple random sample of n = 100 with a sampling fraction of 10%, we would expect by chance alone that we would only get 10 and 5 persons from each of our two smaller groups. And, by chance, we could get fewer than that! If we stratify, we can do better. First, let's determine how many people we want to have in each group. Let's say we still want to take a sample of 100 from the population of 1000 users of extension services over the past year. But we think that in order to say anything about subgroups we will need at least 25 cases in each group. So, let's sample 50 large, 25 medium, and 25 small. We know that 10% of the population, or 100 users of veterinary extension services, are belong to medium group. If we randomly sample 25 of these, we have a within-stratum sampling fraction of 25/100 = 25%. Similarly, we know that 5% or 50 users of veterinery extension services are small. So our within-stratum sampling fraction will be 25/50 = 50%. Finally, by subtraction we know that there are 850 large farmers who have used veterinary extension services. Our within-stratum sampling fraction for them is 50/850 = about 5.88%. Because the groups are more homogeneous within-group than across the population as a whole, we can expect greater statistical precision (less variance). And, because we stratified, we know we will have enough cases from each group to make meaningful subgroup inferences.

Multi-Stage Sampling

The four methods we've covered so far; simple, stratified, systematic and cluster sampling are the simplest random sampling strategies. In most real applied social research, we would use sampling methods that are considerably more complex than these simple variations. The most important principle here is that we can combine the simple methods described earlier in a variety of useful ways that help us address our sampling needs in the most efficient and effective manner possible. When we combine sampling methods, we call this multi-stage sampling. For example, consider the idea of sampling Andhra Pradesh residents for face-to-face interviews. Clearly we would want to do some type of cluster sampling as the first stage of the process. We might sample townships or census tracts throughout the state. But in cluster sampling we would then go on to measure everyone in the clusters we select. Even if we are sampling census tracts we may not be able to measure *everyone* who is in the census tract. So, we might set up a stratified sampling process within the clusters. In this case, we would have a two-stage sampling process with stratified samples within cluster samples. Or, consider the problem of sampling students in grade schools. We might begin with a national sample of school districts stratified by economic and educational levels. Within selected districts, we might do a simple random sample of schools. Within schools, we might do a simple random sample of classes or grades. And, within classes, we might even do a simple random sample of students. In this case, we have three or four stages in the sampling process and we use both stratified and simple random sampling. By combining different sampling methods we are able to achieve a rich variety of probabilistic sampling methods that can be used in a wide range of social research contexts.

Cluster Sampling

The problem with random sampling methods when we have to sample a population that's disbursed across a wide geographic region is that you will have to cover a lot of ground geographically in order to get to each of the units you sampled. Imagine taking a simple random sample of all the residents of Andhra Pradesh in order to conduct personal interviews. By the luck of the draw you will wind up with respondents who come from all over the state. Your interviewers are going to have a lot of traveling to do. It is for precisely this problem that cluster or area random sampling was invented. In cluster sampling, the steps to be followed:

- Divide population into clusters (usually along geographic boundaries)
- Randomly sample clusters
- Measure all units within sampled clusters

For instance, let's say that we have to do a survey of town governments that will require us going to the towns personally. If we do a simple random sample state-wide we'll have to cover the entire state geographically. Instead, we decide to do a cluster sampling of five districts. Once these are selected, we go to *every* districts, municipalities and panchayat in the five areas. Clearly this strategy will help us to economise on our mileage. Cluster or area sampling, then, is useful in situations like this, and is done primarily for efficiency of administration. Note also, that we probably don't have to worry about using this approach if we are conducting a mail or telephone survey because it doesn't matter as much (or cost more or raise inefficiency) where we call or send letters to.

Non-Probability sampling

The difference between non-probability and probability sampling is that non-probability sampling does not involve *random* selection and probability sampling does. Does that mean that non-probability samples aren't representative of the population? Not necessarily. But it does mean that non-probability samples cannot depend upon the rationale of probability theory. At least with a probabilistic sample, we know the odds or probability that we have represented the population well. We are able to estimate confidence intervals for the statistic. With non-probability samples, we may or may not represent the population well, and it will often be hard for us to know how well we've done so. In general, researchers prefer probabilistic or random sampling methods over non-probabilistic ones, and consider them to be more accurate and rigorous. However, in applied social research there may be circumstances where it is not feasible, practical or theoretically sensible to do random sampling. Here, we consider a wide range of non-probabilistic alternatives.

Accidental or Haphazard or Convenience Sampling

One of the most common methods of sampling goes under the various titles listed here. I would include in this category the traditional "man on the street" (of course, now it's probably the "person on the street") interviews conducted frequently by television news programs to get a quick (although non representative) reading of public opinion. I would also argue that the typical use of college students in much psychological research is primarily a matter of convenience. (You don't really believe that psychologists use college students because they believe they're representative of the population at large, do you?). In clinical practice,we might use clients who are available to us as our sample. In many research contexts, we sample simply by asking for volunteers. Clearly, the problem with all of these types of samples is that we have no evidence that representative of the populations the sample is where interested in generalization.

Purposive or Deliberate or Judgment Sampling

In purposive sampling, we sample with a *purpose* in mind. We usually would have one or more specific predefined groups we are seeking. For instance, have you ever run into people in a mall or on the street who are carrying a clipboard and who are stopping various people and asking if they could interview them? Most likely they are conducting a purposive sample (and most likely they are engaged in market research). They might be looking for working woman between 30-40 years old. They size up the people passing by and anyone who looks to be in that category they stop to ask if they will participate. One of the first things they're likely to do is verify that the respondent does in fact meet the criteria for being in the sample. Purposive sampling can be very useful for situations where you need to reach a targeted sample quickly and where sampling for proportionality is not the primary concern. With a purposive sample, you are likely to get the opinions of your target population, but you are also likely to overweight subgroups in your population that are more readily accessible. All of the methods that follow can be considered subcategories of purposive sampling methods. We might sample for specific groups or types of people as in modal instance, expert, or quota sampling. We might sample for diversity as in heterogeneity sampling. Or, we might capitalize on informal social networks to identify specific respondents who are hard to locate otherwise, as in snowball sampling. In all of these methods we know what we want and we are sampling with a purpose.

Quota Sampling

In quota sampling, you select people non-randomly according to some fixed quota. There are two types of quota sampling: *proportional* and *non proportional*. **Proportional quota sampling** you want to represent the major characteristics of the population by sampling a proportional amount of each. For instance, if you know the population has 40% women and 60% men, and that you want a total sample size of 100, you will continue sampling until you get those percentages and then you will stop. So, if you've already got the 40 women for your sample, but not the sixty men, you will continue to sample men but even if legitimate women respondents come along, you will not sample them because you have already "met your quota." The problem here (as in much purposive sampling) is that you have to decide the specific characteristics on which you will base the quota. Will it be by gender, age, education, caste, religion, etc.?

Non-proportional quota sampling is a bit less restrictive. In this method, you specify the minimum number of sampled units you want in each category. here, you're not concerned with having numbers that match the proportions in the population. Instead, you simply want to have enough to assure that you will be able to talk about even small groups in the population. This method is the non probabilistic analogue of stratified random sampling in that it is typically used to assure that smaller groups are adequately represented in your sample.

Heterogeneity Sampling

We sample for heterogeneity when we want to include all opinions or views, and we aren't concerned about representing these views proportionately. Another term for this is sampling for *diversity*. In many brainstorming or nominal group processes (including concept mapping), we would use some form of heterogeneity sampling because our primary interest is in getting a broad spectrum of ideas, not identifying the "average" or "modal instance" ones. In effect, what we would like to be sampling is not people, but ideas. We imagine that there is a universe of all possible ideas relevant to some topic and that we want to sample this population, not the population of people who have the ideas. Clearly, in order to get all of the ideas, and especially the "outlier" or unusual ones, we have to include a broad and diverse range of participants. Heterogeneity sampling is, in this sense, almost the opposite of modal instance sampling.

Snowball Sampling

In snowball sampling, you begin by identifying someone who meets the criteria for inclusion in your study. You then ask them to recommend others who they may know who also meet the criteria. Although this method would hardly lead to representative samples, there are times when it may be the best method available. Snowball sampling is especially useful when you are trying to reach populations that are inaccessible or hard to find. For instance, if you are studying the homeless, you are not likely to be able to find good lists of homeless people within a specific geographical area. However, if you go to that area and identify one or two, you may find that they know very well who the other homeless people in their vicinity are and how you can find them.

Procedure for selecting a sample

Defining the Universe : The whole group from which the sample is to be collected is technically called a universe. In order to draw the sample one must have a clear idea of the universe from which the sample is to be drawn.

- Sampling unit: Before drawing the sample we have to decide the unit of sample.

    1. Geographical unit, ex: state, district, city etc.

    2. Structural unit, ex: house, flat etc.

    3. Social group unit, ex: family, school, club, church etc. and

    4. Individuals.

- *Source List*: The list, which contains names of the units of the universe from which, the sample is to be collected.

- *Size of the Sample* : The size of the sample is an important problem to be decided in sampling because the size has direct bearing upon accuracy, time, cost and administration of survey. The sample should be small enough to avoid unnecessary expenses and large enough to avoid sampling defects.

Advantages of Sampling

- *Saving of Time* : As smaller number of units is studied in sampling method, it requires much less time than census method. In certain types of social survey time is most important factor and the result of study has to be declared quite early, in all such type of survey, sampling is the only method, which can be used.

- *Saving of Money* : Survey of smaller number of units not only requires less time but also requires less money.

- *Detailed Study* : When the number of units is large, detailed study is possible. The smaller number of units in the sample permits a more minute observation and detailed study.

- *Accuracy of Results* : The results drawn by sampling techniques are more reliable than other methods.
- *Administrative Convenience*: Small sample is usually more convenient from administrative point of view.
- *Impossibility of use of Census Method* : When the universe is too vast and geographically scattered then every unit cannot be contacted. In that case sampling method is more useful.

Disadvantages of Sampling

- Chance of bias: It may lead to biased selection and there by lead us to draw false generalizations.
- Difficulties of representative sample: Results are accurate and usable only when the sample is representative of the whole groups.
- Need for specialized knowledge: The use of sampling method requires specialized knowledge.
- Difficulty in sticking to sample: Although the number of units in the sample is small, it is not always easy to stick on to it. The units may refuse to cooperate with the researcher, which have to be replaced by other units. This leads to bias in study.

Techniques of Data Collection

Social research is based on various types of information. The more valid is the source of information, the more reliable will be the information received. The type of information to be collected depends upon the type of social research. This requires a wide knowledge of kinds and sources of information.

1. *Primary Data :* Is the actual information received by the researchers for social studies from the actual field of research, usually obtained through questionnaire, schedule, interview and observation methods.
2. *Secondary Data :* Is the information attained indirectly either from published or unpublished sources such as literature, documents, bulletins etc.

Important Techniques of Data Collection

- Observation method
- Case study method
- Interview method

Observation Method

Science begins with observation and must ultimately return to observation for its final validation. Observation is probably the oldest method used by man in scientific investigation. The first knowledge of the universe around man begins with observation. In social research too the earliest method of investigation was probably the 'observation' method. This method involves a careful and systematic watching of facts as they occur in course of time. It could be considered as a basic method of collecting information about the world / phenomenon around us. Observation is important in social sciences including Animal Husbandry Extension as an independent method of research and as an auxiliary to other methods. It is an act of recognising and noting some fact or occurrence. It is a methodical way of looking at the behaviour of others in a specific situation and recording the same thoroughly in some form or the other. Observation method of data collection is the technique for gathering information without direct questioning on the part of the investigator.

Characteristics of Scientific Observation

The accurate and fruitful observation is an art and it demands training and practice, following are criteria of a good scientific observation:

- It must be specific but not haphazard
- It must be systematic but not by chance i.e., 'dropping in' on a situation
- It must be quantitative in nature and
- Scientific observation must be recorded immediately.

Kinds of Observation

Observation can be classified into various categories, depending upon 'type of method' and 'type of control'

Depending on the type of method used observation is classified into

1. *Participant Observation :* When the observer participates with the activities of the group under study it is known as participant observation. Thus a participant observer makes himself a part of the group under study. The observer need not necessarily carry out all the activities as carried out by other members of the group, but his presence as an active member of the group is necessary. Behaviour of the members of the group is observed under natural situation. During this, it is possible to get minute and hidden facts. A participant observer can easily learn a wide variety of intimate details to interpret his observation.

2. *Non-Participant Observation :* When the observer does not actively participate in the activities of the group, but simply observes the group from a distance it is known as non-participant observation. The observer will observe the group in a particular situation as an outsider by standing at a distance. It is a useful method, if an extension scientist wants to find out the real reaction of farmers. During this, it is possible to get complete, thorough, scientific & objective method of data collection is possible.

Researcher behaves like a stranger and stands outside the group so the group may forget about the researcher and they behave in a natural way.

Depending on the Type of Control Observation is Classified into

1. *Non Controlled Observation :* When the observation is made in the natural surroundings and the activities are performed in their usual course without being influenced or guided by any external force it is known as non-controlled observation. The observation may be biased and colored by the views of the observer, because there is no check upon him. For example, observing the farmers when they are carrying out their farming activities in their fields.

2. *Controlled Observation :* Observing the people when they are working under some authority. The purpose of controlled observation is to check any bias due to faulty perception, inaccurate data and influence of outside factors on the particular incident. For example, studying the people working under some administration like in factories.

Guidelines for Data Collection Through Observation

While collecting data through observation method, following are the guiding points to be considered in the field of extension:

- Physical environment of the village such as rivers, valleys, hills, types of houses, building etc.
- Various centres of the community such as marketing centre, recreation center, temples, educational institutions etc.
- Agricultural activities such as farming system, farming types, varieties of crops grown, species of animals reared, pattern of animal rearing, sources of irrigation, farm implements, etc.
- Religious activities such as beliefs, associated with an activity.
- Relationship pattern among the livestock owners and with other community members.
- To whom more respect is given whether to high adopters or literates or wealthier persons.
- What are the different groups such as castes existing in the village? What are its different activities? What kind of relationship exists between and within the groups?
- What is the pattern of leadership existing in the community?
- What are the problems faced by the livestock farmers?

Case Study Method

Case study research involves studying individual cases, often in their natural environment, and for a long period of time. The method of exploring and analyzing the situation in depth of any social unit is called as Case Study. It is studying 'every thing about a single thing'. For example, study about any institution, any phenomena and about any successful persons.

Characteristics of Case Study

- It studies the whole unit in totality.
- It often studies a single unit – one unit is one study.
- It perceives the respondent as an expert, but not as a source of data.
- It studies typical cases.
- It employs several methods primarily to avoid errors and distortions.

Steps in Conducting Case Study

- Introduction about the unit under study.
- Explanation about the existing situation.
- Course of events
- Cause and effect relationship.
- Validation of diagnosis.
- Follow up of case.
- Summary of findings.
- Conclusion and generalization.

Uses of Case Study

- To gain more information about the structure, process and complexity of the research object especially when relevant information is not available or insufficient.
- When the research content is complex for survey studies and when the researcher is interested in the structure, process and outcome of a single unit.
- To assist with formulating hypothesis.
- To illustrate, explain and to offer more detail findings.
- To test the feasibility of quantitative studies.

Interview Method

Interview is a technique of face-to-face contact used to watch the behaviour of an individual to record the statements, to observe the concrete results, and group interactions. It may be regarded as a systematic method by which a person enters more or less imaginatively into the life of a comparative stranger. This method helps in securing special information by getting in contact with the respondents.

Purpose of the Interview

- To bring the scientist and layman into direct contact, so that both may know and recognise the needs of each other.
- To gather first hand, factual, tactful and useful information.
- To provide chance for the extension worker to observe the behaviour of livestock owners.
- To find out the views, reactions, attitudes and beliefs towards the livestock farming activities and various developmental programmes.

Kinds of Interviews

Structured Interview : Also known as controlled, guided or direct interview. In this a complete schedule is used and the interviewer is asked to get the answers to those questions only. Interviewer generally doesn't add anything from his own side and does not even change the language. Interviewer can only interrupt or amplify the statement wherever necessary.

Unstructured Interview : Also known as uncontrolled, unguided or undirected interview. No direct or predetermined questions are used in this type of interview and are generally held in the form of free discussion or narrative story type. The interviewee is asked to narrate the incidents of his life, his own feeling and reactions and the researcher has to draw his conclusions from it.

Focused Interview : The interview generally focused on a particularly issue of social, economic, cultural, psychological aspects of the society. For example, study on reaction of livestock farmers towards an animal husbandry programme broadcasted in television. The researcher tries to focus his attention to the particular aspect of problem and tries to know his experiences, attitudes and emotional response regarding the situation under study. The interview is also in the form of free story or narrative type.

Repetitive Interview : When it is desired to note the gradual influence of some social or psychological process Interview then, it is repetitive in nature. To know the social change that have a far-reaching influence upon the people and it is some times desired to know the effect of such factors in time sequence. For example, when veterinary college adopts a village, naturally it will have its own influence upon the livestock and farmers owning the livestock. There will be betterment in the production of livestock, and also in the economic status, and standard of living of farmers. In order to study this influence in time sequence the study can be conducted through repetitive interviews.

Guidelines for Conducting an Interview Method

Conducting an interview is a technical matter & requires a thorough knowledge of the methods and the principles underlying it. Following are the different steps to be taken in conducting an interview.

- *Establishing Contact* : Interviewer has to greet the respondent & introduce himself. He should explain the purpose of selecting him and also how the information supplied by him is going to benefit the farmers.

- *Starting an Interview* : A beginning may be made from the general discussion of the problem. The interviewer should see that the respondent does most of the talking and the interviewer should listen to him attentively by guiding and directing the respondent towards the subject matter.

- *Securing Rapport* : The interviewer should be tactful and expert in creating friendly atmosphere and gain confidence of the respondent.

- *Recall* : At times respondent is full of emotional feelings so that he drifts away from the main interview. Then after some narration he/she generally lapses into silence. In such a case the researcher should wait for some time, and should try to draw the attention of the respondent to the actual point carefully by probing some questions to continue again.

- *Encouragement* : The respondent has to be encouraged from time to time during the course of interview by complementing for his expressions and also by supporting him wherever necessary.

- *Recording* : The entire process of interview should be recorded which can be used for future references.

Tools of Data Collection

Questionnaire and interview schedule are the two important tools used for data collection in the field of extension.

I. Questionnaire

Questionnaire is one of the instruments used for collecting data from the respondents. It is a device for securing answers to questions by using a form in which the respondent fills himself. The use of questionnaire is very common in social sciences. They are administered to the respondents personally or by mail.

Types of Questionnaire

Structured Questionnaire : Contains definite, concrete and pre-ordinate questions with additional questions limited to those necessary to classify inadequate answers or to elicit a more detailed response.

Non-structured Questionnaire : Often known as interview guide. It is used for focused, depth and non-directive interview. It contains definite subject matter areas, the coverage which is required during the interview. The interviewer is free to arrange the form and timing of enquiry.

Types of Questions

Closed From : The questions having only fixed alternative answers are called, as closed type of question and the respondent has to choose one among two or more alternative questions.

Dichotomous Questions : are those, having only two alternative answers. For example, do you own crossbred animals? Yes/No.

Multiple-Choice Questions : are those, having more alternatives as answers. For example, what are the reasons for owning a dairy enterprise? (a) Small scale enterprise (b) For livelihood (c) A household enterprise (d) Profitable enterprise

Open-Ended : Are unstructured questions, which provide free scope to the respondent to reply with their own choice of ideas and wording. For example, what are the reasons for not adopting the livestock production as major occupation?

Principles of Preparing Precise Questions

The wording of a question is extremely important. Researchers strive for objectivity in surveys and, therefore, must be careful not to lead the respondent into giving a desired answer. Unfortunately, the effects of question wording are one of the least understood areas of questionnaire research. Many investigators have confirmed that slight changes in the way questions are worded can have a significant impact on how people respond. Several authors have reported those minor changes in question wording can produce more than a 25 percent difference in people's opinions. Several investigators have looked at the effects of modifying adjectives and adverbs. Words like *usually, often, sometimes, occasionally, seldom,* and *rarely* are commonly used in questionnaires, although it is clear that they do not mean the same thing to all people. Some adjectives have high variability and others have low variability. The following adjectives have highly variable meanings and should be avoided in surveys: *a clear mandate, most, numerous, a substantial majority, a minority of, a large proportion of, a significant number of, many, a considerable number of,* and *several.* Other adjectives produce less variability and generally have more shared meaning. These are: *lots, almost all, virtually all, nearly all, a majority of, a consensus of, a small number of, not very many of, almost none, hardly any, a couple,* and *a few.* Besides, this following may be opted for preparing precise questions.

- Terms that could easily misinterpreted should be defined.
- areful in using double negatives
- Inadequate alternatives should be avoided.
- Double-barreled questions should be avoided.
- Important words should be underlined.
- Unwarranted assumptions should be avoided.
- Questions should give same meaning to any respondent.
- Questions should be designed in such a way that will get complete response.
- Enough provision should be made for quantification of responses.

Structure of the Questionnaire

Whether the questionnaire is administered personally or by mail it has to include three main elements, such as:

The Covering Letter : To make familiar the topic to the respondent and to motivate the respondents to participate in the study, answer all the questions and to assure them confidentiality. Covering letter includes main objective and significance of the study. It also specifies the time allotted for answering the questionnaire and also the conditions to fill it. It also specifies the reasons for selecting him/her as a respondent.

Instructions: How to fill the questionnaire and how to state their answers etc.

Main body: Includes the questions that are to be answered. This part of document should be worked out care fully which enable the researcher to collect the data required for the study.

Criteria of a Good Questionnaire

- Questions should be brief, attractive and clear.
- Questions should be arranged in predetermined categories.
- Questions should be objective, comprehensive and easy to answer.
- Simple questions must be arranged first and there after complicated questions.
- There should be clear-cut instructions to fill up each item.
- Avoid the tendency of putting many ideas in one question.
- Avoid delicate and embarrassing questions.

- Questions be framed in such a way, so that the data generated be easy to tabulate and analyze.
- Questions should be pre tested on small scale before the actual use.
- Leading questions should be avoided.

II. Interview Schedule

Schedule is the one, which contains some questions and or blank tables, which are asked and filled in by an investigator while gathering information in a face-to-face situation from the respondents. Schedule provides opportunity to establish rapport, to explain the purpose, and to make meaning of items clear.

Purpose of Schedule

1. To provide a standardised tool for observation and interview in order to attain objectivity.
2. The data received can perfectly be comparable.
3. It acts as a memory guide to the investigator or otherwise, in the absence of any schedule the investigator may not get the same information and he/she may forget some important aspects of the problem.
4. It facilitates the work of tabulation and analysis.

Kinds of Schedules

Observation Schedule : Are used for observation purposes. They contain specific topics upon which the observer has to concentrate and the nature of information has to record. This makes observation more pin pointed and accurate by pointing clearly what is to be observed and how it is to be recorded.

Rating Schedule : Used for social or psychological research. Used in those cases where the attitude or opinion is to be measured.

Document Schedule : Used for recording data from written documents like autobiography, case history, diary, or official records maintained by the Government. Such schedules are used for preparing the source list of collecting preliminary information about the universe.

Interview Schedule : Used for interview purposes. They contain standard questions the investigator has to ask, and blank tables that he has to fill up after getting information from the respondents.

Essentials of a Good Schedule

Accurate Communication : The basis of accurate communication is proper wording of questions. The respondent has to understand the questions in the same sense that they are expected to convey. The questions should be so worded that they may clearly carry the desired sense with out any ambiguity.

Accurate Response *:* It can be achieved when response contains the required information, which is unbiased and true.

Procedure for Framing a Schedule

- Clear understanding of the problem under study and identifying different aspects of the problem.
- Each aspect should be divided into sub parts and information required for valid generalisation under each aspect should be elicited out.
- Framing of actual questions: This is the most vital part of schedule. Care is taken to see that the questions convey exact sense, easily followed by the respondents and they will be willing to supply information with out any hesitation or bias.
- After framing the actual questions, proper arrangement namely layout of the schedule is important. A well-planned and attractive schedule will get good responses.
- Lastly the validity of the schedule has to be tested on a sample population to find out any discrepancies. Thus the schedule should first of all be prepared on trial basis.

Techniques of Interviewing

1. ***Preparatory Thinking*** *:* It includes
   - Introduction of investigator and the respondent
   - Time and place must be convenient to the respondent.
   - Politeness in gestures towards the respondents
   - Avoid backdoor approaches and encroachments.
   - Time of interview should not be more than 30 minutes.
   - Interview the local leader first and then ensure the cooperation of the others.

2. ***Approach to the Interview*** *:* It Includes
   - Friendly exchange of greetings with the respondents
   - Explain the purpose of interview
   - Give the importance of interview
   - Use informal approach with the respondents
   - Build up rapport with the respondents.

Advantages

- It is possible to get answers to all the questions

- Probability of getting real and correct information is more.
- Interviewing technique is very useful, when livestock farmers are ready to talk.
- Doubts can be cleared.
- Cross-questioning is possible.
- Possible to go to the depth of the problem.
- Possible to ascertain the reasons for the behaviour and feelings of respondents.

Limitations

- It takes lot of time
- It costs more
- It demands good abilities on the part of investigator
- Chance of investigators bias

Size of a Sample

Size of a sample is an important aspect to be decided. This is because the size has a direct bearing upon accuracy, cost and administration of the survey. Large samples are generally hard to manage and are unfit for detailed study but they may be essential for representativeness. According to Parten "An optimum sample in survey is one which fulfills the requirements of efficiency, representativeness, reliability and flexibility. The sample should be small enough to avoid unnecessary expenses and large enough to avoid intolerate sampling over.

Factors affecting size of the sample :

The size of the sample depends upon a number of factors the chief of which are stated below:

1. *Homogeneity or heterogeneity of universe :* If the universe is comparatively homogeneous; a smaller size of the sample may be sufficient. If all the units were exactly alike one single unit could serve as example, but if the universe is heterogeneous than the sample has to be essentially larger.

2. *Nature of study :* The size of the sample will also depend upon the nature of the study. If an intensive study is to be made a large sample is unfit for the purpose, as it will require very large finance and other resources. Thus in the case of opinion surveys, people have to be contacted for knowing their opinion about the problem under study, the

interview is quite short and a much larger number of units may be included in the sample. Similarly in the case general study large number of cases may be taken, but if the study is of a technical nature a large number of cases may become difficult to manage.

3. ***Practical Consideration*** : Practical consideration of availability of finance, time, number of trained field workers etc., may also be taken as important factors in deciding the size of the sample. The limitation of these resources necessarily limit the size of the sample. Although these practical considerations do weigh heavily in determining the size of the sample it should never be done at the cost of accuracy. Any account of money, however small, spent on an unrepresentative sample is a waste and must be avoided at all cost.

4. *Standard of Accuracy :* It is generally considered that larger the size of the sample, greater the standard of accuracy or representativeness. However largeness of size is no guarantee for representativeness. A small but well selected sample can give better results.

5. *Size of the Questionnaire or schedule :* The size of the questionnaire, and the nature of question to be asked is also a limiting factor for the size of the sample. Larger the size of schedule more complicated the questions to be asked, smaller is the size of the sample suitable for proper administration.

6. *Nature of Cases to be Contacted :* If the cases are geographically scattered a small sample is suitable. On the other hand if the refusal rate is likely to be heavy or losses of cases likely to be quite big a larger sample has to be selected.

7. *Type of Sampling Used :* If absolute random sampling has been used a much larger sample is required. Random sampling is reliable only when there is a sufficiently large number of cases so that the law of statistical regularity properly works and every class of units get a chance of being selected. On the other hand if a stratified sample is selected reliability can be achieved in a much smaller size. But in stratified sampling th essential condition is that stratification must be proper. If stratification is improper a large number will only add to bias in the sample.

These are some of the broad considerations that are to be kept in mind. In fact no rigid number can be prescribed for an optimum size of the sample. The nature of the problem is the only deciding factor in this matter.

Sample Error

A most significant feature of the sampling procedure is that any estimate based on a fraction of the individuals will not usually be equal to the estimate based on data measured from all the individuals constituting the population by following the identical measurement. There is bound to be a difference between the sample estimate and that of population value. Ex: if a population consists of N individuals and a sample x is taken from it and the average value of the characteristic per individual is given as follows:

Population Value	Sample value
X	n'
N	n

These two values n' and n will not be identical values each and every individual constituting the population has the same value. The difference between n' and n is usually denoted as sampling error. The sampling error is an inevitable part of any sampling procedure. In other words no estimate based on sampling however accurately planned and conducted will be free from error. It does not mean that the sample estimate cannot be useful. Its utility will be judged by the magnitude of sample error. Obviously, larger value of the sample error, larger will be the risk in using it. It is therefore always attempted to design the sampling enquiry in such a way that the error is as small as it could possibly be obtained within the resources at the disposal of the researcher.

Since in calculation of sampling error, the quantity XN, which is unknown and in fact is the quantity to be estimated, the knowledge about the sampling error is difficult to obtain. This difficulty is however overcome by utilising certain mathematical and statistical laws. The quantity arrived at on the basis of mathematical and statistical laws is generally called 'standard error' (S.E). Sampling error is usually defined for the individual sample while standard error denotes the average of sampling error for all the samples. The statistical laws which are used in determining the value of sample error are applicable only to the situation where the sample estimate is based on individual drawn according to the law of chance and not according to individuals judgment. Thus for a judgment sample no sample error can be calculated which is so essential in sampling. This is the basic reason that the statistician firmly advocates the use of a probability sample for studying the characteristics of population on sampling basis. (In other words, sample error which is so essential for judging the utility of a sample estimate can be calculated only from a sample selected on the basis of probability law).

Variables

Variables are the ones, which may change magnitude from one situation to another or symbol to which values are assigned.

Classification of Variables

1. (a) *Dependent :* The variable not capable of causing change in other variables is called a dependent variable – it's a presumed effect.

   (b) *Independent :* The variable capable of causing change in other variables is called an independent variable – it's a presumed cause.

2. *Dichotomous:* which has two variables.

   Polytomous : which has more than two variables.

3. (a) *Qualitative:* which can not be manipulated (or) at least difficult to manipulate.

   (b) *Quantitative :* which may take various magnitudes.

4. (a) *Stimulus :* is the one that is created by researchers i.e. knowledge.

   (b) *Response:* any kind of behaviour of a respondent.

5. (a) *Active :* manipulated variables e.g. behaviour,

   (b) *Attribute :* measured variable.

Measurement

Measurement is the assignment of numerals / symbols to objects/ events according to rules i.e what is it to be measured? And how is to be measured?

Four levels of measurements

Nominal level : also known as classification scale. This is the one, which is used simply to classify an object or individual. It consists of two or more categories into which objects are classified.

Properties :

- The relation could be reflexive i.e. $x = y$ for all values of x.

- The relation could be symmetrical i.e. if $x = y$ then, $y = x$

- The relation could be transitive i.e. if $x = y$ and $y = z$ then $x = z$.

Ordinal level : also known as ranking scale. The one, which defines relative position of objects or individuals with respect to a characteristic. The basic requirement is that one be able to determine the order of position of each object or individual with respect to others.

Properties :

- It incorporates not only equivalence but also the relation
- It can be reflexive
- It is asymmetrical
- It is transitive

Interval level : The one which defines distance between positions of objects or individuals. For example, the distance between positions labeled one and two on the scale is equal to the distance between positions labeled two and three. It is used mostly in social science researches.

Ratio scale : The one, which defines the empirical operations between two or more objects or individuals. The basic requirements to establish a ratio scale are : methods for determining equivalence and non-equivalence, rank order, equality of intervals and equality of ratios. It has 'true zero point' as the origin; where multiplication, division, geometric mean, variance and all other statistical tests are possible.

Validity

It is the extent to which a test measures correctly. It can be expressed by square root of true variance.

Internal validity : It expresses the logical relationship of two variables. For example, if A = B and B = C then, A = C

External validity : It expresses the statistical relationship of two variables. For example, number of children and supply of food should tally with respect to number.

Reliability

It is the accuracy on precision to measure an instrument. The reliability of test is its ability to yield consistent results from one set of measures to another. For example, if the same test is administered more than once and if we get the results correctly, then, the particular test is considered to be reliable and if the test score is stable, the test score is said to be reliable score.

Relative Reliability : means the extent to which individuals in a group maintains relatively consistent position / scores, when two sets of measures are obtained and correlated using same test or two equivalent forms.

Absolute Reliability : meaning standard error, is an estimate of the deviation of a set of scores obtained from their true scores.

Methods of Computing Reliability Coefficient

1. *Test-test Method:* Same test is administered to the same individuals twice or more number of times.
2. *Parallel Method:* Two separate tests but of equivalent forms are administered to the same individuals separately.
3. *Split of Method:* The test is splitted off into equal proportions and then administered to the individual.
4. *Rational Equivalence Method :* Two forms of tests which are equivalent are administered into two different individuals.

Reliability is satisfactory when reliability coefficient is large which in turn depends on:

- Nature of the test
- Size and variability of the group
- Purpose of the test
- Reliability coefficient should range from 0.90 to 1.0
- Standard deviation of test distribution must be known

Hypothesis

It is a proposition, which can be put to test to determine its validity. It is one of the important steps in scientific method of analysis.

Sources of Hypothesis :

- General pattern of culture and its understandings
- Analogies
- Personal experience
- Scientific theory about society and human behaviour

Functions of Hypothesis :

- It gives orientation to theory
- It gives direction to research problem
- It helps in selecting pertinent data and methods
- It helps; in drawing specific conclusions
- It seems as a working instrument of theory : its application or a new tentative theory.

Steps in Testing of Hypothesis

- Anticipating all the outcomes of research
- Deciding on the procedure that can assess the outcomes
- Deciding in advance the basis for rejection or acceptance of a relationship or event
- Deciding on the significant levels
- Conducting the experiment and observe the outcome
- Making decisions whether to reject or not to reject the hypothesis
- Examining the relationships of the tested outcomes with the theory
- Making a theoretical model out of generalisation

Item Analysis

When we want to study any aspect of an individual, whether it may be knowledge / attitude etc; we select the particular item and the analysis of the selected item is called item analysis. It is based on three aspects: (1) Item selection, (2) Item difficulty and (3) Item validity.

Table 16.1 Difference between social survey and social research

Social Survey	Social Research
1. Concerned with specific problems and situation	1. Concerned with general and abstract problems
2. Practical in nature for immediate needs	2. Theoretical in nature for long term
3. Results in change in social reforms removes social evil	3. Results in formation of new techand sniques
4. Basis for some hypothesis	4. Develops the hypothesis
5. Hypothesis is not necessary	5. Hypothesis is essential
6. Conducted on professional basis	6. Not neccessarily so
7. Improves the man on all aspects	7. Improves the knowledge of man

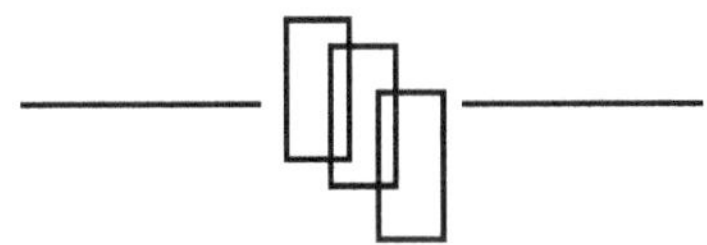

CHAPTER - 17

Participatory Rural Appraisal

Introduction

It is crucial to understand the intricacies of the socio-economic and ecological environment within which resource-poor farmers operate and their problems can then be resolved by collaborating with them according to their needs and priorities. While at the other end, the development professionals tend to operate on a basic presumption that they are the best judges of the welfare of rural masses. Hence, with the best of knowledge, resources and benevolent motives they remain oblivious of the fact that the majority of their rural clients are passive spectators of their technology of rural development. There exists sea-gap between rural perceptions and perceptions of these developmental professionals involved in planning and implementation of conventional models of rural development. In the sphere of agriculture and rural development, efforts were "production centered" as well as concentrated in the hands of rich farmers, while the problems of resource-poor farmers trickled up over the years rather than trickled down.

Proper understanding of the farmers, their situations, problems and solutions from their point of view has always been the challenge before social scientists and extension workers. The conventional approaches are generally time consuming and tedious and where the farmers play a passive role. Thus, new approaches incorporate flexibility of methods, peoples' participation and elements of both qualitative and quantitative methods in different proportions to reflect better field level realities. In the early 1970s, RRA (Rapid Rural Appraisal) was started, which is a systematic and structured activity conducted by a multidisciplinary team with an aim to acquire new information quickly and efficiently about the rural life and resources. PRA is a social science approach which emerged in the late 1970s, which is a growing combination of approaches and methods of enabling local people to analyze their living conditions and to share the knowledge. PRA is a flexible and adaptable need assessment technique.

Features of PRA

1. *Triangulation* : This is form of cross checking by changing the team composition, the sources of information and the techniques applied. There is need that each activity or phenomenon is considered from different viewpoints and studied using different techniques. The diversity in the PRA scientists; team and also in villager group having different interest groupings help in triangulation and thereby in higher reliability of information.

2. *Learning in the Community* : PRA is an approach for learning from, with and through members of the local community. The PRA scientists team should make all possible efforts to see the situation through the eyes of the affected people. To achieve this the team members must become friendly with the villagers by taking part in their every day activities and by listening from them.

3. *Appropriate – Imprecision :* In a conventional survey, efforts are made to collect information on all possible aspects very exhausitively and most of the time only a part of it is used. On the contrary in PRA approach the information is generated only on the aspects which are directly required in accordance with the aims and objectives of the study. Further, an effort is made to avoid more precision than required during the process of collection and analysis of data.

4. *Use of Appropriate Tools / Techniques* : The selection of tools / techniques needs to be done carefully. Tools should be simple, clear, self evident and appropriate to local conditions. The team should also be ready for any modification in the technique that might be suggested by the residents of the study area. Appropriate means and local materials may be taken for describing quantities and for qualities and setting up classification schemes.

5. *On Site Analysis* : The team should meet each evening to jointly discuss and analyse that day's findings and also to plan the work to be done on the following day. The repeated on the spot analyses enable them to understand the existing situation, concentrate on the problem areas and lead to growing understanding and accumulation of knowledge. The results should be evaluated by the team before departure from the area and should be publicly presented and discussed with members of the community.

6. *Offer Setting Biases, Self-Critical Awareness* : This process should try to have involvement of those who would otherwise never get a

chance to speak, the poorest people, women, disadvantaged groups in remote areas etc. and it is important that the team must refrain from any value judgments about others.

7. ***Multi disciplinary Team*** : The scientific team conducting PRA must have a fairly broad base, meaning thereby; inclusion of scientists of all important disciplines relevant to the area of study. It is also important to have women scientists in the team so that rural women could be effectively involved in the appraisal exercise.

Principles of PRA

- Congenial attitudes and friendly behaviour with villagers.
- Outsider (scientist) as listener and villager as teacher
- Facilitating
- Sharing
- Reversal of learning
- Learning rapidly and progressively
- Seeking diversity
- Traingulation
- Self Critical awareness and commitment

Approach of PRA

L Listen to what they say

E Encourage them to join the discussion

A Ask question (do not make judgements)

R Review what has been discussed

N Note what has been discussed

Steps of PRA

1. ***Direct Observation*** : Direct observation has no subsitute as a technique to collect reliable data from the primary source since it provides proof in itself. The team members, therefore, must always act as obervers throughout the PRA exercise. Valuable hints and information should be collected by critically observing the objects and acts beginning right from entering into the area. Each observation must be recorded as early as possible. The critical analysis of these observations/notes will not only provide the additional information but will help in triangulation.

2. ***Rapport Building*** : The first important requirement of PRA is rapport building and environment creation for free and frank interaction with the

villagers. The best way to achieve this condition is that the team members; during the study period to stay and eat with local people. The team should listen and take part in every day activities.

3. **DIY (Do-It Yourself)** : The team members should make efforts to perform the activities that the villagers are doing such as sowing the seeds, transplanting, harvesting, threshing, feeding the animals, milking, chaffing the fodder etc. This is related to participant observation and helps not only in rapport building but provides practical ideas and feelings of complexities of activities.

4. **SSI (Semi Structured Interview)** : This is the most important and critical step of PRA process. The individuals are asked relevant but informal questions. No structured schedule or questionnaire is prepared for interviewing. Various roles, such as team leader, process recorder and environment controllers are divided among the members of PRA team. These should be prepared well before entering into the village.

5. **Focus Group Interviews** : Group interviews are always better than individual interviews. It provides exhaustive information with all possible triangulation and does not leave any chance of missing information. Focus group interviews are arranged under special circumstances depending upon the specific objectives of the study. Focus group interviews provide specific and pointed data for the specific group without any intervention, hindrance and hesitation from other villagers not belonging to the focus group. Separate focus group interviews may be arranged for women, for weaker sections, for paddy growers, for sugarcane growers, for dairy farmers and so on.

Guidelines for Designing PRA

1. Distinguish clearly the different levels, topics, sub-topics, questions and indicators to avoid confusion during field work.
2. Start with something easy to make the informants relaxed.
3. Review secondary sources, interview key informants and use your own knowledge and logic to identify topics, key issues and to formulate hypotheses.
4. Think of analysis early. Use analytical tools throughout the PRA.
5. Think of ways of involving community members in the ananysis of information

Types of PRA

PRA can be of Different Types :

1. Deductive PRA
2. Topical PRA
3. Exploratory PRA
4. Planning & Implementing PRA
5. Monitoring and Evaluative PRA

1. *Deductive PRA* : Deductive PRA is used when the problem requires tact and cannot be directly handled such as ethnic violence, communical riots and corruption etc. These problems can be studied through relating them through indirect probing.

2. *Topical PRA* : For topical PRA, the sessions have to be on a particular topic. For instance, if one has to study the crop or animal diseases, PRA can help in exploring different dimensions of diseases. A topic can be probed across different groups in a village, across gender, caste etc. Its purpose is to generate as much information as possible on a selected topic. It is intensive in its approach and can help in analysis of a particular topic at great length.

3. *Exploratory PRA* : The exploratory PRA are used to reveal different aspects of rural life and also suggest important issues to be probed. If can form a basis for further probing into selected areas which appear important and useful. A theme for further analysis can emerge out of such exploration which can help in undertaking either a PRA exercise or help in framing a survey questionnaire on realistic lines.

4. *Planning and Implementing PRA* : PRA can be done for designing project and implementing them. It can make projects more adaptable to local needs. If village communities are involved in a project right from its beginning, the chance of success of the projects increase through people's participation. PRA seeks commitment from the people as partners in executing the project.

5. *Monitoring and Evaluative PRA* : PRA sessions in the village communities can be done for monitoring and evaluation of projects and development of projects can be better informed and solution to any problem can be sought for. In case of problematic situations, monitoring through PRA can bring different groups in the villages together for solving the problems.

PRA Tools

Four patterns are normally chosen to reveal the key functional relationship that determine the properties of an agro-ecosystem. These are space analysis, time analysis, seasonal analysis and matrix ranking. These analyses provide an understanding about the properties and the process of human management of agro-eco system.

Mapping

The spatial patterns are mostly revealed by maps and transects. The functional relationship can be depicted through various maps (social map, resource map, occupational map etc.), while transects are useful in defining a system boundaries and in identifying problem areas.

Steps

- Ask the villagers to draw the map of their own village at a suitable place which is convenient for all the villagers (dry, visible and spacious place so that many people can look at it easily)
- Allow them to draw the map in their own way and don't suggest.
- Don't interrupt from outside.
- There could be 5-6 or more numbers of villagers preparing the map of the village at a time. More the participation of villagers better the quality of map.
- Triangulate the information indicated in the map. Ask some of the villagers to point out his/her house, crop field, orchards, etc. on the map.
- Ask others if they would like to add anything to the map.
- Transfer the map on a paper. Write down the names of all the village mappers.

Types of Mapping

Social Map or Social Profile : The social map is made by the villagers to describe locations of different castes and communities in the village, the conditions of households, roads and rivers etc. The social map forms a useful base for identifying problems of different households, their number and their characteristics (Fig. 17.1).

Resource Map or Land Use Pattern : The resource map of a village can show different kinds of natural resources. In a resource map, the villagers depict different kinds of rivers, ponds, trees, crops, animal species and the land utilisation pattern. The problems can be discussed on the basis of a resource map. This can also form a basis for a transect walk. (Fig. 17.2).

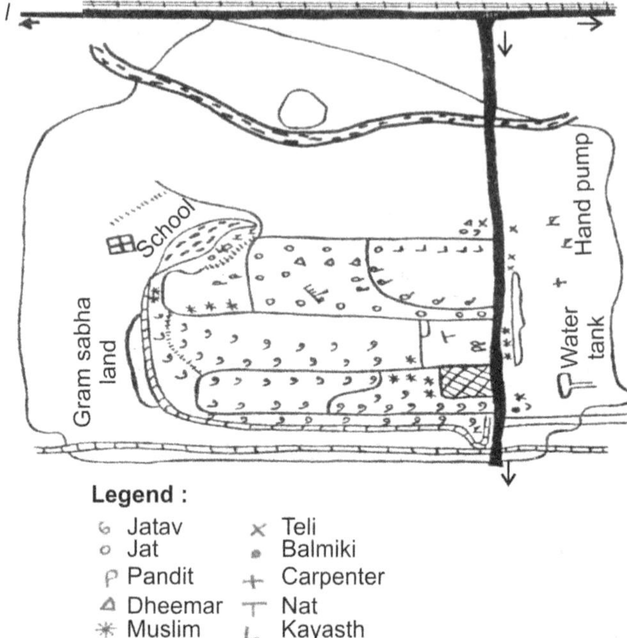

Legend :

ϧ	Jatav	✕	Teli
o	Jat	•	Balmiki
ρ	Pandit	+	Carpenter
△	Dheemar	⊤	Nat
✳	Muslim	∟	Kayasth

Fig. 17.1 Social map.

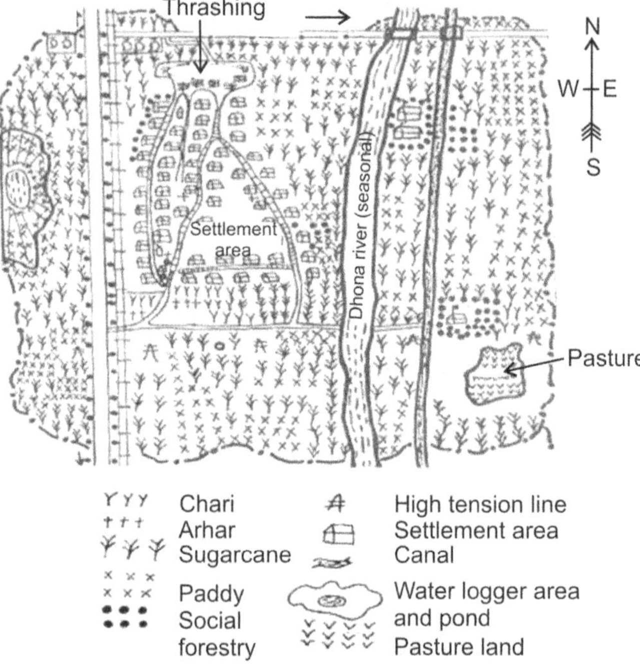

Ɏ Ɏ Ɏ	Chari	A	High tension line
+ + +	Arhar	⊟	Settlement area
Ɏ Ɏ Ɏ	Sugarcane		Canal
✕ ✕ ✕ / ✕ ✕ ✕	Paddy		Water logger area
• • •	Social forestry		and pond
		⌄⌄⌄ / ⌄⌄⌄	Pasture land

Fig. 17.2 Resource Map.

Occupational Map/Enterprise map : This map indicates the diversity of earnings. This type of map can be used for studying the income sources, expenditure pattern etc. (Fig. 17.3).

Fig. 17.3 Enterprise Map.

Transect Walk

A transect walk by PRA team across the village or its surroundings with villagers help to develop a clear idea on the natural resources, land use, vegetation, topography, cropping pattern, etc. It is a useful method to know rural ecological conditions. A walk from one point of the village to another enable the outsider to observe difficult aspects of rural ecology and to discuss with rural people about soil conditions, land use pattern, crop, livestock, agro-ecological zones, water flows and water sheds etc. (Fig. 17.4).

Land	Slop	Low level	Settlement	Highlevel	Low level
Soil fertility	Third grade	First grade	First grade	Second grade	Third grade
Crops	Paddy, Wheat Lenfil	Paddy Wheat Surgarcane Lenfil, Lehta	Sugarcane, Paddy, Wheat Lentil, Lehta	Surgarcane, Wheat, Chari, Lenfil	Water logged area, Wheat
Soil type	Sandy, Loam clay loam	Laom, sandy loam	Loam and sandy loam clay, Hard	Sandy loam	Clay
Irrigation	Irrigated by boring	Irrigated by boring	Irrigated by boring	Irrigated by boring	Irrigated
Fodder Production	Chari, Berseam	Berseem	Berseem	Chari	-
Wild animals	Antilop, deer, rabbit Jaikal	Antilop	Antilop	Antilop rabbit	Antilop deer,
Trees	Shesham	-	Neam, pakad Banyam	Sheesham, mulbery, sagon, mango Guava, jamin	Mango, Shesham
Shrubs	Pawar, khairti	-	-	Pawar, Guma khairti, labhera	Large makry pir grass

Fig. 17.4 Transect walk of a Village.

Steps :

- Select a group of villagers having good knowledge on physical resources of the village.
- Discuss with these selected villagers on the physical map of the village and mark the routes that they would like to follow during transect walk.
- Distribute responsibilities to the team members.
- Observe and record the minute details of all important things that you can who.
- All team members must complete the full transect walk
- Collect the indigenous as well as scientific practices followed by villagers.
- Travel slowly and patiently.
- Record the names of informants.
- After the completion of walk, sit down at a suitable place and record the information on paper.
- Draw an illustrative diagram of the transect using the collected information.

Transect walks can concentrate on specific areas. Transect for agriculture would observe the land use pattern, crops, soil, types, crop insects and pests, weeds, irrigation sources, cropping pattern, etc. The transect would have more emphasis on walking in the fields rather than in the village lanes. On the other hand transect walk for animal science would concentrate visits on animal species in the village houses and lanes. It may not be exactly straight from east to west or from north to south but will concentrate on animal herds. It may include fields also for pasture and fodder situations.

Time Analysis

Time analysis show the quantitative and qualitative changes over time and can be used for many variables such as area and yields of different crops, livestock population, farming practices, variety, breeds, use of fertiliser and other inputs etc. In this method of elderly people are chosen as key informants. They narrate the history on different parameters. Triangulation may be done with the secondary data obtained from records.

Time Line

Time line is a method of knowing history of major recollected events in a community. This method is also known as historical time-line of a community. Time-line illustrates diagrammatically the past events which the members of a community remember as being significant to their life. The historical time-line is narrated by elderly villagers in details with respect to the major changes, that have taken place in their society and surroundings has impact on their economic, social and agro-ecological life. (Fig.17.5).

<table>
<tr><td colspan="2" align="center">Historical Time Line</td></tr>
<tr><td>Year</td><td>Events</td></tr>
<tr><td>1500</td><td>Village settled</td></tr>
<tr><td>1915</td><td>Cholera epidemic in village</td></tr>
<tr><td>1935</td><td>First cycle was purchased in the village and first-pucca house contracted</td></tr>
<tr><td>1949</td><td>Primary school started</td></tr>
<tr><td>1950</td><td>Pucca mosque constructed</td></tr>
<tr><td>1954</td><td>Power substation established</td></tr>
<tr><td>1957</td><td>Co-ordination of land holding was sone for the first time</td></tr>
<tr><td>1972</td><td>Flood came in the village</td></tr>
</table>

1974	Brick road was constructed
1980	Village was electrifield
1982	Second time consolidation of land holding was done
1985	First TV set was purchased in the village
1995	First gobar was plant was installed

Fig. 17.5 Time line.

Steps :

- Develop rapport with elderly villagers by explaining them the objectives of the visit clearly.

- Select such focus group from the village which could give correct information.

- Ask the group about major significant events of various kinds that have affected their life pattern in past.

- Ask the group about the impact of these events on their past and present life style.

- Arrange the narrated major events and its impact in logical sequences and in tabular form.

- Each of these events should be triangulated/cross checked by two to three other villagers.

Time Trends

This tool of PRA is used for depicting quantitative changes over a time in different aspects of rural life. By applying this method one can see the increasing or decreasing trend of hardship in the life of rural people (Fig. 17.6).

Steps :

- Obtain data from the focus group on a time frame about development in particular field

- Verify the provided data with other villagers

- Arrange this information in a tabular form and analyse clearly the increasing or decreasing trend

- Draw the conclusion of given data

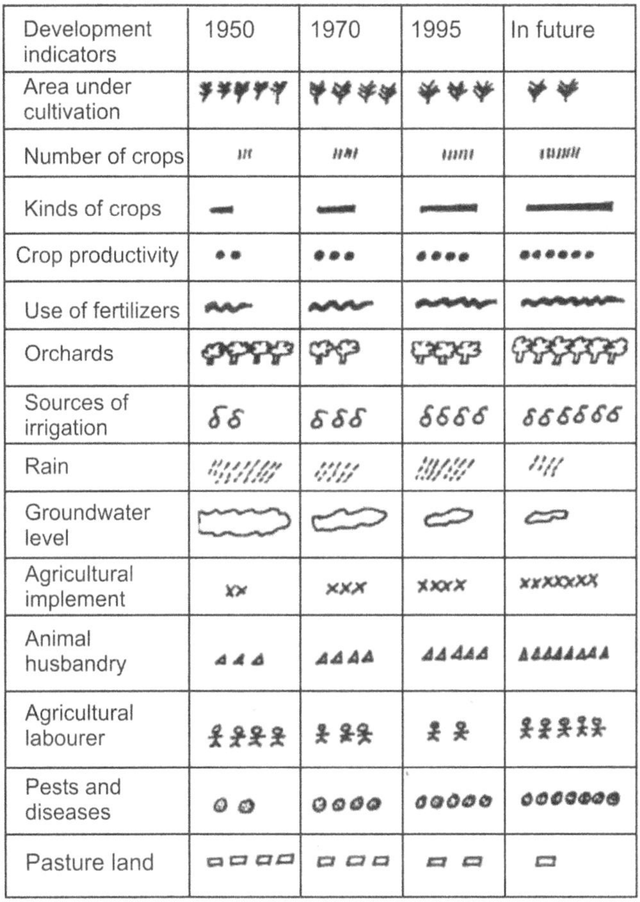

Development indicators	1950	1970	1995	In future
Area under cultivation				
Number of crops				
Kinds of crops				
Crop productivity				
Use of fertilizers				
Orchards				
Sources of irrigation				
Rain				
Groundwater level				
Agricultural implement				
Animal husbandry				
Agricultural labourer				
Pests and diseases				
Pasture land				

Fig. 17.6 Time Trends.

Seasonal Analysis

Seasonal analysis helps in understanding seasonal cyclic variations in selected parameters such as rainfall, temperature, humidity, cropping systems, crop rotations, labour availability, fodder availability, disease occurrence etc. Seasonal analysis may be done through drawing of seasonal diagrams through PRA. The community members are asked to draw the diagram for a given year preferably starting from the first month to; say; kharif season i.e. July and up to the month of June. If required each month may be divided into two fort nights or four weeks. Each activity or event is shown with the help of a bar drawn across the month. The diagram drawn by the community members on the ground is then transferred onto the papers.

Steps :

- Brief the villagers or the focus group about the item with regard to which you require information. For instance, occurence of livestock diseases and fodder availability in a calendar year.
- Do all necessary preparations required for drawing a diagram.
- Tell them to enlist the crop or livestock diseases that have affected their crops or livestock in previous years.
- Prapare a table and write down the name of the diseases on the first column on different rows.
- Write down the names of each month/season on the first row in different columns.
- Show this table to the focus group and ask them to depict the months in which a particular disease occur in a year.

Seasonal analysis helps in drawing out information about the months/ seasons in a year when the villagers face the hardships/problems. Such analysis helps the researchers and policy makers in formulating projects to solve rural problems. The seasonal analysis can be done for rainfall, seasonality of diseases, seasonwise work load in agriculture and animal husbandry, season wise work fodder availability etc. (Fig. 17.7).

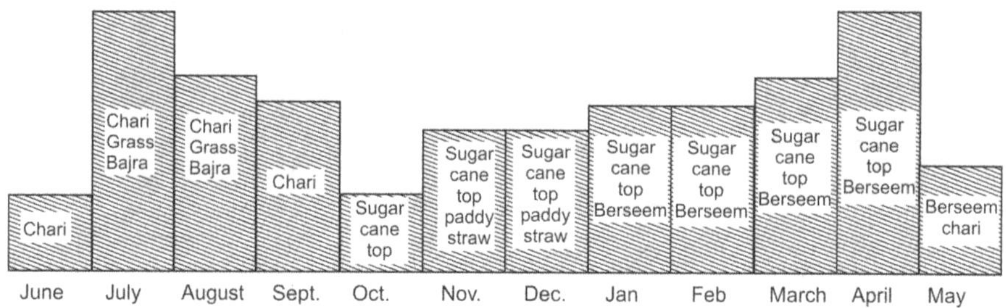

Fig. 17.7 Seasonal Analysis.

Matrix Ranking

Matrix ranking helps in comparing two or more objects in respect of perceived attributes or criteria by the villagers. In this exercise, different attributes and criteria are listed which are ranked either on the basis of fixed scoring or free scoring according to their relative importance. It helps in understanding different criteria against which the villagers rank any particular object/item, ex. crops, varieties or crops, livestock species, breeds of livestock, fodders and their varieties etc. (Fig. 17.8).

Steps :

- Decide the topic(s) on which you would like the villagers to score.
- Have an initial discussion with the interested villagers and set the climate for discussion.
- Ask the villagers to list different alternative objects/items such as breeds, varieties species etc. Ask them to write on horizontal line to have columns.
- Ask them the possible criteria for ranking or preference and to write them on the left side column in veritical order to make rows.
- Draw the table of matrix by drawing horizontal lines dividing the possible criteria and vertical lines dividing the different objects / items.
- Start talking about the first criteria and ask them to score out of 10 for each object / item. The most preferred object may have the highest score on the criteria.

Kinds of animals / Criteria	Cattle production	Buffalo production	Goat production	Poultry production
Use in farm operations	•• ••	•• •	–	–
Milk yield	•• •	•• ••	••	–
Meat production	–	–	•• ••	•• •
For manure	•• •	•• ••	••	•
For eggs	–	–	–	•• ••
For drafting	•• •	•• ••	–	–
Productive age duration	•• •	•• ••	••	•
Additional income	•• •	•• ••	•	••
Ranks	II	I	IV	III

Fig. 17.8 Matrix Ranking Livestock and Poultry Production.

- Move to the next criteria and ask them to score each object. Against that criteria. Complete for all the listed criteria.
- Traingulate the information with other villagers.
- Also ask the villagers for their best choice of the object/item and record for its vertification.
- Transfer the information onto the sheet of paper.

- Calculate the total score for each object/item and triangulate with the best choice of object by the villagers.

Livelihood Analysis

Livelihood analysis helps in obtaining information about decisions and problems and also the coping strategies of different households in relation to socio-economic factors. This type of analysis can be used for studying the household size, composition, income and expenditure pattern, income from different sources etc (Fig. 17.9).

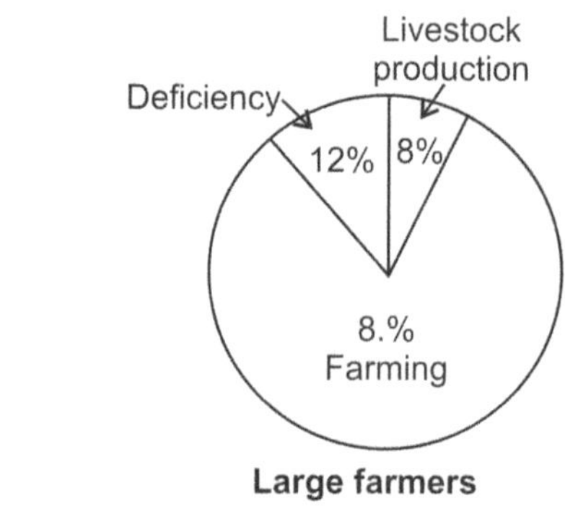

Fig. 17.9 Livelihood Analysis

Steps :

- Decide on the topic which respect to which livelihood analysis is to be carried out such as income and expenditure pattern for different types of household.

- Decide on the classification of household in terms of socio-economic categories, caste categories etc.

- Choose a representative household from different socio-economic categories.

- Ask the head of the chosen household for different sources of income and ask him to show it in the form of a pie diagram by showing the income from different sources.

Village Institution Linkage Map (Venn or Chapati Diagram)

Venn diagram is a visual method to represent the role and relationship of individuals and institutions, their degree of importance and relationship with other individuals and institutions. The individuals or institutions are depicted by circles and their importance by the size of circles. The relationship is depicted by connecting lines and distance of placement. Different aspects of decision making relationship can be studied by Venn diagram. The villagers are asked to enumerate various institutions contributing to a specific action, project or decision making. They are asked to place them in the form of circle, the size on the basis of importance and the distance indicating the relationship among each other. (Fig. 17.10).

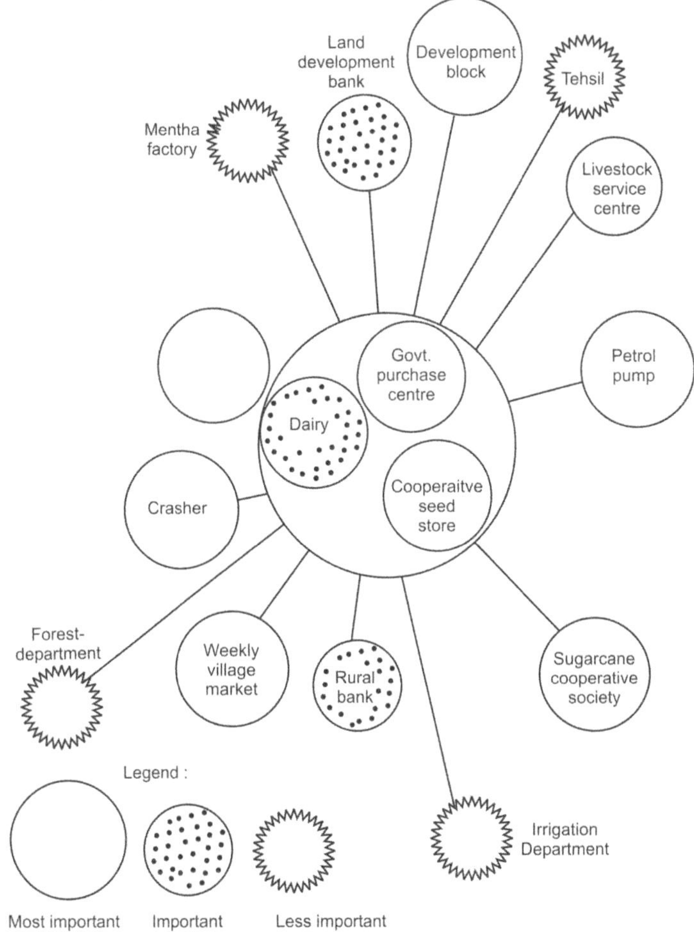

Fig. 17.10 Venn diagram.

After sharing and analysis of all the information collected from the villagers using different PRA tools, the process of participatory planning may be initiated. In this exercise, efforts are made to involve the villagers in developing a plan for action using their resources in best possible manner. The villagers come up with their own views in planning their own future activities and offer suggestions. The villagers are asked to identify and prioritise their problems and suggest possible solutions / interventions.

Participatory Planning

Steps :

- Display all the information collected through PRA exercises in at presentable form.
- Arrange meeting with villagers at a convenient time and place.
- Allow the villagers to look and discuss their own analysis and drawing prepared by themselves during PRA exercises.
- If possible arrange a presentation of results by villagers to a large group.
- After sufficient discussion on existing situations, ask the villagers to identify the problems of their village. Ask one villager to write each problem on a separate piece of paper.
- The villagers arrange the problems (by putting the pieces of papers on the ground) according to their priority / severity.
- Allow sometime to prioritise the problems by villagers.
- Initiate the discussion on the first problem and ask the villagers to indicate the possible causes.
- Keep the problem in the center and its causes around it and connect with the help of arrows.
- The socio-economic causes may be placed towards left, while bio-physical causes may be placed towards right of the problem.
- Ask the tertiary causes also in the same manner and complete the problem cause diagram for each problem one by one.
- Ask the villagers to identify the causes, which could be removed / tackled / managed by the villagers within the available resources.
- Prioritise the causes in respect of managing them easily.
- Ask the villagers to suggest solutions or the way to manage the cause. Alternative solutions may be noted.
- Have discussions on the alternative solutions and let the villagers decide on action.
- Enlist each action plan and consolidate the entire plan along with the responsibilities fixed for each action.

Have second meeting of villagers and arrange presentation of action plan by the villagers before the large group (Fig.17.11).

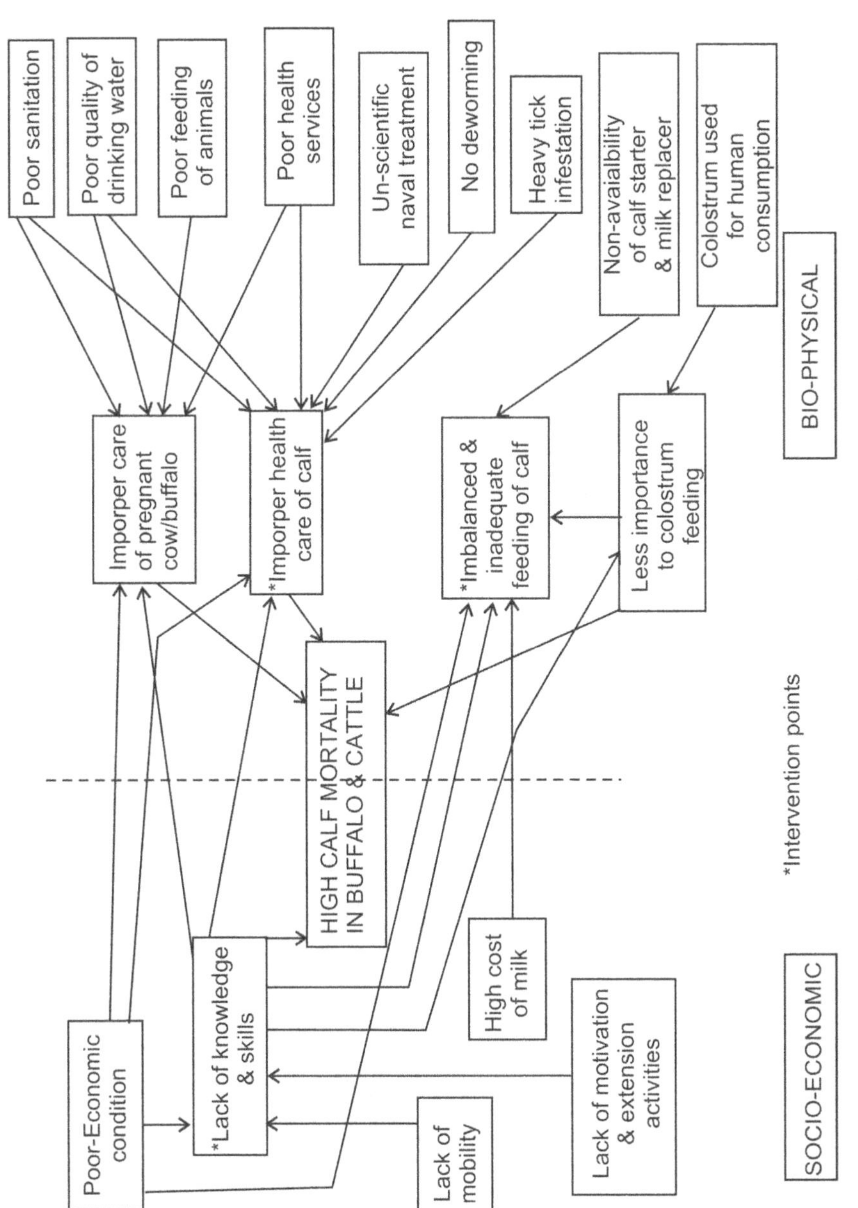

Fig. 17.11 Problem – Cause Diagram

Why use Diagrams?

Many people find images easier to understand and remember than text (i.e., they are "visualisers" rather than "verbalisers"). Drawing a diagram shows the relationships or linkages between different concepts or variables more clearly and immediately than is possible with text. Establishing the relationship of new information to that already assimilated is considered to be one of the most important cognitive processes in learning. Drawing diagrams therefore stimulates thinking about a situation. When diagrams are constructed as a group process, they aid brainstorming, analysis, communication and a common understanding of a situation. Most communication is mediated by language. Often, concepts are difficult to explain or appreciate in another language, but can become clearer when visualised as an image or diagram. When working with people of different cultures or different languages, diagrams can therefore help overcome the language barrier. Also, people that are considered to be "illiterate" (because they cannot read or write) can often show considerable analytical capability when they can express themselves in diagrams in an environment and using materials with which they are familiar. Getting people to analyse the current situation as expressed in a diagram is a good way to initiate a discussion of what could be changed, the impact these changes would have, and hence what sort of future could be possible. Diagrams thus facilitate *"visioning"* of alternative futures by rural people. In analysing systems – where an analysis of the *interrelationships* between different elements (stakeholders, entities, processes, etc.) is paramount, rather than the more reductionist analysis of the elements themselves – different types of diagrams are some of the main analytical tools used. Mostly, the diagrams are used to identify where a study needs to be focussed: i.e. to identify which elements (stakeholders, processes, problems) of a complex situation form the *relevant system* for further analysis, with the ultimate aim of designing an improvements to that system that can be implemented by stakeholders involved.

General Guidelines in Drawing Diagrams

In this learning resource, several diagram types are described. It is also possible to combine objectives and types, to make a hybrid diagram, but it is probably best to decide on the objective of a diagram and choose the best diagram type for this objective. Also, it is usually better to make several different diagrams to convey understanding of a situation, rather than try to include lots of different sorts of information in one complicated picture.

Simple diagrams (5-10 elements) are easier to understand, although complex situations often result in complex diagrams. Keep the clutter to a minimum. Avoid elements that are not strictly relevant. If words are necessary to explain the working of a diagram, try to use single words or short phrases rather than sentences; a more complete explanation can be given in accompanying text. Give every diagram a title, to show its purpose.

Before starting a diagram, decide on the main purpose and – more importantly – who will be involved in drawing and interpreting the diagram: if the main purpose of a diagram is to facilitate analysis by rural people, then diagrams that use materials that are more familiar (e.g. using a stick in the ground) will allow people to feel more comfortable and hence facilitate their participation (the diagram can always be photographed or redrawn later). In this situation, the use of drawings or symbols is usually preferable to words in the diagram. If the main purpose of the diagram is to convey information in a report to other development professionals or decision makers, it may be better to use symbols or text. If text is used, single words or very short phrases should be used rather than long sentences. If using a computer to write a report, the excessive use of clip-art should be avoided, as this can clutter up and distract, rather than aid understanding.

When diagrams are being used to facilitate team analysis and understanding, the value is mainly in the process of drawing itself rather than the finished drawing. It is therefore a mistake to wait for a complete understanding of a situation before starting to draw a diagram, as the act of drawing will identify areas where such understanding is incomplete. Few diagrams reach a "final" form without redrawing several times. As with all teamwork or facilitation of group interviews or participatory activities, you should be aware of *process* issues as well as the content or output. In particular, you should be wary of forcing a consensus on a particular point – there is room in a diagram for different points of view. These differences of perspective are often highly significant.

Preparing a diagram is not a quick process, if the associated analysis is to do more than "scratch the surface". Managing discussion and disagreements takes a lot of time, but is necessary to build common understanding within a team, between rural people and development professionals, or between an analytical team and decision makers. Take your time, get feedback from others, and make sure the diagram is understandable. Ask others what they understand when they "read" the diagram. If your diagram is being used to convey meaning in a report, make sure that it contains only the information necessary to convey the meaning you intend.

Concept Maps

"Concept Mapping" is a way of representing and relating knowledge, by showing important elements (concepts) and the relationships between these, usually around a central theme. It was developed by Joseph Novak of Cornell University in the 1960s, based on the theory that new knowledge is more easily assimilated when it is related to existing ways of thinking. Concept maps are often advocated as way of conveying information in a form that is facilitates learning. (Fig. 17.12).

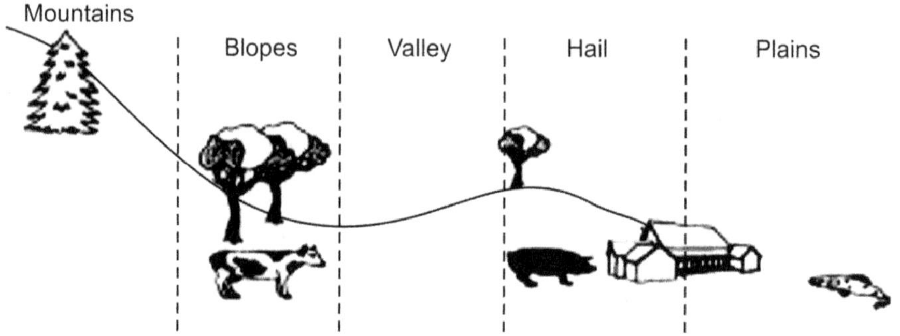

Fig. 17.12 Concept map of village.

When to use Concept Diagrams

As can be seen from the above introduction, concept maps (and the related diagram types) can have several uses. These include: personal brainstorming (organising and exploring one's thoughts); group brainstorming; exploring and recording many ideas. Making notes for personal use, many people find it easier to retrace their thoughts, pick out important points, recognize what they know or have learned, if they record and present this information in the form of a concept map, rather than sequential text. Making notes in an interview. when conducting an informal interview (i.e. one where there is no formal, fixed "questionnaire"), the note taker will often find it quicker to take notes in the form of a concept map, rather than text (especially after a little practice). It also forces the note taker to make sense of what he/she is hearing and so encourages a more active listening. Interview notes recorded as a concept map on a large sheet are also a good way for both interviewer and interviewee to see what has been discussed: i.e. to "map" the interview. Seeing what has been discussed, and what remains to be discussed (by identifying all the main issues first) helps keep the interview focused and on track, avoiding the common problem of going of at a tangent and discussing something relatively inconsequential (Fig. 17.13).

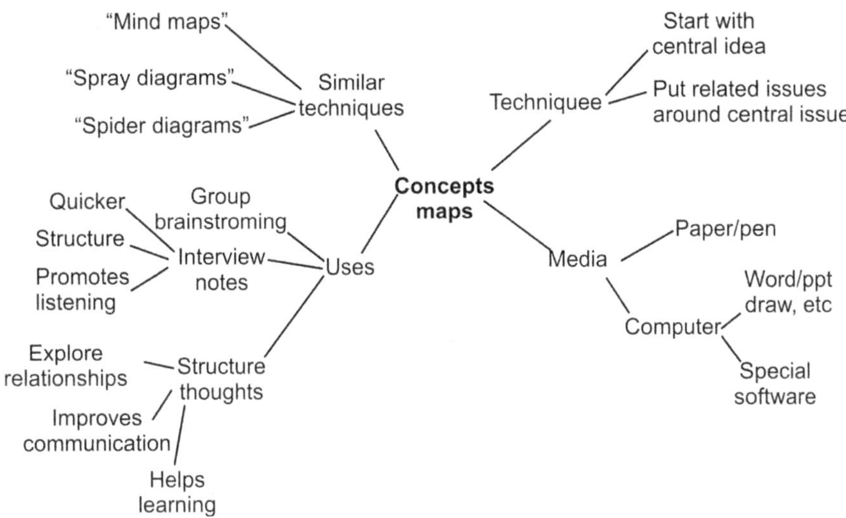

Fig. 17.13 Which is the easier to follow – the text in the above sections, or the diagram? Concept mapping.

Steps :

Usually, the place to start is the central idea, or topic of interview, etc. This can be put in a circle (as a "blob") on the page, or just written down as a single word or concise phrase (no long sentences!). Then place around this the (4-10) main elements or issues or questions that are related to the central issue, also as words or concise phrases. Connect the secondary issues with the central issue by a line. Explaining the line – putting in the nature of the relationship usually helps, but may not always be necessary. Once the main secondary issues or elements have been identified, the "tertiary" elements or issues can be similarly arranged around the secondary issues. In this way, the "map" is built up – usually from the centre outwards. Of course, it is sometimes difficult to know where to stop. But usually 3 or 4 "layers" are sufficient to explore or summarize a particular situation, or identify elements and/pr relationships that merit more detailed analysis.

Flow Diagrams

Flow diagrams emphasise the flow of something within or between systems. In agricultural or rural livelihood analyses, these flows are usually of materials, nutrients, labour, cash or information. When a particular system is depicted with boundaries, this type of diagram can also be referred to as an "input-output" model. This type of diagram is sometimes also referred to as a "flow chart", although flow charts are often used to depict a *process* with a defined starting and end point, and showing the activities and decisions at different points within this process.

When to use Flow Diagrams

Flow diagrams show how scarce resources are deployed within a system. When the boundaries to the system are shown, they can also show clearly the inputs and outputs of a particular system (e.g. a "farm system") and how interrelated this system is with other systems (e.g. other farms within the same area, with communal resource systems, with marketing systems, etc.). Making explicit the flows between components of a system facilitates discussion about potential changes to those components and the likely impact on the system (for example, what would happen if this farm type stops growing sorghum and instead grows tomatoes?). Flow diagrams can be stylised or pictorial. (Fig. 17.14). Pictorial diagrams are easily visualized and can, promote communication between development professionals and rural people. Use of such diagrams facilitates "visioning" where farmers can explore or model ideas that involve exploiting new agroecological niches, adding new elements (crops, livestock) to their activities, changing resource flows, etc.

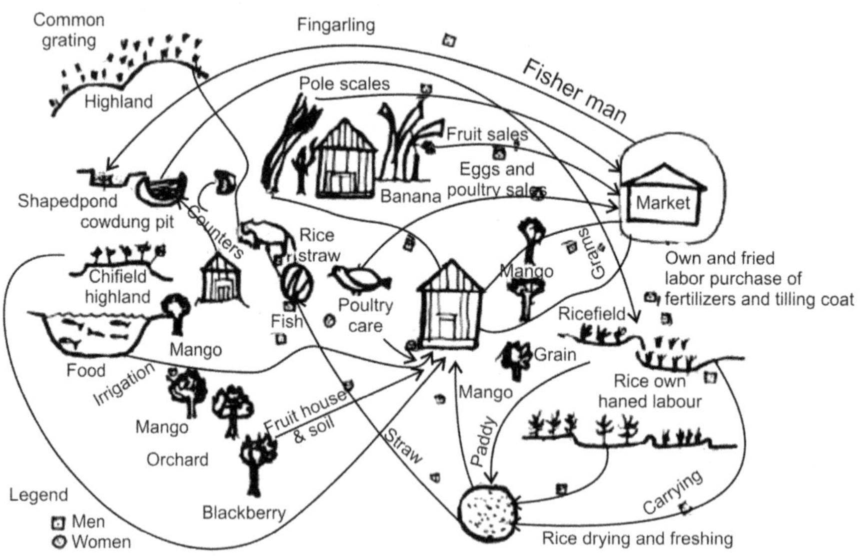

Fig. 17.14 Pictorial flow diagram of a Farm.

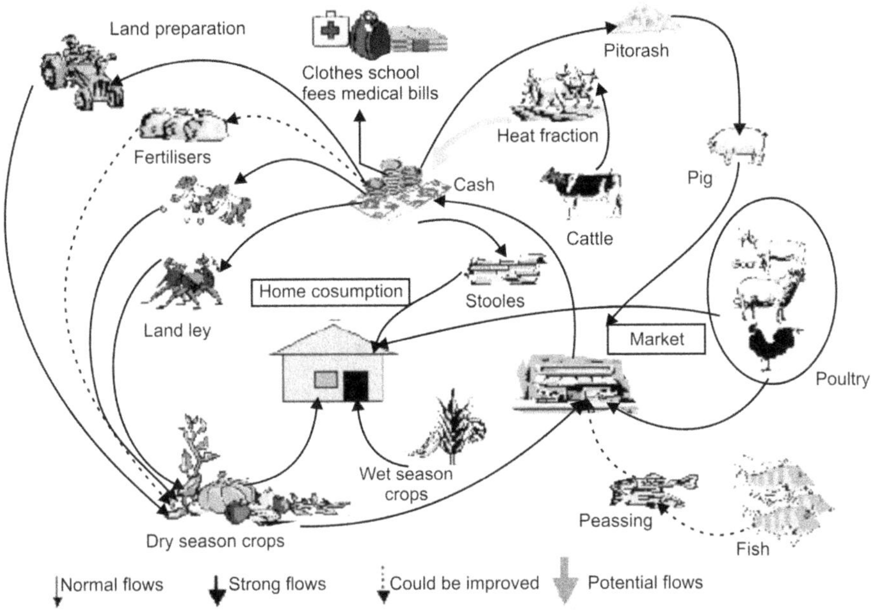

Fig. 17.15 Economic flow diagram of an integrated system.

Steps :

- First, decide on the purpose of the diagram: is it to facilitate analysis and visioning with farmers what sort of flow is to be modeled, according to the need for analysis. Diagrams that try to combine different sorts of flow (eg materials, labour and cash) can turn out to be too cluttered – it is often better to draw separate diagrams to show these different aspects.

- Next, use boxes, icons or drawings (depending on who will be involved in drawing or reading the diagram) to depict the main elements of the system being modeled. Be sparing of "clip-art", (this usually clutters diagrams for little additional gain in understanding). Show the main flows with arrows between these components, labeling the arrows for clarity if needed. Do not use double-headed arrows to depict flows (e.g. information) in both directions between two elements, as the flows are not identical (use two separate arrows and distinguish what *type* of information is being transferred). It is usually helpful to distinguish the boundaries of the system that is being considered (e.g. farm, village), so that inputs and outputs are more clearly identifiable (e.g. communal resources such as grazing lands, forests, rivers, wells, etc.)

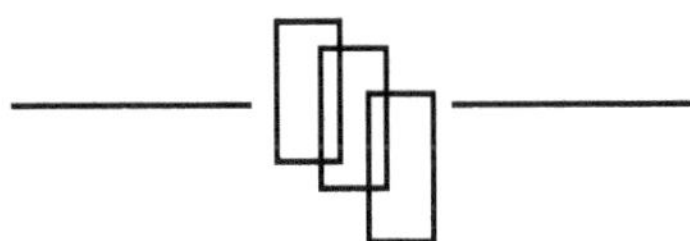

CHAPTER - 18

Entrepreneurship

Introduction

Our knowledge of entrepreneurship, like our knowledge of management, comes to us from many different disciplines. Anthropology, sociology, psychology, and economics have all contributed to our understanding of the entrepreneurial phenomenon. However, because of the diversity of approaches to the study of entrepreneurship, there is difficulty in defining just what entrepreneurship is and identifying just who is an entrepreneur.

Even Webster's dictionaries disagree among themselves. Webster's New American Dictionary defines an entrepreneur as "one who undertakes an enterprise" or "an employer of workmen". Webster's *Third International Dictionary* offers a more complete definition: "an organizer of an economic venture, especially one who organizes, owns, manages and assumes the risk of a business". The difference between the definitions is one of scope and emphasis. The first definition is very broad and can include organising a football team or hiring a contractor to fix the roof. The second definition is more useful because it narrows the focus of entrepreneurship to its important qualities: economic activity, ownership, venturing and risk.

Some writers on entrepreneurship such as economist Joseph Schumpeter, have stressed "innovation" as being the key factor in entrepreneurship. Innovation, the creation of new products, markets and services, sources of supply or forms of industrial organisation, is viewed as the dynamic force that moves the capitalist system. Entrepreneurs take capital from less productive sectors of the economy and invest it in new, growing, more profitable industries. This "creative destruction" of capital makes the entrepreneur the linchpin of a strong free enterprise economy.

Another key element of entrepreneurship is risk-taking. John Stuart Mill, a nineteenth-century British economist, saw riskbearing as the distinguishing factor between entrepreneurs and managers. Entrepreneurs stand to lose or gain for their own purse as a result of their own efforts. Managers are merely office holders and they bear none of the risks or liability of loss. Today, however, with employee stock ownership plans, managerial stock option incentives and large pension fund investments, managers and employees are often important risk-bearers in their organisations.

Sociologist Max Weber offered a significant insight into our understanding of the entrepreneur; as an ultimate source of power and authority in an economic organisation. In his analysis of bureaucracy, Weber saw no role for personal power in managerial positions. Only the entrepreneur possesses this. However, recent commentators have noted that managers do act entrepreneurial in certain situations. We will look at these cases later in the chapter. But first, we will examine what makes an entrepreneur from a psychological and sociological perspective. Then, the discussion will cover how entrepreneurs create new ventures and what entrywedges they employe as competitive strategy. Since many ventures become small businesses, we will offer a managerial perspective of small business operations. The chapter ends with a look at corporate venturing and the role of innovation in larger firms.

Psychology of the Entrepreneur

Psychologists have investigated the personality characteristics of entrepreneurs in an attempt to understand why some people become entrepreneurs and others don't. This is known as the trait approach. Among the traits thought to be related to entrepreneurship are: flexibility, extroversion, independence and aggressiveness. In this section, we examine three traits which investigators have shown to be entrepreneurial: the need for achievement, the internal locus of control and a tolerance of ambiguity.

The **need for achievement**, or **n-ach**, is one of the traits studied by David McClelland (1997) in his book *The Achieving Society.* McClelland developed the theory that individuals, indeed whole societies, which possess n-ach, will have higher levels of economic wellbeing than those that do not. McClelland's work indicated that there are five major components to the n-ach trait: responsibility for problem solving, setting goals, reaching goals through one's own effort, the need for and use of feedback and a preference for moderate levels of risk-taking.

The individual with high levels of n-ach is a potential entrepreneur. The person views the economic undertaking as a problem to be solved and enjoys solving problems. Solving the entrepreneurial problem requires setting goals with hard but achievable expectations. The potential entrepreneur also believes that these goals can be achieved by his own efforts – that he possesses the skills, abilities and resources. Along the way, the entrepreneur is open and responsive to feedback. This feedback may come from other individuals, organisations, the environment or even the entrepreneur himself. The entrepreneur uses positive feedback for encouragement and negative feedback for correction. Lastly, the achievement of the goals and the possibility for failure are perceived as moderate risks. The n-ach individual neither plays it safe nor takes unnecessary chances.

A trait related to n-ach is locus of control beliefs. People who do not believe that their efforts in a business venture will be related to its success or failure are unlikely to become entrepreneurs. These people are known as "externals" since they believe that outcomes are determined by forces external to their own behavior. "Internals", on the other hand, believe that they have the ability to influence the outcome of an event. These people are potential entrepreneurs. A third important trait is tolerance of ambiguity. Ambiguous situations are defined as novel, complex or unsolvable. These conditions parallel entrepreneurial activity and therefore, becoming an entrepreneur requires the ability to tolerate these factors, achieve a goal or impose their own will on the situation. People who are intolerant of ambiguous situations are threatened and uncomfortable in this context. While the trait approach is useful for understanding the psychological makeup of potential entrepreneurs, it doesn't offer a model of the person who actually goes into business for himself or herself. For this we turn to a sociological explanation.

Sociological Appraoch

While personality characteristics may provide evidence of entrepreneurial tendencies, the actual activity of starting an enterprise is a social process and it begins with a life plan change. The potential entrepreneur may have just experienced a displacement, be between things, or have a positive force encouraging entrepreneurial activity. For example, the negative displacement of being a refugee helps explain overseas Chinese entrepreneurship and Jewish entrepreneurship worldwide.

However, perceptions of desirability must also be positive if the process is to continue. Some cultures have a disdain for entrepreneurship and generally speaking, find alternatives to venture creation. Although there are always exceptions, most Irish immigrants found assimilation in the United States through civil service, as policemen and teachers. In addition, family tradition sometimes makes commercial activity seem undesirable. Many families have traditions of professional work (medicine or law), the family firm, or strong labour and union ties. These tend to decrease the desirability of entrepreneurship.

Perceptions of feasibility must also be present. Entrepreneurial activity must appear do-able. This means there must be models of successful venturing and resources available for venturing. Only when all the conditions of the entrepreneurial event process are met can new company formation be predicted.

Are they any other differences between the characteristics of entrepreneurs and non-entrepreneurs? A recent survey by the Gallup Organization conducted for *The Wall Street Journal* provides us with insight and information. The survey tends to confirm some of the psychological and sociological theories. Entrepreneurs are younger on an average than executives and less likely to be with others. These findings, coupled with the findings on changing jobs, indicates that entrepreneurs are most likely to be displaced, unsettled, or between things. The need for achievement is evident in that entrepreneurs tend to begin their business activity while still in school and most before they finish college. Overall, from both theoretical and empirical perspectives, entrepreneurs are a different breed of executive.

New Venture Creation

It is the goal of almost every entrepreneur to start a new business someday. For the entrepreneur this challenge can lead to personal development and satisfaction as well as wealth, power and status. Furthermore, in democratic capitalism the primary role of the entrepreneur is new venture creation. A healthy, thriving capitalist system needs new ventures for jobs, productive capacity, innovation and creativity. It is a fact that while larger, older firms in steel, autos and chemicals are slowing down and becoming smaller, new ventures in computers, services, robotics and genetic engineering are creating jobs and growing.

The nucleus of the entrepreneur's attempt to create a new venture is the **business plan**. This is the document that describes the who, what, where, when and how of the venture. The plan and the planning process serve a number of important functions. Primarily, the process of developing the business plan informs the entrepreneur of all the potential threats, opportunities, problems and strengths facing his venture. It is an educational process. The entrepreneur, who may be a specialist in an area, must learn how to be a generalist and understand all of the factors, which affect his creation. Even if events don't occur as the entrepreneur has envisioned, his responses are likely to be more effective if he has gone through the planning process and prepared himself for contingencies.

The major barrier for the entrepreneur is raising the money for the venture. The business plan serves as the document that will help the entrepreneur finance the business. All potential sources of capital will want to familiarize themselves with the business plan before investing: friends, relatives, business associates, banks, small business investment companies and venture capitalists. A highly stylised form of the business plan is the initial public offering prospectus, which is required by the Securities and Exchange Commission when stock in a venture is sold to the public at large.

Table 18.1 Elements of a comprehensive business plan.

S.No.	Elements	Sub Components
1.	Entrepreneur's philosophy and goals	
2	Venture's mission statement	
3	Strategy of the firm	1. Entry wedges 2. Generic strategy 3. Strategic posture
4	Feasibility of the venture and requirements	1. Market analysis and competition 2. Location analysis 3. Operations policies 4. Personal policies 5. Financial projections
5.	Form of business	1. Legal form 2. Tax strategy
6.	Schedules	1. Checklists of mile stones 2. Time tables
7.	Key Executives	1. Resumes 2. Compensation Plans

The elements of a business plan will vary from venture to venture but certain types of information are always (or should be included). Table 18.1, provides an outlie of a comprehensive business plan.

The elements outlined in Table 18.1 indicate the comprehensiveness needed for the business plan and the generalist qualities required of the entrepreneur who authors the plan. While much of the outline is self-explanatory, a few items dealing directly with the entrepreneurial nature of the venture plan elaboration. These are mission statement, the firm's strategies and the feasibility study.

The mission statement describes the new venture's products and services, the markets to be served and the technology, which will be employed to (a) make the product and (b) deliver the product. This key element of the plan answers the question: "What business are we in? This is not as trivial or as simple as it may seem. The entrepreneur has a great deal of discretion in terms of product, market scope and technological configuration. Consider an entrepreneur about to open a computer store.

The mission that promises to do everything for everyone will have as difficult a time succeeding as one where the market is defined so narrowly that there aren't enough customers to be profitable. In addition, a mission statement should include the nature of the entrepreneur's competitive advantage. It answers the question: "Why should anyone do business with us?" Again, this is not trivial and if no answer can be found, the potential entrepreneur may wish to reevaluate his options.

Table 18.2 Entry wedges.

S.No.	Wedge	Description
1.	New product	Invention or commercialization of previously unavailable product, service or product type
2.	Parallel competition	Entry into already established industry or market based on a minor variation of what is offered and/or how it provided
3.	Franchise	This employs a proven product or service without variation, but in a new geographical area under licence
4.	Partial momentum	Exploiting the momentum of established enterprises, e.g. by filling a supply shortage or tapping underutilized resources
5.	Customer sponsorship	Entering at the behest of the customer, e.g. with contract in hand or as a second source
6.	Parent company sponsorship	Creation of a new firm assisted by a larger firm with existing resources, e.g. joint venture, licence agreement or spin-off
7.	Government sponsorship	Creation of a new firm at the behest of the government or to meet the government market, e.g. due to a favoured purchasing agreement or a rule change (set aside)

The strategies of the firm describe how the mission will be accomplished. It answers the questions of how the business will be entered, how a sustainable, competitive strategy will be maintained, and what tactics will be employed to implement the strategy. This section details the strategic management of the new venture. While a complete analysis of strategic management is beyond our scope here, of special importance to the entrepreneur are the entry wedges used to overcome market entry barriers.

Entry wedges are the firm's initial advantages. Karl Vesper has developed a typology which includes three major competitive wedges – new products/services, parallel competition, franchising – and a number of partial wedges. Table 18.2 describes these strategies.

Any major change in a society's lifestyle is mutually interrelated with changes in the economy; as a consequence, human resources and working conditions may determine some aspects of the technology transfer protocol. In the absence of responsiveness to socio-economic changes and to emerging new individual lifestyles, institutional mechanisms that intended to facilitate the acquisition of technology may instead act as constraints and barriers. Further, for any technology transfer project to be successful, instabilities in the existing political and policy-making systems, internal economic conflicts and institutional constraints have to be dealt with.

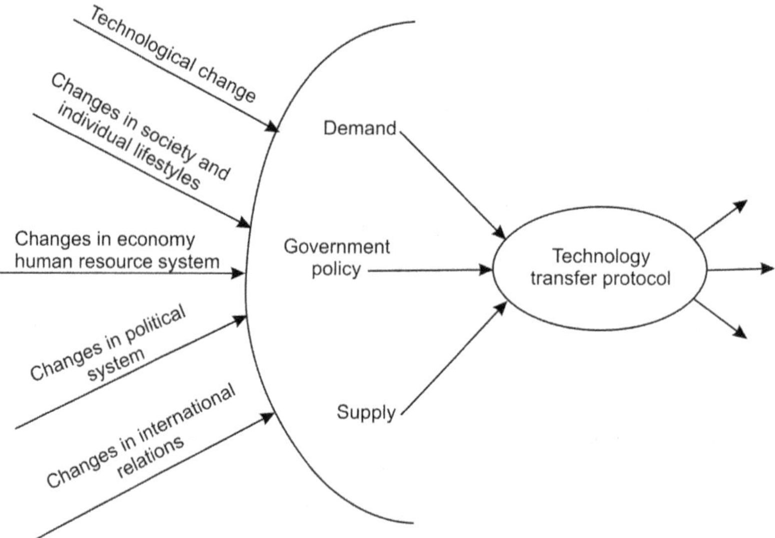

Fig. 18.1 External and internal forces affecting changes in protocol.

A Systematic Approach to the Process of Technology Transfer

At this point, it is useful to recapitulate briefly the approach to technology transfer that we have outlined until now, in order to lay the groundwork for the next stage in our analysis. We contend that in order to arrive at the optimum practices, a thorough review of all the aspects involved in the technology transfer process is required, and this forms the basis of the approach developed here.

The major elements in this conceptual system include : (a) the home country (supply side); (b) the host country (demand side); and (c) a transaction that involves each side along with its relevant variables. The transfer process itself has been analysed into two main stages, each further divided into minor stages or components. The first of these major stages involves 'technology policy planning and assessment'; the second concerns the 'implementation and operational process' of technology transfer. With this background, it is possible now to consider how the transfer process essentially works, how it is affected by the various factors involved, and which groups of variables play key roles in the assessment of candidate technologies.

A number of other questions have to be answered before choosing a technology and implementing the transfer process. For instance, the choice of appropriate technology for transfer is a crucial issue: how can the assessment of technology for transfer be carried out. Evolving a rigorous approach to technology assessment requires us to explore the internal characteristics of the process of transfer. We shall now undertake an in-depth analysis of this nature. In chapter four we analyse the major steps involved in the implementation of technology transferees, with a view to addressing the crucial issue of improving technological capability, the main objective of any technology transfer.

Technology Transfer Pyramid

Technology development is not an autonomous process, but is intimately associated with the needs and pressures created by environment and society. Technology transfer, in our view, is about *continuous* technological improvement of technical know-how and information viability, organisational and managerial ability, technological capability, and human skills and knowledge capacity, brought about through the utilisation of imported technology for the achievement of sustainable capability. (Fig. 18.2). Achieving technological sustainability involves passing through a series of stages includes:

- Technology assessment and selection
- Technology acquisition
- Technology adaptation
- Technology absorption and assimilation
- Technology diffusion and
- Technology development

Personnel

If the small business is relatively large (may be 75 employees or more), it may be big enough to have a personnel department and a personnel manager. Until the firm reaches sufficient size, the entrepreneur is the decision maker. The personnel responsibilities include staffing, the determination of salaries, wages, fringe benefits, training job; description and performance appraisal. These activities together are known as human resource management (HRM). The proper execution of the personnel function in the small business is similar to that in larger firms. But there are some differences.

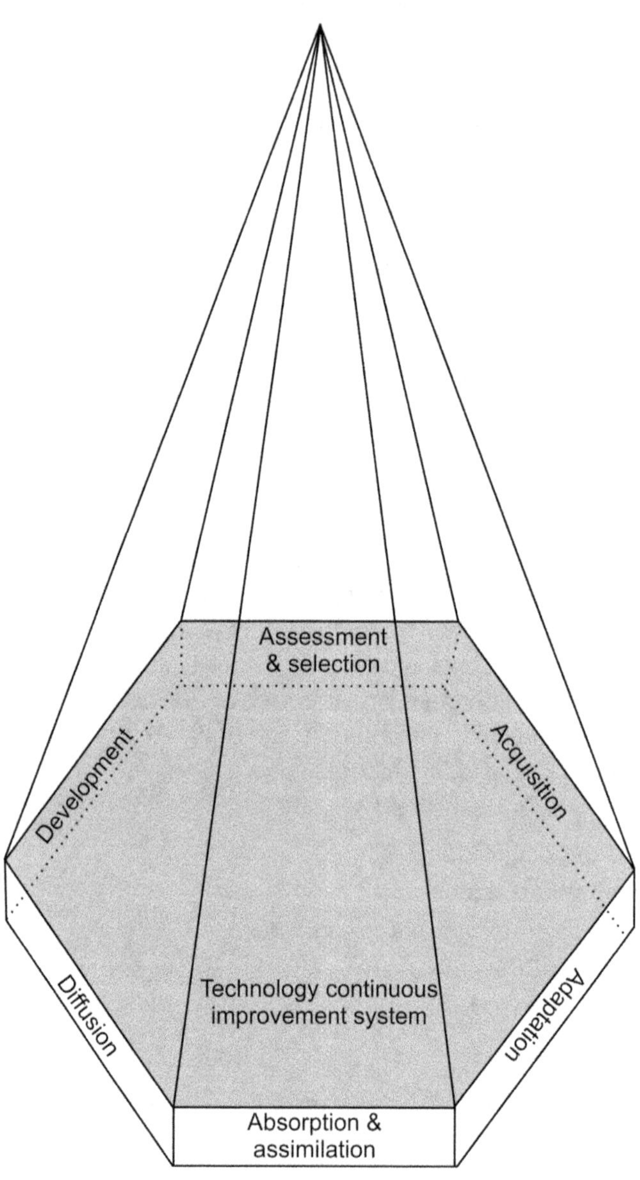

Fig. 18.2 Technology transfer pyramid.

One major difference is the role of the entrepreneur. Entrepreneur occasionally have trouble delegating responsibility and authority for getting the job done. Their tendency to do it all themselves, which was of paramount importance in starting the business, interferes with the normal delegation process once the business is operating. As a result, the supervisory style of the entrepreneur may be authoritarian. Many small businesses remain small because the owner can't let go and allow others to help make the business grow. Successful entrepreneurs learn they must enable and encourage others to do their jobs without interference.

The second major problem faced by smaller firms is the lack of career paths. Employees of smaller firms often find that because of the size of the firm, there are few opportunities for advancement and promotion. If a small firm grows, this problem may take care of itself. If the firm is stuck at a slow rate of growth, people without opportunities may seek employment elsewhere. This is unfortunate for the small business because recruiting, training and developing good workers is an expensive and time-consuming function. While it would be nice to say there is a solution, there really isn't. The consequences of high turnover in small firms can be mitigated, however, through job enrichment and job enlargement programs.

Finance

The entrepreneur must be the firm's chief financial officer (CFO) during the initial capital formation phase and often remains the CFO for the firm. In the simplest terms, the CFO is responsible for managing the balance between the sources and uses of funds. The objective of this balancing act is to provide the firm with enough capital to meet its needs and minimize the firm's cost of capital. These responsibilities include securing equity for the firm, arranging for short and long-term debt financing, analyzing the financial statements and making capital budgeting decisions.

Financial management today will be different from past decades. First, the institutional structure has undergone probably its greatest change during the last decade. Second, changes in interests rates, governmental regulations and development of new ideas (popularly known as creative financing) have raised interesting issues for financial managers. In the recent past, a group of corporate raiders have developed techniques for raiding even the largest corporations through tremendous debt financing. These raiders contend that the value of the common stock of many of these corporations is much less than the value of their individual assets and thus the shareholders have not been effectively represented by the operating management. This treat of takeovers and actual mergers results in shifting the attention of top managements from improving operations and marketing to techniques for protecting themselves through defensive financial measures. Thses new developments are forcing top management to give special attention to financial management at a time when foreign threats are made in operations management, especially in quality and in productivity. As a result, financial management as a scholastic endeavour cannot be separated as easily from general management and its operational and marketing functions. However, the single most damaging factor in small business failure is under capitalisation. This means that the firm is always short of capital and is unable to obtain enough to meet its financial obligations and expand. Early bankruptcy is often caused by under-capitalization because the entrepreneur underestimates how much money it will take to bring the firm up to break even. Later bankruptcy may be caused by a firm's expanding too quickly and trying to grow without proper financing. In either case, inadequate attention to the finance function can spell doom for a small business.

Marketing

Marketing consists of the purchase, merchandising, pricing, distribution and selling of goods and services. It is not just sales. Sometimes the activities of marketing are characterized by the four "p's": product, price, place and promotion. The entrepreneur, who may have exceptional skills as a promoter, needs to manage the active marketing function if the small business is to profit and grow.

Operations

The management of operations is crucial. While all the previous managerial functions help the small firm operate efficiently and effectively, they are at the periphery of either production or service. The others are staff functions and their goal is to support operations. Operations are the line function – it makes the product or delivers the service. Operations are the technical core, the heart of the firm.

Unlike the staff functions of small business management, operations management and the management of the line require a particular orientation. The management of the operations of a small business depends almost completely on what kind of business it is. Accounting, marketing, finance and personnel activities are each a bit different depending on the type of firm, but each has a universal set of techniques, which can be applied to any type of small business. While on operations management and production summarizses the universals, it doesn't teach how to manage or produce anything in particular. Only personal knowledge, experience and on the job training can provide that the following information summarises the most important topics to be considered when forming a new enterprise.

1. Accounting and legal Form of organisation (propreitorship, Partnership, Corporation)

    - Tax treatment

2. Personnel

    - Staffing

    - Compensation

    - Traing and development

    - performance appraisal

3. Finance

    - Source of capital (Debit & Equity)

    - Use of capital

    - Financial statements

4. Marketing

    - product

    - Price

    - Place

    - Promotion (advertising, publicity, personnel selling)

5. Operations'

    - Hands - on experience

    - close to the customer

Outside Assistance

In many cases, "smallest" firms (under 10 employees) to "just-small" firms (between 100-250 employees) need expertise that lies outside the knowledge of the entrepreneur and his immediate company. In the previous section, we encountered the problem of choosing a legal form of organization, and it was recommended that the entrepreneur seek the advice and counsel of an accountant and an attorney. These professionals can be thought of as external functional managers. An external functional manager is a person who doesn't belong to the small business organisation but is hired (or volunteers) to offer counseling, consulting or technical expertise. In addition to accountants and lawyers, there are others who serve as external functional managers. Many of these professionals operate under the banner of consultant. Thus, there are personnel consultants, financial consultants, marketing and market research consultants and consulting engineers.

All of these areas have professional orientations and the consultants should be members of their respective professional organizations before they are employed. Professional certification is often a requirement before a consultant in a particular field can practice. Entrepreneurs are advised to check certification and references carefully before hiring a consultant. Furthermore, all paid consulting engagements should begin with a written proposal of the work to be done, a timetable and schedule of fees. The use of hired external functional mangers to augment the entrepreneur's technical sophistication is an expensive proposition. There are other alternatives. Three of these are: establishing a board of directors, using government sponsored counselling and networking with other entrepreneurs and small business owner-operators.

Large firms have boards of directors composed of people who have fiduciary and oversight responsibility for the management of the corporation. Most often times managers are directors, but almost all corporations have outside directors, too. These outside directors are hired (large firms pay high directors' fees) from the pool of captains of industry, political figures; ex-office holders and respected public servants.

Small businesses can also have boards of directors. For nominal fees, the small business owner-operator can create a board of knowledgeable experts who can help advise and manage the firm. Prospective candidates for a small business board would include: accountants, lawyers, bankers, Chamber of Commerce staff members, local political figures, college Professors and other business people in similar or related businesses. A motivated, skilled and enthusiastic board can provide the small business with insights, technical expertise and perspectives that are hidden to the entrepreneur. It also generates valuable goodwill among important local community leaders. There are also a number of govenment ignored programmes designed to offer management councelling and abundance to mall business.

Networking is a term used to refer to making business contacts, finding common problem areas and offering support to people within the network. For the entrepreneur, networking can be a source of technical and managerial assistance, as well as emotional support. Networks can be built from Chamber of Commerce committees, caste groups, philanthropic organisations NGO's and local political activists. By meeting people from different parts of the business environment and cultivating these people as advisors and confidants, the entrepreneur can create a team of external advisors for his small business.

Corporate Entrepreneurship

So far we have discussed entrepreneurship and entrepreneurs in the context of new venture creation. Recently, however, the importance of entrepreneurship has been recognised within the corporate environment. Corporate entrepreneurship can be defined as the development of new products, business lines, services and innovative technology within a large, mature organisation. Sometimes this is referred to as **intrapreneurship** indicating that the activity takes place within an existing business.

In the corporate environment, entrepreneurship is not an all-or-nothing phenomenon. Parts of the corporation may be mature, stable and oriented toward professional administrative management; however, other parts of the firm, new product development groups, venture teams, applied research task forces, may be quite entrepreneurial. These parts of the firm possess the entrepreneurial mindset and try to recreate for the corporation the environment that stimulates innovation and creativity. The entrepreneur's thought patterns define the entrepreneurial process by asking the following questions:

1. *Where is the opportunity?* Entrepreneurial teams are sensitive to changes in market conditions and buyer preferences. They monitor improvements in technology and product quality characteristics. These teams are empowered by the corporation to take action to mobilise the firm.

2. *How can we take advantage?* The action-oriented venture group must be able to act quickly to gain an advantage once opportunities are identified. These groups are risk-takers. Corporations must be willing to tolerate and accept failure since many, if not most, products never prove to be commercially successful. But the payoffs can be large, too. Members of venture groups understand that their jobs are not on the line with every risk they take.

3. *What resources do we need?* Entrepreneurial teams make imaginative use of limited resources. They *borrow* people, equipment, material and money from other parts of the corporation where these resources exist. Project development for venture groups is a multistage process. Commitment of major resources may be appropriate only as progress to commercialisation is made.

4. *How do we control our resources?* When entrepreneurial groups receive their *borrowed* and formal appropriations from the firm, the emphasis is on the results that can be obtained. This allows the new venture team more flexibility in how to use their resources. Since the team has limited resources and may exist on a temporary basis, entrepreneurs avoid owning equipment or hiring people. This is valuable

because the teams can have greater resource specialisation (they can get exactly what they need) and avoid the risk of obsolescence.

5. ***What structure is best?*** Entrepreneurs within a large corporation prefer a flat organisational structure with multiple lateral networks. These lateral networks are groups of colleagues who can help the entrepreneurs get things done. They facilitate coordination and provide a buffer between the administration's need for stability, hierarchy and orders and the entrepreneur's challenge to hierarchy and desire for independence.

Bernard A. Goldhirsh, (1985) has said that an "entrepreneurial revolution is taking place." It has come into being because society has come to recognise the value, importance and productivity of the entrepreneurial business unit. The awesome capabilities of small to mid-sized companies to adapt to changes, to bring new products to market, to generate jobs and to create wealth have been recognized. Goldhirsch gives a number of reasons for this revolution. The rise of two-family incomes in the under-50 age bracket has created accumulated capital, which enables one of the partners to go out and take a risk. The shift to a service economy poses low capital requirements for small but emerging growth industries. Furthermore, college graduates are redefining the relationship between work, risk and fulfillment. He notes that college graduates are thinking in terms of risk and of risk and the entrepreneur alternative looks more and more promising.

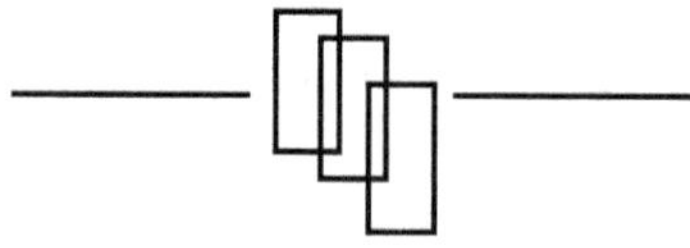

CHAPTER - 19

Private Extension: Indian Experiences

What is Private Extension / Privatisation?

The definition of agricultural extension varies from simple transfer of information to facilitating the process of total human development. The services are mainly funded and delivered by government in the Indian context. But, there are private players who also fund and / or deliver extension services. This process of funding and delivering the extension services by private individual or organization is called Private Extension. But, privatisation is the act of reducing the role of government or increasing the role of private sector in an activity or in the ownership of assets. Very often Private Extension and privatisation are viewed similarly. Private Extension is the act of private individuals or organisations where the decision of privatisation solely rests with governments and is implemented in liaison with Private Extension Service Provider (PESPs). Hence the two are different. And also Private Extension and privatisation need not necessarily involve cost recovery or be feebased. NGOs being PESPs provide extension service free of cost and involving NGOs in the process of privatisation need not end in costrecovery always. Hence, these two concepts are not necessarily always linked to priced services.

Stake holders in Private Extension

In the process of privatisation, public extension is always a part. In Private Extension, players are

1. Agricultural Consultants.
2. Agricultural Consulting Firms.
3. Progressive Farmers.
4. Farmer's organisation / co-operatives.
5. Non-governmental Organizations (NGOs).
6. Krishi Vignana Kendras (KVKs).
7. Agri-business Companies.
8. Input Dealers.
9. Newspapers.
10. Agricultural Magazines.

11. Private Television Channels

12. Private Sector Banks.

14. Donor Agencies.

The role played by these players has significant impact on Indian Agriculture. But no academic attempt is made to assess the strength of private extension in indian agriculture though such an attempt will help the government to regulate the activities of PESP also to formulate the strategy to utilise their services through privatisation and to reduce the burden on the public exchequer.

Indian Experiences

India has the largest extension system in the world with 1,17,603 paid agricultural extension personnel catering to the farming and allied needs of over 90 million farm families. Among these an overwhelming majority were small and marginal farmers with an average land holding of 1.63 ha, scattered and fragmented over different agro-climatic zones. There are five well-specified extension systems prevalent in our country. They are First-Line Extension Education System (ICAR/SAUs), National Agricultural Extension Service / T&V System, Special Extension Programme on specific crops, Rural Development Programmes and Extension Programme of Non-governmental Organizations, each having its own mandate on poverty eradication. The First-Line Extension System of Ministry of Agricultural Research (ICAR) and State Agricultural Universities (SAUs) have only limited jurisdiction in coverage, while compared to if, the extension system of Ministry of Agriculture has a broader network. The non-government organisations engaged in extension activities include mostly voluntary agencies (partly or fully funded by government for extension activities), business houses, agricultural processing firms, producer's co-operatives, input agencies and private consultants.

Privatised extension involves any personnel in the private sector delivering advisory services in the areas of agriculture and is viewed as an alternative to public extension. In India and many of the developing countries extension is not merely transfer of technology and information. It takes care of the broad perspective of human resource development of its clients. Speaking of the Indian context, the stress on two crucial issues need to be considered. First, the ability of the farmers to pay for the extension services and secondly, how to demarcate the benefits of extension as 'private and 'public'. So the scope of privatisation or cost sharing for extension services in Indian situations is limited owing to poor socioeconomic conditions of the majority of farmers. More than two-thirds of the farmers in India manage resource poor situations and to expect them to pay for not too remunerative advices needs a lot of thinking. Even at present big farmers are buying advice and other extension services from private consultancy firms in different parts of the country. The public extension system can re-orient its strategies in such areas and such high value cash crops for partial withdrawal and cost recovery. Commercial agencies concentrate their activities on areas possessing extremely favourable physical environments and they normally won't invest in rain fed, unfavorable and resource-poor environments where there is the least possibility to make profits. Agricultural development in India has attained a stage where more inputs have to be made on rainfed / dry land areas for national food security, correction of nutritional and regional imbalances and generation of widespread rural employment opportunities in hinterlands. These situations therefore necessitate the need for more public support to extension services in those areas and to put it briefly, the time is not yet ripe in India to think of the complete withdrawal of public support to extension.

Extension services have been traditionally organised and delivered by the public sector all over the world, which led to a situation wherein whenever one refers to extension, it has denoted public extension service. Similarly, whenever one refers to private sector, there is a tendency to consider only the corporate sector in that category. Private sector also includes the consultancy firms, contracting firms, producers' associations, non-governmental organisations, media organisations; NGOs and the like. Thus private extension has a broader canvas in it including in it all relevant groups; than the narrow canvas of the corporate sector. The growing investment needs of the agricultural sector in the present context; especially commercial agriculture and the influence of the reform processes in the economy have necessitated the need for new initiatives to provide effective extension services for agricultural development. One of the key initiatives attempts to involve private sector in providing extension services to farmers. The new initiatives in extension management designed and developed by the private and voluntary sector have proved to be more successful, and these are getting readily accepted by the farmers; Thus, private extension service, as a strategy for providing effective extension support to the farmers is gradually becoming popular among farmers in many countries and India is no exception to this general trend. Privatisation of extension services does not aim at substituting private sector for public extension service. It essentially aims at a reduction in the role of public sector to pave way for an enhanced role of private initiatives in agricultural extension services.

As a strategy of privatisation, if public sector extension has to be restricted, it could be done only in crops and areas where farmers' associations or producer's co-operatives exist. There is considerable scope for initiating paid extension services in agriculture for high value crops and resource-rich farmers. However, there are technical, legal and institutional changes that are needed for promoting privatisation of extension services. Many public extension organisations have a narrow view of extension wherein they see it as a process of supplying information to farmers on demand and of introducing new technologies in agriculture which they consider to be desirable, rather than the one of promoting farmers' development and independence. However, with the emergence of many private initiatives, especially consulting services, the perception had changed a lot. Many young technically qualified professionals have entered into consulting services and at present, there are quite a large number of firms in the market with the potential to offer their technical expertise and services for a fee which the farmers also welcome.

Privatization of extension service in India has adopted a variety of forms involving different stakeholders. However, their roles are mutually non-exclusive and complementary. A simultaneous participation of all the private groups is needed for effective extension service. Corporate sector participation covers the involvement of three categories of private groups. Input firms, consulting firms and contracting firms. Most of the agro-input firms perform the function of marketing, in which the marketing personnel also oversee extension related functions. Farmers join together to avail the services of consultants in the case of cash crops like coffee, tea, etc. There are farmers who have come together to have professionals on a retainership basis. There are many farmer organisations with facilities for paid consultancy. With the spurt in agricultural exports, many overseas consultants are providing their services to Indian farmers and companies. The various farmers' associations organise different types of services for themselves, including input supply, credit, technical services and marketing arrangements; activities that would increase their productivity and time, and at the same time, reducing their dependence on the public sector. Agricultural service is also offered through voluntary organisations, which is presently being encouraged by the public sector to a large extent. There are many NGOs, which can even compete with the public extension system in offering various services to farmers. The role of private media organisations in technology transfer is also commendable. The need to increase food security while at the same time conserving the resource base requires profound changes in the extension approach, which necessitates real farmer participation. It has been proved beyond doubt that the top-down approach which the public extension agencies adopt tend to be much less effective than the participatory, farmer-centered approaches adopted by the private extension agencies. The public extension system views technology transfer as a typical delivery mechanism, while the private extension service views it as an acquisition system, wherein the farmers are empowered to take control of their own agenda. The private extension system aims at helping the farmers to own and operate their own extension services. With ownership and responsibility lying with clients, the basis for more demand-driven as against a supply-driven extension service is established. Food security has at least three dimensions: (a) the production of food through crop husbandry, animal husbandry, fish cultivation, etc.; (b) the marketing and distribution of food so that people have access to food; and (c) the entitlement to food.

Public extension services mainly focus on the first dimension, while private extension services concentrate on the second dimension, which deals with storage, transport, processing, wholesale and retail activities. The contracting firms also do a lot of services in relation to the second dimension, while NGOs play a vital role in relation to the entitlement to food. The success of an extension service for farmers depends on the effectiveness of planning at four levels; policy, programme, projects and strategy. Policy and programmes should be decided by the public extension system. Projects and strategy can be undertaken by the private extension organisations. When private extension organisations get involved in providing extension support to farmers, there will be competition among the various extension providers, which will take care of at least two elements; (a) the need to subserve consumer welfare and public interest, and (b) the need for competitive advocacy a competition culture. As a result of competition, the private extension services become more efficient, especially in the context of liberalisation and globalisation. Both technical and allocative efficiency, which are basically economic in nature are taken care of by the private extension agencies, resulting in cost minimisation, profit maximisation and optimal use of resources, which are much needed in a competitive environment. Public extension service often views development as a zero-sum game. When there is a change, conditions become better for some and worse for others. Many socially successful public extension activities fit this pattern, which is a fall out of the "induced" development by the professional change agents in the public extension service. Public extension service is frequently involved to protect specific interest groups, with no explanation of how and why the interest of this group transcends all others. Private extension organisations can attempt to bring about changes in all the sections of the society through "spontaneous" development, which will be more sustainable. Public extension service often views sustainability of programmes in terms of continuity. Sustainability is different from continuity. The private extension agencies, especially NGOs and media organisations provide valuable service in ensuring sustainability of programmes for farmers by ensuring conservation of the resource base. The private extension system in India offers the following services for farmers in terms of sharing, augmenting and supplementing the private extension efforts, besides offering unique and innovative initiatives, which the public extension service can emulate. Such services are:

1. Cost sharing by farmers' groups
2. Cost recovery on selected services offered to farmers
3. Contracting services to private initiatives
4. Paid extension service for high value crops/favored regions
5. Value addition of produces by agro-processing firms
6. Problem solving consultancy services
7. Privatised service centres for farmers
8. Self Help Groups of farmers
9. Human Resource Development through need based trainings
10. Information support through media organizations

Existing type of Extension and Need for Change

In India we have seen to a number of approaches putting extension into action. Starting from simple informing people (1950's), now we have come to reaching and teaching people with location and farmer specific technology. Different states of India have been adopting various types of extension activity aiming to help farmers of all categories. Whatever the approach may be; the basic philosophy; of the Government remains to improve the condition of farmers emphasizing much open shifting. The adoption of methodology and approach to achieve this end has been controversial and put to trial and error methods for a quite long time. It is a matter of common observation that the extension machinery of the state uses a number of methods/media & approaches to show the result of new technology, but it is not a fact that these technologies are evaluated in terms of relative advantage, compatibility, complexity, triability and communicability before recommendation. The approaches do not stick to the principle of extension education. This has warranted modification and refinement of the technology; what ICAR & SAU institutes are working on at present. We can broadly state that extension education has been converted to extension service in many of the states; ignoring the concept of education. This has resulted in three types of extension; mainly knowledge driven, input driven & market driven. Out of these three systems the knowledge driven one is different , since it deals with the motivational aspect of human being, which is normally avoided by the extension machinery. Application of inputdriven extension results in lapses in application of technology, resulting in pollution and less productivity of technology, once well defined by the scientists.

Consequences of Privatization of Extension Service

The chain reaction of a social change creates consequences in different directions. It is difficult to induce planned change when basic factors are not homogenous. It is presumed that the privatisation of extension service will lead to the following:

- Social distance
- Social conflict
- Social dis-equllibrium
- Higher gap between haves & have nots
- Sidelining of farmers
- Selectiveness of technologies for adoption
- Commercialisation of human value
- Shortage of essential food grains required at rural village level
- Temptation to act as middleman rather than food producers.

Public Sector Extension System

Public institutions are governed by financial and administrative rules of the government, as they are dependent on government funds. The transaction cost of bureaucratic regulation is very high. Efforts are underway to reduce the transaction cost by decentralising the systems. But nothing serious is done to improve the efficiency of public organisations like research and extension systems. These are still in the public domain. There is a need to make these organisation more autonomous and decentralised for greater efficiency.

Private sector extension systems

Private organisation have more flexible functional structures. Their focus is on development of usable technologies and there are strong linkages between technology development and dissemination. In most cases research, commercial production and marketing are vertically integrated. Usually, there is a tieup between a research organisation and marketing firms. Under these organisational structures there is tendency to quickly respond to client's needs, Production of proprietary materials and appropriation of research benefits largely govern the strategic response of private organisation to changing market force. It is unlikely that the private sector will cater to needs of small-scale farmers operating under various constraints. Private sector extension will focus their attention on high value species; value added and exportable products. The education component will be less cared for by the private extension services. It is argued that public sector extension should continue to serve the small-scale farmers and with draw their operation from the areas where for example, fish farming is being carried out on a commercial scale. Farmer's co-operatives and other types of farmer associations also provide extension to specific commodity or member target groups. The concepts about privatization emphasises three aspects which are :

- It involves extension personnel from private agency/ organization
- Clients are expected to pay the service fee (private extension may not expect fee from clients. e.g. NGOs
- Act as supplementary or alternative to public extension service

Target Group : Private extension mostly concentrates on big farmers, farmers growing commercial crops, areas having favourable environment like high fertile soil, irrigated areas. They are not interested in investing on small, marginal and resources poor farmers, because private agencies are concerned about their profits. Farmers in rain fed areas, farmers with less per capita incomes and in subsistence agriculture will not be able to pay for the extension service.

Clients : In private extension system, clients are more committed and careful about extension services, because they are paying for the service. Clients make best use of private extension workers; time.

Technologies Transferred : Private extension agency transfers the location specific and demand-driven technologies. Technologies become specialized and profitable. Private extension ensures timely input supply.

Organizations Involved : Private organizations such as, agricultural consultancy, commercial firms, agro-based industries, input agencies organizations etc., will enter in the area of extension service.

Extension Personnel : Private extension personnel become more accountable to clients and highly motivated because they are getting remuneration from their clients. They become professionally sound and will put efforts to upgrade their knowledge and technical know-how.

Funding : Private extension service gets funds from farmer's contribution and developmental agencies.

Extension Service: Advisory nature of service, extension becomes purchased input and it generates new income to farmers.

Methods: Private consultancy mostly employees personal contact methods, because group approach will reduce their chances of getting consultancy fee.

Role of Public Extension

- Technology transfer for Socio-economic transformation of rural area
- Safe guarding national food grain production.
- Concentrating on environmental issues
- Sustainable agriculture

- Soil and water conservation measures
- Emphasising integrated nutrient and pest management
- Human resources development in agriculture
- Gender issues, such as Empowerment of women farmers; Training programmes for women and TOT for farmwomen
- Training for present and prospective farmers
- Providing advice for off-season employment to rural poor
- Coordinating efforts of different developmental departments for rural development.

Role of Private Extension

- Farm advisory services for profit maximisation of clients
- Timely inputs supply for better production
- Providing market information and market intelligence
- Processing the clients; produce
- Marketing the clients; produce
- Providing credit facilities for farmers
- Providing infrastructure facilities ex: transport, storage; etc.
- Privatisation of extension demands changes in agricultural research and teaching

Strategies for Privatizing Extension

- Commercialisation of extension services; complex, demand driven technologies in the public extension system should be provided for particular cost.
- Introducing "Contract Extension System"
- Public extension system can enter contracts with registered private agricultural consultancy agencies to transfer technology.
- Private/Public extension agents are provided with remuneration in the form of share crop. It will increase the extension personnel's accountability and commitment to the service.
- Giving partnership rights and more responsibility to private sector and NGOs.
- Private sector and NGOs are entering in a big way in recent years to provide agricultural consultancy. They may be given more responsibility in agricultural technology transfer.

- Gradual withdrawal of public extension system can be done in two ways: area wise or commodity wise.
- Creating and strengthening farmers; groups and cooperatives

Recent developments such as financial constraints of government, disappointing performance of public extension system and commercialisation of agriculture will make privatisation of agricultural or livestock extension service inevitable. Good decisions can be made by weighing the assumptions of privatisation. Intervention of private extension alongwith public extension service will increase the efficiency of agricultural production system. Private extension will make extension service more decentralised, farmer participatory, flexible and provides ownership rights to clients. Co-existence of public and private extension in the new millennium provides the appropriate extension approach. To make Indian farming in the 21st century become more efficient and effective, it is essential to introduce changes in agricultural research and teaching and also the intervention of private extension along with public extension for the overall development of the country.

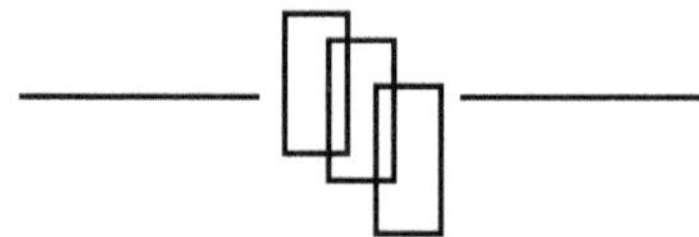

CHAPTER - 20

Transfer of Technology

The transfer of technology from developed to developing nations has grown vastly within the last three decades (Frame 1983). The first reference to technology transfer in modern times was made in 1967 (Cairncross 1967). Indeed, it was only in the early 1960s that the subject of technology transfer began to receive serious attention from scholars and the academia. This was followed by an explosive growth in the literature on technology transfer from the mid 1960s through the 1970s and into the 1980s. But several questions still remain unanswered. There exist three major issues. (a) the lack of a single accepted definition of technology or of what constitutes its transfer; (b) the absence of a rigorous formalisation of the process of technology transfer to, and its impacts on, developing countries, and (c) the lack of universal, comprehensive measures of volume or value by which to aggregate relevant data when examining the different mechanisms of technology transfer.

Technology Transfer : Traditional economic analysis can be dangerously misleading in the study of technology transfer. The problem arises because the term involves two complex and multidimensional concepts 'technology' and 'transfer'. A broad definition of technology transfer is provided by H. Books (1966): transfer of technology is 'the process by which science and technology are diffused throughout human activity.

Technology Transfer in Different Disciplines

Technology transfer is a complex concept, as well as a process that is difficult to operate or manage. The reason is that the circumstances surrounding the interaction between technology development sources and technology receivers differ from transfer to transfer, even within the boundaries of a single organisation.

A. From an economists' point of view, technology is embodied not in aggregate capital nor in particular factors, but in the whole economic process that extends from factory suppliers on the one hand to marketing outlets on the other. It involves the harmonious meshing of a number of subsystems. This makes technological transfer and diffusion a function of the *ability* to change processes that require system adaptation.

B. From an anthropologist point of view, it takes place only in the context of cultural evolution. According to Merrill (1972), a technology is adopted when people or groups find it desirable and possible to change what they are doing in ways that involve particular uses of that technology. Such people and groups are the active, initiating elements in the change in technological practices. Beyond that, it is the circumstances of the society involved, and how they work which determines the wider effects of a technology on cultural evolution. In this, both differ from the economists who are concerned primarily with the (tangible) goals of technology transfer.

C. Technology transfer may be interpreted as a 'chamaeleon' process that may be harnessed for differing applications, environments, participants and problem areas (Kleiman and Jamieson 1978). Transfer involves four major issues.

- Transplantation of technology from within one set of well-defined conditions to another set in which at least one key variable may differ; how the recipient applies the technology may vary greatly from the donor's mode.

- A sense of opportunism, whether justified or not, that prevails in the technology transfer phenomenon.

- A rich variety of mechanisms and relationships between recipient and donor. These may vary from a routine, people-less, passive transfer to a turnkey contract.

- The critical dependence of the success of the transfer upon the nature of the transferred technology and how it is transferred.

Technology Classification

Thompson (1967) sought a technology typology that was general enough to deal with the range of technologies found in complex organisations and systems. Thompson's technology classification comprises three types of technologies: (Fig. 20.1).

1. ***Long-linked Technology:*** Involving serial interdependence in the sense that act X can only be performed after successful completion of act Y, which in turn rests upon act Z, and so on. Long-linked technology is characterised by moderate complexity and formalisation, and should be accompanied by planning and scheduling. For example, mass scale production

2. ***Mediating Technology:*** Which links clients on both the input and output sides of the organisation. It is coordinated most effectively through rules and procedures. For example, computer services, telephone utilities, banks etc.

3. *Intensive Technology:* Which draws upon a variety of techniques in order to achieve a change in some specific object, but the selection, combination and order of application are determined by feedback from the object itself. This form of technology is found in work dealing with humans. Intensive technology, with its high complexity and low formalisation, requires mutual adjustment. For example, technologies dominant in hospitals, research laboratories.

In order to overcome the limitations of the earlier paradigms, we may consider technology as the means of transforming natural resources into produced resources and then, technology may be regarded as the key factor converting certain inputs into desired outputs. Sharif (1998) suggests, that the efficiency of this transformation (or productivity) is influenced by the national economic and socio-cultural factors, including the science and technology climate.

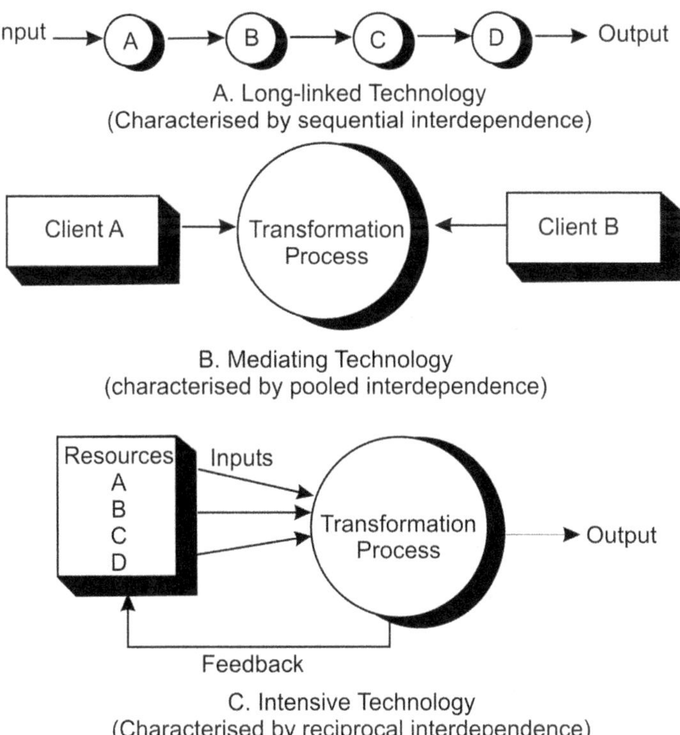

Fig. 20.1 Thompson's three types of technology.

Inputs : Natural resources and produced resources (semi finished products and goods).

Outputs : Produced resources (semi finished goods, consumer goods, capital goods).

Production activity : The mode of converting inputs into outputs.

Technology : The transformer and the core of the transformation facility. (Fig. 20.2).

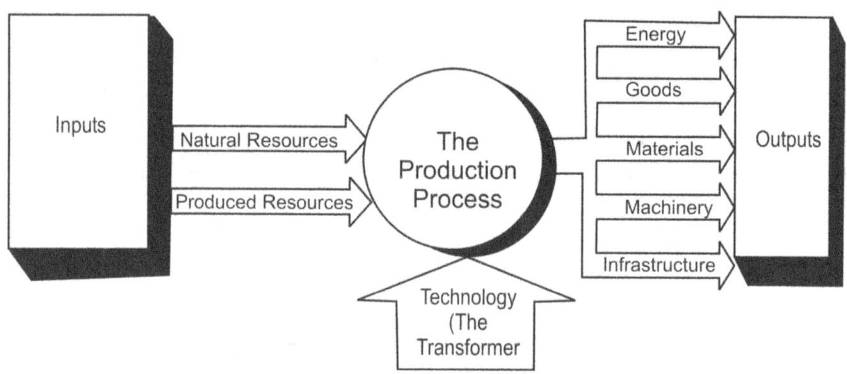

Fig. 20.2 Technology and Transformation.

Form this perspective, wherein technology is viewed as a complex process for the transformation of resources, four basic components of technology may be identified (Kahen & Griffiths, 1995 and Kahen, 2001). (Fig. 20.3).

Technoware: It's an object-embodied technology. It forms the core of any technology (i.e. a transformation facility), which is developed, installed and operated by humanware.

Humanware: It's a person-embodied technology. It represents a key element of any transformation operation that is, in turn, guided by infoware.

Infoware: It's a document-embodied technology, which stores information accumulated for time compression by individuals in learning and doing. It is generated as well as utilised by humanware for the process of decision making and for operations.

Organoware: It's an institution-embodied technology. It acquires and controls infoware, humanware and technoware in order to effect their operation. Organoware consists of the activities of planning, organizing, activating, motivating and controlling the operation.

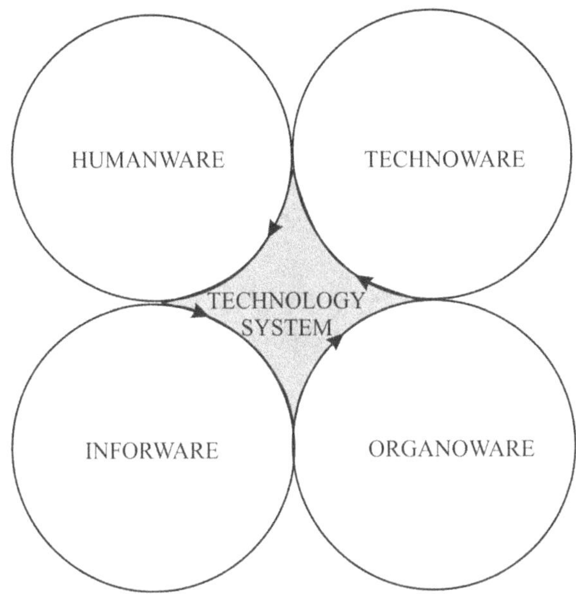

Fig. 20.3 Dynamically interacting components of technology.

Mechanisms of Technology Transfer

New technology may be acquired in either of two basic ways: by *developing* it or *importing* it (Sharif, 1986). Developing countries import technologies for two fundamental reasons: because little or no R & D investment is needed, and because technical and financial risks are very low since the new technology is usable immediately. From the supplier's point of view, three good reasons exist for selling technology, such as:

- The sale of technology may imply increasing returns on R&D investments;
- The transferred technology may have no immediate use for the supplier;
- The technology may have already been utilised up to its limit.

Two kinds of transfers of technology occur: vertical and horizontal (Brooks, 1966). Vertical transfer refers to transfer of technology along the continuum from the more general to the more specific. In particular, it includes the processes by which new scientific knowledge is incorporated into technology, by which 'state-of-the-art' technology becomes embodied in a system, and by which the confluence of several different and apparently unrelated technologies lead to a new technology. On the other hand, horizontal transfer occurs through the adaptation of a technology from one application to another that is possibly wholly unrelated to the first. The transfer of technology has been described as taking place through three main mechanisms: first, through the mobility of personnel; second, sourcing (which is the agreement between two producers to manufacture fully compatible products); and third, cross-licensing agreements (Rada 1982). This classification, however, does not take into account the education and training needed to use a particular equipment as a technology transfer channel.

A combination of these mechanisms is needed for successful technology transfer to occur. The mechanisms most frequently used in developing countries include.

- Purchase or licensing of already developed technology in the form of patents, products or know-how packages;
- Direct purchase of specialised machinery embodying needed technologies;
- Use of foreign experts as technology transfer agents;
- The transfer of technology through the establishment of foreign industrial enterprises, often multinational corporations or joint ventures, in a developing country.
- The training of personnel abroad;
- Establishment of specialised centers for technology transfer, mainly for the training of personnel.
- Establishment of institutions of education, research and development, and agricultural extension services in the developing country; and
- Maintenance of a suitable economic and social climate for innovation and technological change.

Modes of Transfer

According to the International Code of Conduct on the Transfer of Technology, a number of distinct operations may be identified, as follows:

- Assigning or granting of industrial rights,
- Handing over technical or non-technical know-how in the form of documents, plans, diagrams and so on,
- The communication of technical or other know-how in the form of supply of services,
- Providing a combination of services with a view to commissioning an industrial complex and
- Providing technical services related to the selling or leasing of machinery.

Many authors have attempted the following means by which technologies can be transferred.

- Transfer of published material (such as journals, books);
- Purchase of machinery, equipment and other intermediate goods;
- Transfer of data and personnel;
- Granting of licences and trademarks;
- Collaborative agreements;
- Direct foreign investment by transnational corporations (this procedure entails a combination of mechanisms);
- Technologists' mobility (movement of technical personnel from one organisation to another);
- Technological entrepreneurship (technologists moving out of established firms to set up their own business); and
- Interpersonal communication (gatekeeper networks).

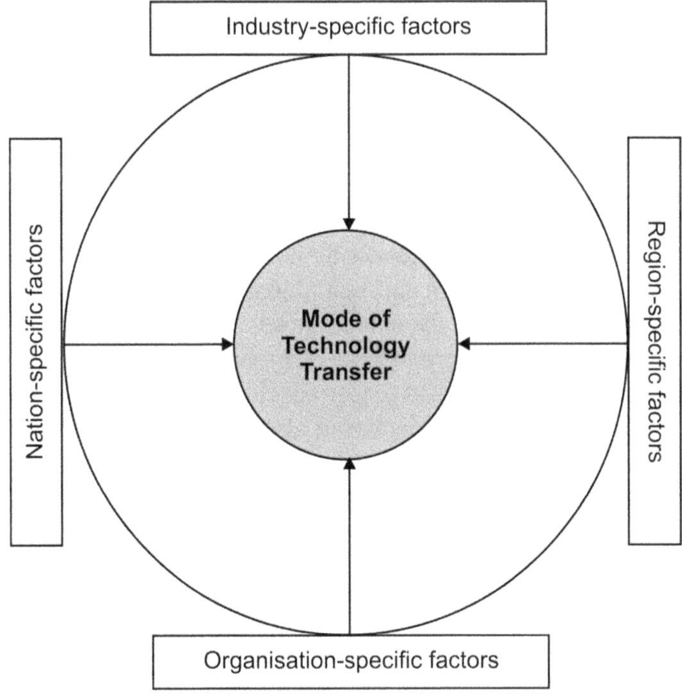

Fig. 20.4 Major factors in mode of transfer.

Whatever the mode of transfer, multinational corporations, as the main suppliers of technology, dominate the process (Buckley and Casson 1976). The mode of technology transfer is affected by the following four groups of factors, which play a crucial role. (Fig. 20.4).

- Industry specific factors, such as product and structure factors
- Region specific factors, such as cultural aspects
- Nation specific factors, such as political aspects
- Firm specific factors, such as management and technical knowledge.

Transactions and Channels for Technology Transfer

Beyond the choice of technology itself, manufacturing firms face other choices when the transfer know-how for use in foreign plants, such as the channels to be used for transferring proprietary technology. Three mutually exclusive channels are available:

- Sale of the know-how to an unrelated party;
- Use of the technology in a facility partially owned by the technology owner; or
- Use of the know-how in a facility wholly owned by the technology owner.

A technology licensing agreement may be employed in all three channels. The first channel of transfer of proprietary technology is often referred to as an 'arm's-length licence', or simply a 'licence'. The second channel is known as a 'joint venture', and the third channel is referred to as a 'wholly owned subsidiary. Despite the emergence of new forms of transfer, technology transfer through direct foreign remains the dominant model, and the supply side of technology still remains very oligopolistic in several industries (Lall, 1982). Direct investment can result in a substantial loss for the investing enterprise. Licensing, in contrast, almost always brings some return to the owner of the technology. Clearly, there is no evidence that any one channel is ideal for all developing countries in all situations. Certain characteristics of the country to which the technology is being transferred also affect the decision about the transfer channel. The international transfer of technology takes place through a number of formal and informal organisational modes. These modes can involve governments, academic institutions, companies and individuals, and may range from direct contact with foreign sources to indirect contact with such sources. Simon (1991) identifies five basic organisational modes:

- The international technology market, which is made up of independent buyers and sellers;
- Intrafirm transfer, wherein organisations (such as multinational corporations) do not resort to the market but transfer technology either through an internal venture or through a wholly owned subsidiary;

- Agreements or exchanges directed by the government, where the counterparts can be either public or private actors;
- Education, training and conferences, through which information is disseminated publicly for consumption by either a general or a specialised audience;
- Pirating or reverse engineering, through which organisations obtain access to technology without resorting to the market but at the expense of the property rights of the owners of the technology.

Thus, the type of transaction adopted and the procedures followed are affected by many factors. Given local market conditions, and depending on the economic situation, level of industrial development and political circumstances of the host country, the receiving enterprise may be a joint venture or may be totally independent. Indeed, there is considerable overlap between these forms. Lack of necessary levels of skills also affects the conditions surrounding the transaction. In many cases, the receiver of technology may not possess the ability to repair and maintain, modify, adapt or develop technology, or to design and produce new equipment or products.

Stages in the Technology Transfer Process

The process of technology transfer to developing countries involves a complex series of stages. The precise nature of the process in each case depends on prevailing local conditions and on the type of technology that is to be transferred. Two major stages may be identified in general; each stage consists of related minor stages deals with technology policy planning and decision making (that is, technological planning and technology assessment), the second revolves around the implementation and operational processes of technology transfer. The operational stage of the technology transfer process includes five different, but coordinated, parts. The steps in this sequence may be summarised as follows: (Fig. 20.5).

- Assignments, including sale and licensing agreements covering all forms of industrial property including patents, inventor's certificates, utility models, industrial designs, trademarks, service names and trade names.
- Arrangements covering the provision of know-how and technical expertise in the form of feasibility studies, plans, diagrams, models, instructions, guides, formulations, service contracts and specifications, and/or involving technical, advising and managerial personnel and personnel training as well as equipment for training.
- Arrangements covering the provision of basic or detailed engineering designs, and the installation and operation of plant and equipment.
- Purchases, including leases and other forms of acquisitions of machinery, equipment, intermediate goods and/or raw materials, insofar as they are part of transactions involving technology transfers.

- Industrial and technical cooperation agreements of any kind, including turnkey agreements, international subcontracting as well as provision of management and marketing services.

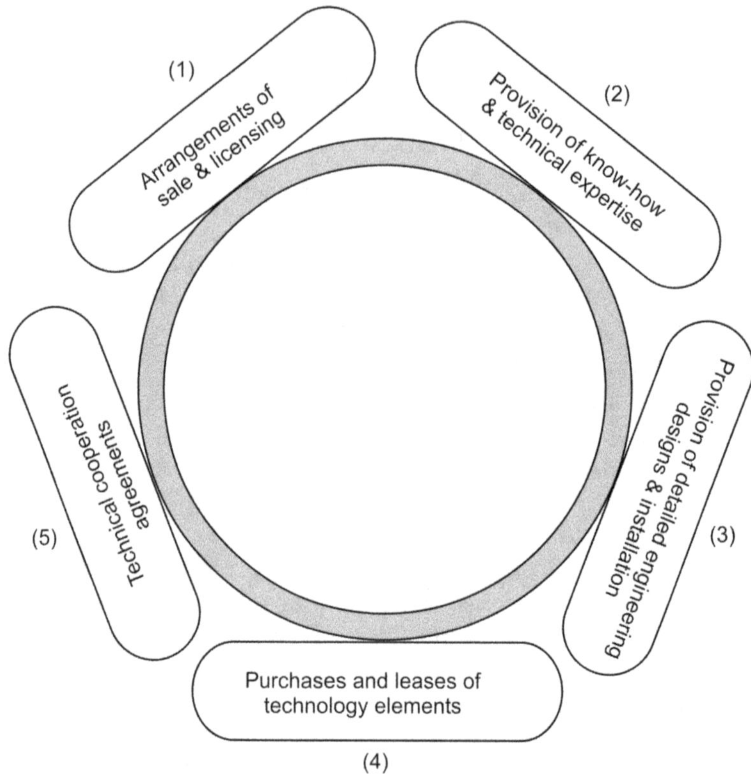

Fig. 20.5 Operational sequences for technology transfer.

Technology is not a homogeneous phenomenon. There are different types of technology, each posing fundamentally different problems and demanding different solutions in the international transfer process. It is now accepted (Kolds, 1981) that adhoc negotiations on a case-by-case basis are often useful in the proprietary technology sphere.

External and Internal Forces Affecting Technology Transfer Protocols: The effectiveness or success of any transferred technology in a developing country depends on five kinds of forces:

- Changes in indigenous technological capabilities
- Changes in society and individual lifestyles (that is, socio-cultural changes)
- Changes in the economy; human resources systems and working conditions

- Changes in the political system
- Changes in the international relations

Together, these five kinds of forces affect the supply of and demand for technology, and influence government policies on technology transfer, or in other words, the national protocols on technology transfer. Technological changes or breakthroughs occur continuously, and have implications for all areas of the host country's techno-economic system (from facilitating research and development and improvement of the services and manufacturing sectors, to encouraging the expulsion of foreign technology). The introduction of foreign technologies in these situations should involve careful management and evaluation. Further, technological changes wrought by transferred technology require a re-examination of all arrangements for the initial and ongoing training of professional staff, and have a profound effect on hitherto accepted working practices.

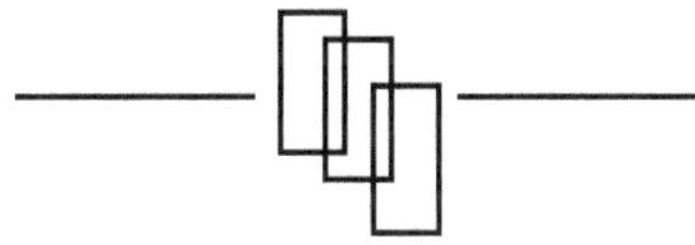

CHAPTER - 21

Fisheries Extension in India

History : Kautilya's Arthasastra, one of the oldest Indian texts indicates that fish culture activity in India dates back to 2000 years. Perhaps it started when the human settlement moved away from the riverbanks to the hinterland. Paddy fields and low lying areas in flood plains and those connected to the estuaries and estuarine creeks became the cradles of aquaculture where inundation, caused either by monsoon rain or by tidal water, brought the natural seed of finfish and shellfish which got automatically trapped after the water receded. That eventually gave rise to the operation of "trapping and holding" of fish seed and raising them to table size and thus this marked the beginning of aquaculture in India.

Indian Fisheries - Capture and Culture

During the last 50 years of fisheries production, India has become sixth largest producer of fish and occupies the second position in inland fish production in the world. Similarly, the contribution of fisheries sector to the GDP has increased considerably compared to agriculture in recent years. The fisheries sector also contributed greatly to the export earning of the country, while adding considerably to the fish availability to about 8 kg/per capita.

Extent of Rural Aquaculture : As the old saying goes "as you sow so shall you reap" appears quite apt for aquaculture. The whole system of production is a continuum and it is very difficult to strictly divide in to different categories like extensive, semi-intensive and intensive culture systems mainly on extraneous feed supply or off-farm agro-industrial inputs. Similarly, it is difficult to segregate rural aquaculture from entrepreneurial aquaculture, which normally concerns with intensive cultivation. In fact, many farmers who have been involved in subsistence level production increased their production over the years, with more inputs and better management skill, resulting in enlarging their resource base and gradually becoming entrepreneurial. In fact, it is most desirable that the rural aquaculture should be evolve ultimately into entrepreneurial aquaculture and make the resourcepoor farmers entrepreneurial farmers. Such an evolution is already taking place in rural India particularly in the freshwater sector. Shrimp aquaculture in India shows two distinct types of entrepreneurship, one comprising of resourceful farmers and small entrepreneurs another of large entrepreneurs or commercial enterprises. The former practices shrimp farming at a low level of intensity which is not very capital intensive or input intensive but the latter undertakes complex commercial aquaculture requiring extensive facilities and management skills with reliable supply of seed and feed to achieve high production. Very high stocking density needs a considerable amount of freshwater and oxygen. It also requires sophisticated infrastructure to ensure that the products reach the markets in the form and quality required by the consumers. Thus, fresh water aquaculture and shrimp farming under traditional and extensive systems only should be strictly viewed as rural aquaculture in India, (Edwards and Demaine, 1997).

Fisheries National Policy

The national policy on fisheries development in general stresses on optimal utilizsation of natural resources through their rational exploitation, centeres round the concept of giving a better deal to socially backward communities, more and better employment opportunities, and increasing aquatic productivity on a sustainable basis. The national development programmes include organisation of necessary infrastructure for fish production, storage and distribution, and mobilisation of manpower at various levels for production and marketing, thereby accruing maximum assured benefits to the producer and consumer, reducing middle level involvement to the minimum.

Long-Term Objectives

To progressively raise the subsistence level aquaculture activities to the level of an organised industry, following plans were reiterated:

- Establishment of an effective mechanism by which the advances made in the technology of fish culture could be disseminated for the purpose of increasing fish production in small water areas such as ponds and tanks.
- Adoption of an integrated approach to the problems of development, comprising of filling up the gaps in the technology, organising training of manpower and provision of necessary inputs.
- Assigning priorities for securing a greater flow of institutional finance for fish culture programmes.

Medium Term Objectives

- Progressive adoption of improved technology of fish production in certain selected areas
- Brackish water fish culture with particular emphasis on shrimp
- Fish seed farms in each district and possibly in each sub-division.

- Developing technique of culture of frogs, mollusks, sea weeds, etc.
- Developing running water carp and catfish culture.
- Further research on nutrition and reproductive physiology of cultivated species of fishes and crustaceans so as to have better control on their growth and breeding. Formulation of suitable feeds.
- Further research to obtain still better fish production through improved methods of aquaculture
- Developing fish culture in the cold water areas of the country.
- Development of culture technology for the non-conventional species of fish, molluscs and crustaceans to bring them into the orbit of culture programmes.

Short-Term Objectives

- The immediate goal was to reach reasonable production levels through substantial development of aquaculture in the country with available technical know-how, and to adopt improved technology of fish production in selected areas under traditional culture.
- Immediate emphasis to be given to the production of cultivable fish seed and their efficient rearing in selected area in each state to provide the stocking material.
- To create and activate the category of fish farmers all over the country by introducing suitable training programmes combined with making water areas and inputs available to carry out fish cultures.
- Making available water areas to the fish farmers on long-range lease, so as to link leasing with production programs.
- To provide basic infrastructure for brackish water fish and shrimp farming by constructing hatcheries for organised production of seed.
- Bringing the financing institutions and fish farmers closer so that they may not be financially handicapped in undertaking scientific fish culture.

Women in Rural Aquaculture

Women form about 48% of the total population in India. About 78% of them are economically active and are engaged in agriculture and allied fields. They are deeply involved in food related activities including food production and thus they are more concerned towards food security of their families by way of producing/procuring food, feed and fodder, maintaining live stock and raising kitchen garden and orchiads. Some recent estimates showed that India possesses about 314.9 million total work force of which 91.4 million are women workers. Of which 81.5 million are in the rural sector and 9.9 million in the urban sector. All the rural work force is not always employed for about 240 days in a year. Hardly about 285.4 million are employed for about 180 days work/year; of which about 16.2% are female. Among them 34.6% are cultivators, 43.6% farm labourers and 4.6% are engaged in sectors like livestock, fisheries, poultry etc. Mechanisation of the agriculture sector with increasing use of tractors, transplanter, weeder, harvestors, thrasher and dehulling mill etc. Has threatened women's employment opportunities. In this context, aquaculture sector opens up the possibility of alternate employment for men and women alike. Analyses of some case studies have been presented below to show the type of labour and gender involved in rural aquaculture. Family labour showed much more womens involvement. Of 18 cases 9 cases showed female involvement and male to female involvement went as high as 40%. However, possibility of their involvement has been considerable. Following are the areas of fish farming, where women could be involved profitably.

- Net making
- Carp breeding
- Hatchery management
- Carp fry rearing
- Composite fish culture
- Paddy cum fish culture
- Air breathing fish culture
- Prawn seed raising
- Prawn culture
- Aquarium fish breeding
- Aquarium keeping
- Integrated fish farming

Fisheries Extension

It is the collective word to describe all organised communication efforts by which an individual or agency tries to bring about changes in the knowledge, attitudes, skills and/or behaviour of a client population, in order to reach one or more objectives that have been established within the framework of an overall development policy.

Prior to 1944, a fishery was a 'transferred' subject. The administration and management of fisheries were the direct responsibility of the erstwhile provinces. Although the Government of India enacted the Indian Fisheries Act in 1897, it was mostly for empowering the provincial governments to frame rules under the Act. The provincial governments concerned themselves mainly with administering the Fisheries Act and issuing licenses for fishing or leasing out Government-owned water areas for fishing. The constitution of a Fish Sub-Committee of the Policy Committee on Agriculture, Forestry and Fisheries in 1944 to review the position of fisheries of the country and the recommendations of that Committee paved the way for the formulation of various programmes for fisheries development. After independence, fisheries development was given enhanced importance particularly in the National Plan Schemes (Five-Year Plans), during the first five Five-Year Plans and intervening three annual plan years (1951-79), the main emphasis in marine fisheries development was on the introduction of mechanized fishing boats and mechanisation of fishing craft. Besides this, up to Fifth Five-year Plan emphasis was given to popularisation of synthetic fishing twine and assistance to small fishermen in the form of supply of fishing requisites, salt for fish curing, etc. The public and private sectors established a large number of ice plants and cold storages during this period. This development, along with the increasing demand for shrimp in the international market, stimulated the private sector to establish facilities for freezing and canning and it also led to an export trade. Institutes for the training of fisheries personnel of various categories to meet the technical and operative manpower requirements of the fishing industry were established. Fishery survey and research programmes were intensified. The construction of fishing harbours at major and minor ports and provision of landing facilities at various sites were taken up. In order to improve the socio-economic conditions of fishermen and marketing, establishment of fisheries cooperative societies and fisheries corporations was encouraged. Various techniques have been employed to estimate the productivity of Indian waters. These include exploratory fishing surveys by Exploratory Fisheries Project (EFP), Integrated Fisheries Project (IFP), Central Institute for Fisheries Nautical and Engineering Training (CIFNET), Central Marine Fisheries Research Institute (CMFRI), National Institute of Oceanography (NIO) and various universities.

Emergence of a Fisheries Extension Service

As mentioned earlier, a fisheries extension service works within the context of an overall development policy. Such a policy may have objectives that are related to an increased production of fish and/or objectives that are related to the increased well-being of the fishing families involved in fishing. In order to attain these development objectives, the Fisheries Agency, or any other organisation, designs a number of developmental programmes. Extension support can be one of the instruments for a successful programme. In this context, policy-makers should be concerned with defining the general outline of programmes, but not with deciding the actual content of the programme. That should be left to the experts concerned, such as extension officers for extension programmes. As mentioned earlier, it is important that policy-makers find an appropriate mix of production- and output-oriented programmes and human resources development programmes. In addition, while identifying the programmes, the following are kept in mind.

- The client orientation of fisheries extension: WHO?
- The problem analysis: WHY?
- Choosing the right extension method(s): HOW?

- Timing the extension programme: WHEN?
- Extension requires resources and has a cost
- Monitoring and evaluation of extension programmes

Working with Clients: The Demand Side

The Principle of the Client's Self-Reliance in Decision-Making : A very important principle of fisheries extension services is that it should help the clients to remain, or to become, self-reliant decision-makers in improving the economics of their fisheries activities. For certain sectors, such as health and education, it is usual for people to become dependent on the basic facilities offered by the government. But when it concerns economic enterprises, and each fishing unit is an economic enterprise, such dependency is likely to be disadvantageous in the long run. Each enterprise requires constant investment (however small), as well as decision-making on how such investment should be made in the most economical way. As a result, the client may become reluctant to think about future investments or make decisions, as they would expect government, at least partly, to take care of it. This can be detrimental to the enterprise, and the responsibility for decision-making should be left as much as possible to the client.

Assessing the real needs and Analysing the Problem : In finding possible solutions to the clients' problems and needs, and information on how to mobilise their resources and, when required, how to obtain such outside resources as credit.

Working with Adults : One of the most important features of fisheries extension is that the clients are usually adults. However, there may be occasions when children are the target group.

Working with Social Groups : Groups have different interests, learning capabilities, income-generating activities, household activities and different cultural values

Establishing a Demand for Fisheries Extension : *As we have seen, the Fisheries Extension Agent is involved in the two sides of fisheries extension: the demand side and the supply side. On the demand side, the FEA assists the clients in analysing their problems, assessing their needs and, if necessary, helping to organize for them extension methods, such as a field workshop. On the supply side, the FEA is responsible for providing the required extension services to the clients, such as arranging for technical information.*

Fisheries Administration

Both the Union Government and States share responsibilities for the development of fisheries. Each of the States is directly responsible for the development of fisheries within the territorial waters of the sea and the inland waters. The Union Government is responsible for the development of fisheries beyond the territorial waters and for fisheries research, although these are shared by the State Governments as well. The Fisheries Wing in the Department of Agriculture under the Ministry of Agriculture, Government of India, is in overall charge of all important matters relating to policy and administration of the fisheries of the country. It is responsible for the formulation of national policies and programmes of fisheries development, fishing harbours, processing and preservation of fish, fisheries education, training, fish trade, etc. It is also responsible for taking all necessary steps for making available timely and adequate supply of inputs and services required; for participating in international organisations, promoting bilateral and multilateral cooperation and collection and maintenance of relevant statistics. It assists State Governments in formulation of policy, plans and projects, and in the setting up of fisheries corporations, and offers technical advice and guidance whenever required. Apart from the Department of Agriculture at the Union Government level, the Ministry of commerce also looks after certain functions concerning fisheries such as export promotion and quality control. While fisheries education and research are the responsibilities of the Indian Council of Agricultural Research (ICAR). In order to plan, coordinate and develop indigenous capability in oceanographic research, an Oceanic Science Technology Agency (OSTA) has been established.

Research and Development Institutes

Central Marine Fisheries Research Institute (CMFRI) : The Institute was established in 1947 by the then Ministry of Food and Agriculture. It came under the control of ICAR in 1967. The overall objectives of the Institute are to conduct short-term and long-term multi-disciplinary researches on the marine capture and culture fisheries of the country. It thus seeks to provide research support for the rational exploitation, conservation and management of marine and brackish water resources, and development support for the growth and stability of the industrial, artisan, and culture fisheries. Its functions include transfer of technology, dissemination of

Information and Education, Training and Extension

Central Institute of Fisheries Technology (CIFT) : CIFT was established in 1957 with its headquarters at Cochin and is now under the Indian Council of Agricultural Research. The aims of CIFT are to develop improved fishing techniques, gear, craft and implements; to develop improved and new technology for the optimum and economic utilisation of fish catch; to develop technology for economic utilisation of wastes from the fishery industry and also fish and shellfish which do not have a ready market for human consumption; to popularise the research results; to provide a forum for feedback to the Institute on technical problems and to conduct short-term refresher and training courses in the improved technology evolved.

Central Inland Fisheries Research Institute (CIFRI) *:* The Institute was established in 1947 at Kolkata, under the then Ministry of Food and Agriculture, Government of India. The administrative control of the Institute was taken over by the ICAR in 1967. The headquarters of the Institute is at Barrackpore near Kolkata. The main objective of the Institute is to elucidate the scientific principles, which can be applied for full utilisation of all available inland waters of the country for maximizing fish production.

This objective entails evolving sound fish husbandry techniques; acquiring understanding of the biology of food fishes; conducting investigations on the hydrology and ecology of different waters; performing research on population dynamics of fish in natural waters, such as rivers, lakes, reservoirs, estuaries, etc. ; formulating artificial feeds and evolving feeding techniques; and developing fishery management techniques relating to both fresh and brackish water environments. The Institute presently has 37 centres spread across the country.

National Institute of Oceanography (NIO) *:* The National Institute of Oceanography was established in 1966 under the Council of Scientific and Industrial Research (CSIR). The headquarters of the institute is at Panaji in Goa and it has three regional centres at Visakhapatnam, Cochin and Mumbai. The institute is engaged in studies of the seas around India with the objective of making proper and judicious use of their resources.

Integrated Fisheries Project (IFP) *:* Until 1973 this was known as the Indo-Norwegian Project. It was started in 1952 as an area development project with the objective of uplifting the fishing communities of Neendakara and Saktikulangara villages in Kerala State.

Exploratory Fisheries Project (EFP) *:* It was known until 1974 as Deep Sea Fishing Station, Mumbai and was established in 1946. There are 12 operational bases at Porbander, Veraval, Mumbai, Goa, Mangalore, Cochin, Tuticorin, Chennai, Visakhapatnam, Paradeep, Kolkata and Port Blair. A new operational base has been recently established at Vizhinjam to conduct exploratory surveys of the Wadge Bank as a part of the Vizhinjam fishing harbour project development. The objectives of EFP are to carry out exploratory work in respect of charting of fishing grounds, fishing seasons, types of fish available for exploitation and suitability of different fishing vessels and gears; to train personnel for modern fishing operations; to test the commercial feasibility of deep fishing; and to make available requisite data and information towards the expansion of the fishing industry.

Pre-Investment Survey of Fishing Harbours (PISFH) *:* It was established with UNDP special fund assistance in 1968, initially for a period of five years. SIDA (Swedish International Development Agency) provided funding for two more years. Thereafter it has been a national project fully funded by the Government of India. The objective of PISFH is to conduct economic and engineering investigations to locate prospective sites for fishing harbours and to prepare project reports.

Marine Products Export Development Authority (MPEDA) *:* It is the successor to the Marine Products Export Promotion Council set up in the Ministry of Commerce in 1961. It was established in 1972 by an Act of Parliament (Act No. 13 of 1972). MPEDA has its headquarters at Cochin, regional offices at Calcutta, Bhubaneswar, Kolkata and Mumbai and trade promotion offices at Tokyo and New Delhi. The main objective is to develop the marine products industry with special reference to Exports.

Training Institutes

Central Institute of Fisheries Education (CIFE) *:* It was established at Mumbai in July 1961 under the Ministry of Agriculture. Its administration was transferred to ICAR in 1979. The objective of the Institute is to train fisheries officers in service in various states in the country by means of a comprehensive course of fisheries education aimed at equipping them with the technical know-how necessary for implementing fisheries development projects. A limited number of private candidates and nominees from the industry and foreign countries are also admitted. It also has a number of subordinate establishments for providing junior level training courses, details of which are given below:

Inland Fisheries Training Unit, Barrackpore

Regional Training Centre for Inland Fisheries Operatives, Agra

Central Fisheries Extension Training Centre, Hyderabad

Central institute of Fisheries Nautical and Engineering Training, Cochin

Marine Fishermen Training Centres and *Fishermen Training Centres* are set up by the states for the training of fishermen in the operation of small mechanised fishing boats. There are at present 30 such training centres - 4 in Orissa, 2 in Gujarat, 4 each in Maharashtra and Karnataka, 5 in Kerala, 6 in Tamil Nadu, 2 in Andhra Pradesh and one each in the Union Territories of Lakshadweep, Goa and Andaman and Nicobar Islands.

Fisheries Cooperatives

The fisheries cooperative system in the country was organised with a view to providing assistance to the actual producers, the fishermen. Fisheries cooperatives are societies governed by a separate set of rules to channel government assistance on the basis of principles of self-help and management. The fisheries cooperative structure in India is broadly three-tiered. It consists of a primary cooperative for a village or group of villages; a district or regional federation; and a state-level apex body constituted as a cooperative federation. The primary fishery cooperative is expected to function as a multi-purpose agency providing credit, supplies (including domestic necessities), elementary guidance and supervision of the utilisation of loans, assembly of the fish-catch and its transportation to marketing centres. Member education and extension programmes are supplementary functions expected to be undertaken by the primary society in collaboration with other concerned agencies. In practice, however, primary fisheries cooperatives are mainly engaged in the provision of loans to the member fishermen. Some cooperatives have organised the provision of supplies and only a very few societies are engaged in activities such as construction of fishing boats and processing. The regional federation is the district-level organisation which is expected to act as a useful and conveniently located intermediate agency between the apex body and the primary cooperative. It is expected to have an important role in marketing and, for that purpose, is located in the larger towns. Most of the other functions of the federation are similar to those of the primary cooperative but on a larger scale and catering to a wider area in providing assistance to the member-primary cooperatives. The apex federation is the state-level organisation and is expected to provide support to the primary and district-level cooperatives. Through its location, size, capacity to command resources of men, material and market, it is expected to provide leadership to the entire movement. All State Governments have full-fledged departments of cooperation. Cooperative banks provide credit, while fisheries departments provide share capital, managerial assistance, subsidies and loans for various approved activities, besides loaning the services of departmental officers to undertake activities on behalf of cooperative societies. The Agricultural Refinance and Development Corporation (ARDC) undertakes various integrated projects through cooperatives and cooperative banks. Besides this, two other agencies have a place in the cooperative set-up; National Development Corporation (NDC) and the National Cooperative Union of India (NCUI) for promoting the development of fishery cooperatives since 1974. It provides financial assistance to State Governments for contributing to the share capital of societies to enable to expand their marketing, supply and distribution activities.

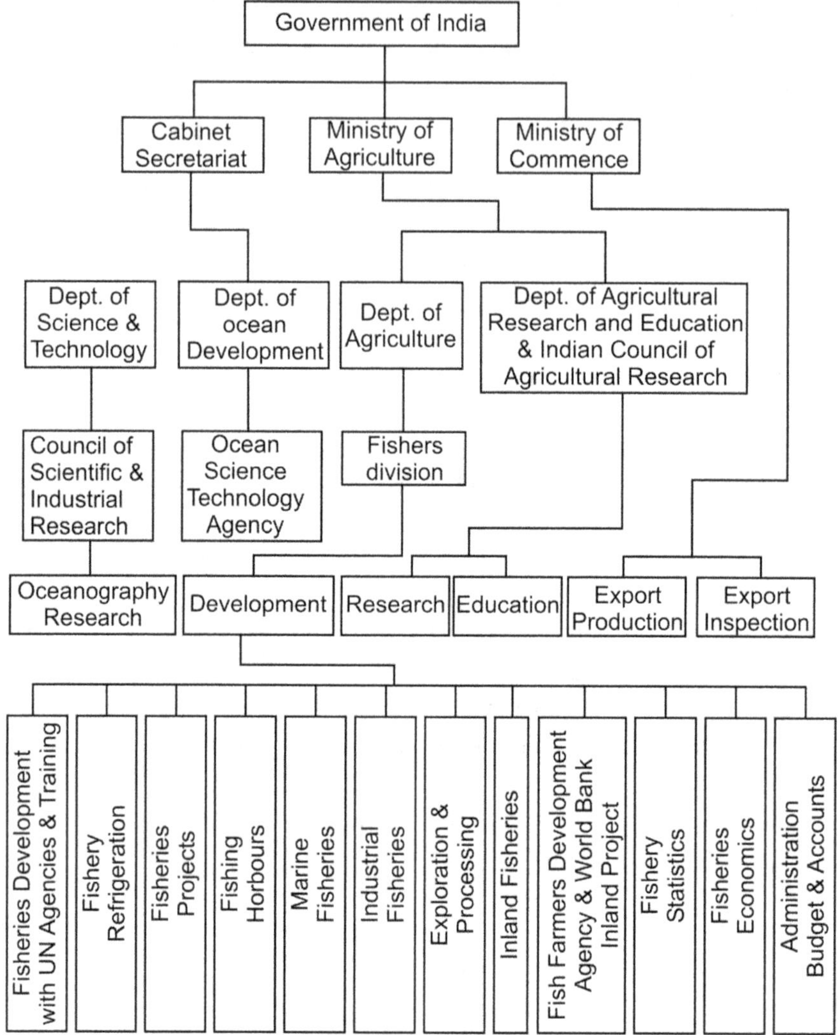

Fig. 22.1 Organizational set-up of the fisheries administration in the Government of India.

Fisheries Development Programmes

FDP is one of the earliest, most attractive and promising Employment and Income Generation programmes. As it involves less investment and more profit, the poor get more interested in this profession.

The Main Objectives of the Programme are:

- To facilitate the access of poor fishers and fish farmers to the public and private water resources;
- To provide credit for employment and income generating activities;
- To introduce ecological aquaculture;

- To improve technical and managerial capacities of the rural poor fish farmers and fishers through training;
- To introduce community-based fisheries management system;
- To increase fish and fingerlings production and raise the nutritional status of the poor and fishers to ensure proper management;
- To ensure women's participation.

Extension and Technical Assistance

Technical assistance along with credit is extended to the group / society members. It is a package of activities that covers feasibility study, follow-up, training (group members and programme personnel), monitoring, evaluation, field demonstration on pond re-excavation methodology, technology transfer on aquaculture, pond preparation, application of feed and organic manure, disease treatment, sampling, harvesting, fish processing, and marketing. Besides this, there is a technical assistance package to the group members to choose from the options and help them to implement projects with the help of the fisheries technical staff. The Global Fishing Industry is suffering from paucity of resources, over exploitation and mounting operational costs. Management and conservation of resources, diversification of fishing effort and economic utilisation of fishing units are some of the chosen directions for the new millennium. When developmental interactives are channelised in this direction, the National Institutes like Integrated Fisheries Project which has acted as a catalyst in the growth of the sector since independence has to change roles and re-equip itself to take up new challenges.

Marketing

It is interesting to note that during 1970s carps from Bangladesh were seen being sold in Kolkata markets but now fish from Andhra Pradesh is said to be sold in Bangladesh. Indian major carps are sold in the U.K markets particularly to those places where Indian population is sizable. However, secondary and tertiary markets are emerging for carps all over India. Yet, strictly speaking poor farmers have to sell their produce in the local market at a very cheap price or to middlemen. proper mechanism is still inadequate to collect and disseminate information about the prices and markets. Wholesalers and commission agents have the key role in marketing fish for the farmers. However, the farmers share some time as high as 88% of the sale price in the local sale though it is much lower in the outstation price. In the outstation sale the producers get about 60% of what consumers pay. It is important that proper infrastructure is created for efficient marketing of aquaculture products. Government should invest more and also encourage private sector investment in market infrastructure development for serving farmers and entrepreneurs. Also creation and expansion of small-scale credit and saving institutions are necessary to facilitate development of aquaculture trading, transportation, and processing enterprises.

Glossary

Abstract : Refers to precise or a pieces of writing over research, that proceeds in a publication.

Action Research : "Small scale intervention in the functioning of the real world, usually in the administration system and the close examination of the effects of such interventions".

Active Adopters : who adopt there an innovation and influence others to do so.

Active Rejecter : who reject an innovation and influence others to do so.

Active Variable : It is a variable that can be changed or manipulated by a researcher.

Administrator : A person with administrative responsibilities.

Adoption : Adoption is a decision to continue full use of an innovation.

Adoption Categories: Adoption categories are the classifications of individuals within a social system on the basis of innovativeness. The five categories utilised are innovators, early adopters, early majority, late majority and laggards.

Adoption Period : Is the length of time required for an individual to pass through the adoption process from awareness to adoption.

Adoption Process : The mental process through which an individual passes from first hearing about the innovation to final adoption.

Adoption Rate: Rate of adoption is the relative speed with which the members of social system adopt an innovation. Rate of adoption is usually measured by the length of time required for a certain percentage of the members of a social system to adopt an innovation.

Adult Education: Education provided for adults for general educational rather than vocational purposes.

Aesthetics : The Judgement of beauty or taste.

Agriculture : The production of plants and animals for the use of human beings. This includes cultivation on soil and the feeding, breeding and managing of crops and livestock. To some extent it deals with preparation of plant and animal products for human consumption and marketing of these products.

Agricultural Economics : A branch of agricultural science, which uses the principles and techniques developed in agronomy, farm business, animal husbandry and economics in a coordinated way for the economic benefits and welfare of a farmer in particular and a region or state or country or the world in general.

Agricultural Extension: Extending knowledge to Agriculturists. Agricultural Extension is concerned with agricultural education assisting farmers to bring about continuous improvement in their physical, economic and the social well being through their individual and cooperative efforts. It makes available to the farming community the scientific and other factual information, the training and guidance for the solution of problems in agriculture including Animal Husbandry, Gardening, Agricultural Engineering, etc.

Aim : This is a broad objective. It is a generalized statement of direction and may have several objectives. It is also said to be an end in view direction to the creative process.

Analysis of Covariance **(ANCOVA)** : Statistical technique used to compare two or more groups on a variable when some other variable may affect the comparison.

Analysis of Variance **(ANOVA)** : Statistical testing technique to determine if variance between the mean on different sets of observations exceeds what may be expected by chance.

Antecedent Variable : A variable, which proceeds the variables under consideration, usually unknown to the researcher at the outset of his study: frequently discovered to be the true causal or independent variable.

Anthropology : A study of man and his works. The discipline has two fundamental aspects: one the study of man as an organism (physical anthropology) and its sub divisions human biology, Primatology anthropromatry and biometrics and two, the study of human societies (ethnology), social anthropology, cultural anthropology, archaeology, enthnography and social psychology.

Aphasia : Inability either to use speech or to understand speech.

Applied Economics : It is that branch of economics, which is devoted to a study of practical problems with the help of concepts, laws and principles provided by Pure Economics.

Applied Research : This type of research is based on the application of known theories and models to the actual operational fields or populations. The applied research is conducted to test the empirical content or the basic assumptions or the validly of a theory under given conditions.

Apprentice : A young person contracted or indentured to be trained over a period of several years in a skilled trade or occupation.

Aptitude : An acquired skill or ability that is assumed to underline and is conductive to an individual's capacity to learn and attain a level of achievement in a specific field.

Assessment: The process of evaluating property for general taxation purposes, also the value as assigned-usually called " the assessed valuation".

Assimilation : Assimilation is a process of inter-penetration and fusion in which persons and groups acquire the memories, sentiments and attitudes of other person or groups and by sharing their experience and history, and incorporated with them in a common cultural life.

Assumption : A fact, which is taken as true.

Atomise : The type of social process by which social groups are broken into fragments or larger social patterns are reduced to small ones.

Attention : Attention may be defined as the process of focussing upon certain phases or elements of experience and the neglecting of others. The relative clearness of perception depends upon the intensity of the attentions process.

Attitude : Refers to predisposition to perceive feel or behave towards specific objects in a particular manner.

Attribute : Refers to basic characteristic of a sensation with the disappearance of which the sensation vanishes.

Basic Data : Records of observations and measurements of physical facts, occurrences and conditions, as they have occurred, excluding the reform any material or information developed by means of computation or estimates as are necessary to present a clear statement of facts, occurrences and conditions.

Behaviour : Refers to how a person known, thinks, perceives, remembers, acts and feels in relation to a given context or idea or experience.

Bias : A conditioned tendency to favour and support a certain point of view or conclusion despite the absence of adequate or even any evidence; a disposition to reject evidence that conflicts with a preconceived conviction. It is frequently used as a synonym for prejudice.

Bibliography : List of books or other written materials placed for reference after a piece of academic writing or appearing as a separate publication.

Body Language : It is non-verbal communication using body movement or posture. May be conscious or unconscious.

Brain Storming : It is a technique of exploring possible solutions to problems through discussion.

Burial Society : A type of friendly society established during the eighteenth centuries for the purpose of life assurance usually on a small scale, to provide a sum of money sufficient to cover the funeral expenses of the diseased where otherwise such an expense might be a serious drain on the resources of a poor family.

Buzz Session : Buzz session is a method of discussion whereby a large group is divided into small groups for discussion of specific problems for a limited period, from 5 to 10 minutes.

Catalogue: A list published or for publicity.

Category : A class, group or type in a classified service. The term is frequently used in the classification of needy persons in order to administer programmes of work, rehabilitation, relief, pension, etc.

Census : A periodic enumeration of the population of a political unit. The data secured ordinarily includes not only the sample of persons, but also facts concerning sex, age, race and a variety of other characteristics which may be very inclusive. The oldest continuous genuine census in the world is that of the United States, which was inaugurated at the beginning of its independent national life in 1690 and has been conducted regularly at ten year intervals ever since.

Character : A quality, trait or sum of traits, attributes or characteristics, which serve to indicate the essential nature of a person or a thing.

Class : A totality of persons having one or more common characteristics, a homogenous unit within a series by which persons may be classified. Class may or may not signify the existence of hierarchical scale of social power. There are nativity, occupational, industrial, social, ideopolitic economic and income classes.

Class Consciousness : The awareness of one's class position. Fundamentally, an awareness of difference existing between one's class position and that of some other individual or individuals. This awareness is also generically accompanied by certain attitudes towards those occupying other class positions. These attitudes may be a feeling of superiority or inferiority towards those who occupy respectively a lower and higher rank; or a feeling of opposition or hostility where a situation of class conflict exists or merely a feeling of aloofness or strongness, because of the difference in folkways, moral and ideology of different classes.

Cognitive Behaviour : Observable behaviour from which it is inferred that an individual is exercising one of the intellectual components of knowing like perceiving, remembering, judging or reasoning.

Cognitive Dissonance : Inconsistent information gathered in or after decision-making.

Cognitive Style : An individual's preferred or habitual style of learning or thinking.

Cohesiveness : The overall attractiveness of a group to the members.

Communication: It is process by which two or more people exchange ideas, facts, feelings or impressions in ways that each gains a common understanding of the message.

Community : Community is more or less well defined group of people with common interests and problems. Such a group may include those within a township, trade area or similar limits.

Community Development : Community development would be defined as the induction and educational management of that kind of interaction between the community and its people, which leads to the improvement of both.

Compatibility: Compatibility is the degree to which an innovation is consistent with existing values and past experience of the adopters.

Competition : When impersonal social forces are in opposition, the struggle is called competition.

Complexity : Complexity is the degree to which an innovation is relatively difficult to understand and use.

Concept : A concept is defined as an idea/mental picture of an object and is often used to denote an idea/convince people.

Conflict : Conflict is the personalized form of opposition, which separates and redefines the relative status of diverse group while at the same time, it cements into closer unity of members within a group.

Content: The term content refers to subject matter, teaching points or learning that enable the trainee to perform the task, duties and jobs that are the terminal objectives of training and development systems. Essentially, content comprises knowledge, habits elements of skill and emotionalized controls.

Co-operation: Any form of social interaction in which personalities of groups combine their activities or work together with mutual aid, in a more or less organized way for the promotion of common ends or objectives in such a way that the greater the success of one party to the interaction, the greater the success of the other party or parties.

Cooperative Society : An organized body of not less than ten persons voluntarily associated on equal terms to work together for their own economic betterment.

Coordination : The working together of groups and individuals in an organized and integrated way.

Cosmopoliteness : Cosmopoliteness is the degree to which individual's orientation is external to a particular social system.

Covert Behaviour : Hidden or concealed behaviour.

Credibility : Credibility is the degree to which a communication source in perceived as trustworthy and competent by the receiver.

Culture : Culture consist of all learned normative behaviour patterns of thinking and feeling as well a doing.

Customs : The accepted ways of eating, meeting folks, giving training to the young ones, supporting the aged, etc. are called customs of the society. when we speak of customs, we think of accepted ways in which people do things together in personal contacts e.g. Kanyadan.

Data : Facts, statistics and things known which make the basis of reasoning on reacting a conclusion.

Debate : In a debate, members have to take-up and maintain sharply "opposed view points". The performance of the speaker in a debate is judged usually by a panel of judges and rarely by audience. The audience, therefore, have a passive role to play in a debate.

Decentralization : Process (condition) of division of some of the powers of a social unit among its parts. Often considered to involve a shift in the geographical location of power from a Central area to a number of outlaying districts but geographical shifts are not synonymous with shift of power.

Decision Making : The process of selecting from alternative courses of action.

Definition : Accordingly to Webster definition is an explanation of the meaning or meanings of a word.

Demand : Refers to the willingness and ability to pay a sum of money for some amount of a particular good or service.

Development : MacIver uses the word "development" to signify an upward course in a process "that is, of increasing differentiation".

Diffusion : Diffusion is the process by which an innovation spreads.

Discontinuance : Discontinuance is a decision to cease use of an innovation after previously adopting it.

Discussion : Is a group deliberation usually carried on through oral discourse under the guidance of a leader, aiming at the cooperative solution of a problem through reflective thinking.

Distance Education : Learning-teaching organized under the non-formal system using multi media, instructional designs and a few face-to-face contact programmes for the learners.

Dogma : Brief or opinion considered as authoritative and true without supporting evidence in data or personal experience.

Dominant Values : Those values, which determine the means used to achieve the ultimate end of security.

Dysfunctions: Behaviour which trend to disturb the equilibrium of a social system, acts which lessen social organization and trend to lead to social organization.

Economic Viability : A scheme or project formulated should ultimately earn adequate return on the investment proposed, so as to enable the borrower to repay the loan along with the interest there on be having a reasonable surplus which can be used to maintain and/or increase the productivity of his farm or enterprise.

Educational Psychology : It is the study of educational growth and behaviour in all its aspects.

Educational Technology : Refers to the use of hardware like film, filmstrip and similar audiovisual aids in education and the behavioural engineering involved in designing and regulating instructions. If the design and implementation of systems of teaching or instructions or to the fields of study and practice concerned with such manifest significance.

Emotion : A complex state of an individual in which certain ideas, feeling and usually motor expressions to produce a condition recognizable as such by the individual and frequently by others.

Empathy : The ability to understand and share the feelings of another.

Endogamy : Endogamy enforces marriage within a social group.

Ethics : It deals with the principles of conduct, which help us judge whether a choice or an action is good or right.

Ethos : The sum of the characteristics, or culture traits by which one group is differentiated and individualized from other groups.

Evaluation : This is the measurement of the progress being made on established objectives and goals and an indication of the effectiveness of method employed.

Evolution : A process of change in which each succeeding stage has a connection with the preceding stage, a growth or development involving continuity.

Exogamy: The law prohibiting marriage between persons of the same blood as a prevention of incest has had wide application in primitive society.

Experiment : An experiment is the proof of a hypothesis which seeks to look up two variables into a caused relationship through the study of contrasting situations which have been controlled on all variables except the one of interest, the latter being either the hypothetical cause or the hypothetical effect.

Ex-post Facto Analysis : Research method that examines events that have already occurred in the hope that they will reveal significant generalities.

Ex-post-Facto Research : Dealing only with available data concerning events, which have already occurred.

Exposure : Process of subjecting a photosensitive surface to radiation, as in a photographic or video camera.

Extinction : The process whereby response to a given stimulus is lessened over a period until there is nil response.

Extrovert : Refers to one having open and willing nature who easily fits in with any situation, quickly establishes relationships and ventures confidently and unhesitating into unknown situations without consideration of possible problems.

Fact : Facts are defined as what has really happened. Facts and data are probably the most frequently referred to term in scientific writings, yet these terms are among the most difficult to define.

Factorial Design : Factorial design is the structure of research in which two or more independent variables are juxtaposed in order to study their independent and interactive effect on a dependent variable.

Family : The family is a unit of interacting persons related by ties or marriage, birth or adoption, whose central purpose is to create and maintain a common culture, which promotes the physical, mental, emotional and social development of each of its members.

Family Matrillineal : In a family where the reckoning of descent takes place along the female line.

Family Matronymic : In a family if the offspring inherits the mother's name.

Family Patronymic: A type of family in which offspring inherit the father's name.

Family Polyandrous : A type of family in which a women marries more than one husband.

Family Polygynous : A type of family where a man marries with more then one wife.

Farm Family : A farm of sufficient size to provide employment for its operator and the members of his family, who themselves perform the entrepreneurial and manual labour on the farm.

Farm Income : Income obtained from operations. The sum of receipts from sales, appreciation of inventory and value of living obtained from the farm represent gross farm income. After expenses are deducted, the remainder is called net farm income.

Farming System : The entire complex of resource preparations, allocations, decisions and activities, which within an operational farm unit or a combination of such units, results, in agricultural production. The harvesting, drying, processing and marketing of the products are also directly related to the system that produces them.

Folk Education : Refers to education concerned with popular and traditional culture.

Folkways : These are the recognized or accepted ways of behaving in society. They include conventions, forms of etiquette and the modes of behaviour, men have evolved and continue to evolve, with which to go about business of social living e.g. in the Northern India, while smoking a Hukka in a group, the first puff is always given to the elderly man.

Forum : It consists of a question period in which members or the audience may ask questions or make brief statements. The forum provides an opportunity for the audience to clear up obscure points and to raise questions for additional information. It also gives an individual an opportunity to state briefly their understanding of a point and see whether he has interpreted correctly the material presented. It is primarily a means of understanding information.

Fundamental Research : These researches deals with the fundamental principles of sociology. They may be conducted either for the verification of some old theory or establishment of a new one.

Goal : This is a distance in any given direction, proposed to be covered in a given time.

Group Dynamics: Refers to the study of the behaviour of groups of people and of the interaction of behaviour of individuals as members of a group.

Habit : An acquired attitude or tendency to act in a specific way, which has became, in a measure largely unconscious and automatic e.g. customs is sometimes referred to as group habit.

Habitat: To designate the natural setting of human existence-the physical features of region inhabited by a group of a people, its natural resources, actually or potentially available to the inhabitants, its climate attitude and other geographical features to which they have adapted themselves.

Human Capital : Human capital can be defined as the skill, capacity and ability possessed by an individual, which enables him to seek a career or profession to earn his living.

Human Resources : Traits or qualities within people that are instrumental to reach goals. Human resources includes cognitive, affective, psychomotor and termporal traits or characteristics of people and other resources that cannot be used independently of individuals.

Hypothesis : A proposition, condition or principle which is assumed perhaps without belief in order to draw out its logical consequences and by this method to test its accord with facts which are known or may be determined.

Ideals : An ideal is a standard, often of perfection, relating to persons, trait, objects or ideas that is accepted by an individual or group. Examples are standard of managership, leadership and craftsmanship.

Ideology: A system of beliefs explaining the nature of man's relation to the cosmos and accompanied always by a system of observances based on these beliefs is a universal aspect of culture.

Illiteracy : Refers to the inability to read.

Illiterate : The term illiterate is used to designate a person who cannot read, write and figure at the level of first grade and has had no formal schooling or its equivalent.

Illusions : Refers to false perceptions of reality, whose study provides information about the process of perception.

Imitation : Imitation is the process by which specific bits of knowledge, skill, symbolic behaviour are taken over from a social object in the interaction process.

Income : Income is the flow of money or goods of an individual, group of individuals or firm over some period of time. It is originated from the sale of productive services (as wages, interest profits, rent, national income) or it may simply reflect a gift, e.g. a legacy from a will or income from a trust, fund or transfer payment; for example, old age pensions, etc.

Independent Samples : Population samples in educational and other statistical research that do not interact in any way to influence the variables under study.

Independent Variables : Refers to the variable whose chances are considered as not dependent upon transformations in other specified variables.

Individual : Those human beings comprising of any social aggregates when they are considered from the point of view of the characteristics, which make each one different from the others.

Information : Statement with no attached direction for action (illusionary purpose) or value orientations to modify/change or affect the behaviour of the receiver of the information.

Innovativeness : It is defined as the degree to which an individual is relatively earlier in adopting new ideas than the other members of his social system.

Instinct : The inherited, standard way of behaving to meet minimum requirements for survival.

Instruction : A process or a series of (learning) steps involving communication and interaction by which an individual is helped to move from one level of behaviour (t entry) to a higher level of behaviour (learning out-come) gradually and assuredly in respect of receiving, perceiving and using information appropriately.

Instructional Methods : Forms of teaching like the lecture, demonstration, etc.

Instructional Models: Suitable learning-teaching methods selected and subsequently combined to inculcate particular learning strategies (like inductive learning, formulating concepts, drawing inferences developing team spirit problem, identification and solving etc.).

Integration: That social process which tends to harmonize and unify diverse and conflicting units, whether those units be elements of personality, individuals group or larger social aggregations.

Intelligence Quotient (**IQ**): It is the ratio of actual mental age, as measured by intelligence tests, to the mental age that is normal for a particular chronological age. The ratio is multiplied by 100, thus giving as average IQ.

Intensive Farming: Farming in which comparatively large amount of capital and labour is utilized per cultivable acre of farm land. Dairy, poultry, vegetable and fruit crops are examples of intensive enterprises.

Interference: Refers to the inhibition of one piece of learning by another considered as a major cause of forgetting.

Interview : In simple terms interview means "conversation with a purpose".

Job Analysis: Job Analysis is the process of collection, tabulating, grouping, analyzing, interpreting and reporting data pertaining to the work performed by individuals who fill operative or supervising positions.

Job Satisfaction: Refers to the extent to which a person is pleased to satisfy by the content and environment of his work or is displeased or frustrated by inadequate working conditions and tedious job content.

Joint Family: Collection of more than one primary family on the basis of close blood ties and common residence.

Joint Family, Matri lineal: A joint family where inheritance and reckoning of descent takes place along the female line.

Joint Family, Matri local : A type of family where the husband goes to live with family of his wife, and the family line goes with the mother.

Joint Family, Matromymic: A joint family where offspring inherit the mother's name.

Joint Family, Patrilineal: Refers to a joint family where reckoning of descent take place along male line.

Joint Family, Patri local: When the married couple and their offspring put-up with the husband's family.

Joint Family, Patronymic: A joint family where offspring inherit the father's name.

Knowledge: Refers to collection of facts, values, information, etc. to which man has access through study or experience.

Lab to Land Programme : A planned programme of rural development through induction of technology evolved by the scientist engaged in agricultural research and education. It constitute the core of the programme contemplated for the celebration of Golden Jubilee Year (1978) of the ICAR. The programme formulates integrated approach involving inter-disciplinary team work on large scale for lasting beneficial impact to bring a break-through in the transfer of technology or so called LAB TO LAND PROGRAMME.

Land Holding (Operational) *:* That area of land which is actually cultivated by the farmer and his family. It includes land kept fallow.

Language: It is a means of communication.

Laws: Laws are consciously and deliberately formulated behaviour patterns.

Lay man: A person without technical training or competency in a specified calling such as a science, medicines, or the ministry.

Linguistics: It is the study of language in its widest sense rather then simply as a direct means of communication.

Literate: Able to read and write or to communicate by means of writing. Such capabilities evidence the capacity of a person or people to become articulate in symbols.

Marriage: Marriage is a socially approved way of establishing a family of procreation.

Marriage; Affinal: A marriage with a relative of a spouse.

Marriage; Auncular: Marriage of a man with his sister's daughter.

Marriage; Beena : A man's being required to leave his name and live with his wife's family; the marriage is usually matripotestal.

Marriage by Exchange: Marriage, which involves two men's exchanging daughters or sisters, so that, each man, obtained a wife for himself or for a brother or son.

Marriage by Capture: Getting a wife by abduction, whether genuine, ritual or pretended.

Marriage Child: The marriage of a child under about 21 years old (Boy) and 18 years old (girl).

Marriage by Group : A group of females married to a group of males.

Marriage; Monogamy: A marriage of one man to one woman.

Marriage; Patrilocal: A marriage in which the wife lives with her husband's tribe. It is associated with patronymic system.

Marriage; Polyandry: Marriage of one woman to two or more men.

Marriage; Polygamy: Marriage of one man to two or more women.

Marriage; Preferential: Marriage between people of specific status, whether, it is enjoined or preferred.

Marriage; Primary: A person's first marriage.

Marriage; Secondary: A second marriage contracted after a first marriage.

Marriage; Symmetrical Cross-Cousin: A marriage in which either type of cross cousin is acceptable or sought after as a mate while marriage to the other cross cousin might not be permitted.

Media: Refers to the channels of mass communication, including the press, radio and television.

Media of Communication: These are the means by which information or knowledge is passed from one group or individual to another.

Mental Age: It is the degree of mental development of intelligence of an individual as compared with the average intelligence of others at different chronological age.

Method: The term method applies to any procedure that achieves some rational order or systematic pattern for diverse objects.

Motivation: The process of initiating conscious and purposeful action. Refers to behaviour brought about by psychological or social needs and directed towards goals.

Motor Ability: Refers to skill in organizing and executing muscle actions.

Movement (Social and Political) : The necessary and sufficient definition of social and political movement consist of three elements found in every day social life (i) Organization or structure, (ii) Beliefs or ideals and (iii) Action or behaviours, in which people engage. It is the particular shape of each of these elements that gives social and political movements their own particular cost.

Multimedia: A presentation employing a mixture of media, e.g., tapeslide, motion picture, video and live action.

Myth: Refers to a narrative cultural tradition, that has a number of proven factual basis, about natural events or the deeds of fictional heroes. ii) A traditional story of religious import, especially on account of the activities of supernatural beings or a fictive explanation in narrative form of the origin at religious rites. Social usages, or natural phenomena.

Need Analysis: The analysis to establish the instructional programme needed by a particular trainee to bring him to the standard laid down in the relevant job description.

Need Assessment: To determine the needs and priorities of a person or an organization.

Needs or Wants: The positive driving forces that impel a person towards certain objectives or conditions.

Noise: Unwanted sound or signals in a communication system, especially random electrical energy interfering with audio and video.

Norm: A norm is defined as the most frequently occurring pattern of overt behaviour for the members of a particular social system.

Null Hypothesis : Hypothesis that has no causal relation (or) interrelation exists between two variables.

Objectives: Objectives are expressions of the ends toward which our efforts are directed.

Paradigm: A set of theories, models and methods that serves as an example.

Pedagogy: Learning-teaching organized according to the developmental needs of children (and also adolescent) and their learning style.

Perception: Process whereby the individual organizes and makes sense of his sensory experience.

Performance: Refers to actions of a person or group when given a learning task.

Personality: Personality is the dynamic organization within the individual of those psycho-physical systems that determine his unique adjustment to his environment. Personality is a more or less enduring organization of forces within the individual associated with a complex of fairly consistent attitudes, values and modes of perception, which account, in part for individual's consistency of behaviour. Personality refers to a person's more or less organized predispositions to "perform, perceive, think and feel" in a certain way with regard to somebody or something.

Philosophy: The word "Philosophy" has two common usages. In one, it represents the more or less unspoken assumptions and values implicit in a certain approach to acting and way of acting. In the other, it means an explicit and systematic articulation of concepts and principles that have been tested for inner coherence and for adequacy of reference to the events it is intended to make meaningful. It is concerned with the study of some forms of thought and argument and the grounds of knowledge in other subjects. The search for truth, understanding of causation and the underlying principle of reality through logical and rational processes but not requiring empiric or factual observations is called philosophical enquiry.

Polygamy: Refers to marriage or durable mating of one man to two or more women, or of one woman to two or more men.

Population *:* Total number of people in a definable group. An aggregate of individuals defined with reference to spatial location, political status, ancestry, or other specific conditions, either at a specific time or in a temporal continuum.

Postulates: Statement of the relationship between variables, which are accepted as axiomatic and therefore must be taken for granted.

Prediction: The act or process or forecasting with greater or less probability, the outcome of an event or series of events by inference from a scientific, especially a statistical analysis of known events. In view of the multiplicity of casual conditions affecting social events, prediction in the sociological fields is less precise than in the physical sciences.

Prejudice *:* Mean a hostile attitude towards a whole group of people, or towards one person simply because he is a member of that group.

Primary Data: Data which the researcher collects himself through surveys, on the spot observation and recording or tabulation.

Principle: A principle is a statement of policy to guide decision and action in a consistent manner (Mathews).

Probability: The likelihood of the occurrence of an event. Sometimes expressed as the ratio of the number of equally likely ways in which a particular event can occur to the total number of occurrences possible.

Problem: The term problem came from the Greek word Proballe which means anything thrown toward, a question proposed for solution; a matter stated for examination. R.S.Wood-worth defines problem as "a situation for which we have no ready and successful response by instinct or by previously acquired habit. We must find out what to do".

Programme: A programme is a planned combination of training session taking place over a period of time.

Prospects: Publication produced by a college, university or school on its courses and facilities with the particular purpose of providing information for prospective students.

Publicity: Data made available or common knowledge given for public circulations. The dissemination of information through any of the channels of mass communication, in simpler societies and in neighbourhoods, the media are gossip, public complex societies, the media includes news papers, radio broadcasting television, bill boards etc. These channels are available through advertising and "free publicity".

Qualitative Data: Refers to data (e.g composition, diaries) not presented in a quantative form but that may be analysed or categorized in ways that may allow systematic treatment and presentation.

Race: Refers to human group with common, distinctive innate physical characteristics. A major division of mankind, with distinctive, hereditarily transmissible physical characteristics, e.g., the Negroid, Mongaloid and caucasoid races. It may also be defined as a breeding group with gene organization differing from that of other intra-species groups.

Rating: The process of evaluating by judgement is called rating, judgement involves the collection, correlation and interpretation of fact and impression to arrive at an estimate or an opinion about a person, trait, object, or situation. In order words, the fact do not "speak for themselves", and human interpretation of the facts is essential. Since rating is judgement and judgement is saturated with opinion, rating is always subjective to some degree.

Ratio: It is a quotient of two numbers showing the measures of two similar quantities.

Ratio Scale: Highest form of scale in educational measurement, others being nominal, ordinal and interval scales.

Rationalization: Refers to the process by which a course of action is given ex-post-facto reasons that not only justify it but also conceals its true motivation.

Reasoning: Thought process involving influence or problem solving by making use of general principles.

Religion: The sociological field of religion may be regarded as including these emotionalized beliefs prevalent in a social group concerning the supernatural, plus the overt behaviour, motivational objects and symbols associated with such beliefs. It is the human response to the apprehension of something or power which is supernatural and suprasensory. It is expression of the manner and type of adjustment effected by people with their conception of supernatural. In it, the person perform necessary actions which bind him with the supernatural powers.

Research: Research is a systematic activity directed towards discovery and the development of an organized body of knowledge.

Research Design: Research design will refer to the entire process of planning and carrying out a research study. Research design refers to an entire process of conducting a research study.

Rituals: It consists in the observance, according to a prescribed manner of certain actions designed to establish liason between the performing individual and the supernatural power. The offering of flowers, pouring of water, saying of some mantras or offering of prayer, all are the rituals in worshiping the god or goddesses in the temples and they are to be followed according to the prescribed manner.

Role: The role is the part an individual plays as a result of his status in the society or group.

Role-Playing: Role playing is a laboratory method of instruction that involves the spontaneous dramatization or acting out of a situation by two or more persons under the direction of a trainer. The dialogue grows out of the situation developed by the trainer assigned to the parts. Each person act his role as he feels it should be played. Other trainees serves as observers/critics. Following the enactment, the group engages in discussion.

Sampling Unit: The ultimate unit of assessment or measurement in sample. It may be the individual or any given number of individuals, a given area of ground, etc.

Scale: Set of symbols (usually numbers) that are given to an individual on the basis of measured or inferred behavior. For a given population of objects, the multivariate frequency distribution of a universe of attributes will be called a scale if it is possible to drive from the distribution a quantitative variable with which to characterize the objects such that each attribute is a simple function of the quantitative variable.

Scaling: Refers to the scaling of examination marks, where in the basic principle is that a common test is given to all pupils concerned, which is used as a yardstick against which performance to each may be compared.

Scattergram: Plotting scores of the same individuals on two separate variables giving a picture of the relationship between the variables.

Secondary Data: The data, which have been previously collected by some one other than the researcher. Census records constitute a well used example.

Seminar: It is a form of group discussion. The discussion leader introduces the topic to be discussed. Members of the audience discuss the subject to which ready answers are not available. A seminar may have two or more plenary sessions. This method has the advantage of pooling together the opinion of a large number of persons.

Short Courses: The short course is a method of intensive training in some specific area. It may vary in length from one day to two weeks. The dictionary defines short course as short, meaning brief or limited and course, meaning progress or direction.

Simulated Training: It is a training in a condition simulating real working or operating conditions and giving opportunity for formal instruction.

Simulation: Refers to a technique in which a real-life situation is simulated by substitutes displaying similar characteristics.

Simulator: Refers to physical simulation or the physical means, which a simulation is created.

Skewness: Refers to a sample that is biased or to a distribution curve that is symmetrical.

Skill: Refers to systematic and coordinated pattern of mental and physical activity, involving both receptor process and effector processes and effector processes. Skills may be perceptual, motor, manual, intellectual processes, social, etc. Skills are behavior that require some degree of a facility in the performance of part or all of a complex act. Speed and accuracy are usually required. Skill may be either mental or motor, e.g. Public speaking, repairing, reading.

Social Action: Behaviour intended to influence the action of others.

Social Behaviour: Behaviour influenced by the presence of other persons, *i.e.* controlled by society, or seeking to influence others.

Social Climate: Refers to the overall characteristic atmosphere in a society.

Social Control: Social control is the term sociologist apply to all the ways and means by which any society achieves and maintains a normative social systems. The chief reliance for social control is, of course, upon the internationalization of society's norms, as in the socialization of the young.

Social Development: Social development comprehends some overall movement toward greater efficiency and complexity but with the recognition of the concomitant problems.

Social Distance: It is the degree of intimacy between any two individuals or groups of individuals.

Social Ethos: Value-systems, attitudes and orientations considered desirable and followed by a society.

Social Facts: Social facts are external to the individual and are engendered with a power of cohesion over him. e.g. Public mortality, family and religious observers.

Social Interaction: Refers to interaction of people and institutions within a community or society. exists when two or more persons mutually influence each other's behaviour.

Social Learning: Refers to knowledge or abilities required or acquired by a person in the process of socialization or living in a social situation.

Social Mobility: Social mobility refers to the movement of persons, families or larger groups from one position in the social structure to another. Horizontal mobility – the movement from one group to another of equal standing and prestige constitutes horizontal mobility. Vertical mobility – the passage of an individual or family from affiliation with one another class of higher or lower status is known as vertical social mobility.

Social Order: Social order consists primarily of the economic, religious and Kinship systems.

Social Organization: Is form structure, orderly-relationship and systemic modes of acting together. It is a state of being, a condition in a society, in which the various institutions function in accordance with their implied purposes. It is characterized by the harmonious operation of the different elements in the society.

Social Psychology: It deals with the study of social behaviour through a study of people in their social relationships with others and through studying the person's individual socialization. Social psychology is the scientific study of the behaviour of human beings. It is a behavioural rather than a natural science, yet like the natural science.

Social Research: Social research may be defined as a scientific undertaking which, by means of logical and systematized techniques seeks to discover new facts or verify and test old facts, analyze their sequences, inter-relationship and casual explanations which were derived within an appropriate theoretical frame of reference, develop new scientific tools, concepts and theories which would facilities reliable and valid study of human behaviour.

Social Science: Refers to scientific study of human relationships through detailed, systematic and logical syllabii in the constituent subjects, e.g., economics, psychology, sociology, social administration, etc.

Social Security: Government measures to give financial assistance to disadvantages members of society like the unemployed, the diabled and the elderly.

Social Services: Provision of services concerned with helping those members of society least able to help themselves. Those organized activities that are primarily and directly concerned with the conservation, the protection and the improvement of human resources and includes as social services; social assistance, child welfare, mental hygiene, public health, social insurance, education, recreation labour, protection and housing.

Social Skills: Refers to skills essential for normal social life in a community.

Social Stimuli: Refers to stimuli that are social in nature or origin.

Social Stratification: Social stratification means ranking of people according to certain characteristics that can be placed in rank order.

Social Structure: Institutional arrangements in a society.

Social Studies: Refers to those portions or aspects of the social sciences that have been selected and adapted for use in the schools or in the instructional situations.

Social System: A social system is a population of individuals who are functionally differentiated and engaged in collective problem solving behaviour.

Socialism: Various forms of state of community ownership or control of the means of production, distribution, etc.

Socialization: It is the process of learning roles and expected behaviour in relation to one's family and society and developing satisfactory relationships with other people.

Speech: Verbal language or communication.

Statistics: Statistics is the theory of analyzing quantitative data obtained from samples of observations in order to study and compare sources of variance of phenomena, to help make decisions to accept or reject hypothesized relations between the phenomena and to aid in making reliable-inferences from empirical observations.

Status: Status is the term we apply to each position. The general behaviour expected of the occupant of such a position is called a role. The status is the position of an individual in the group or society which he occupies by virtue of his belonging to a sex, age, birth, being married, physical abilities possessed, achievements and designated duties.

Symbol: Activity or object representing and standing in substitution for something else.

Sympathy: Refers to ability to appreciate the emotional experience of another person.

Syntax: Rules of grammar in a language.

Taboo: It refers to the world used to designate all the restrictions communicated through verbal "Don't" and generally associated with situalistic behaviour, which a member of a rural society has to submit to. It is also said to be an unwritten law of a society. Its purpose is three-fold, productive, protective or prohibitive. Taboos, associated with process of cultivation are designated to be productive. Those like keeping women, children or even man away from certain places, actions and objects are protective. Those which seclude a person or limit contact with him or her, as is done in case of menstruating women when she is not allowed to enter a cow-shed or handle milk etc are prohibitive.

Tabulations: When a mass of data has been assembled it is necessary to arrange the data in some kind of concise and logical order. This procedure is referred to as tabulations.

Task: A task is one of the work operations that is a logical and essential step in the performance of a duty. It is the work unit that deals with the methods, procedures and techniques by which duties are carried out. Each task has the following characteristics. It occupies a reasonable portion of the work time spent in performing a duty. It occurs with reasonable frequency in the work cycle of a duty. It involves very closely related skills, knowledge and abilities and it is performed according to some standard.

Theory (Sociological): A sociological theory is a set of logically related, operationalized sociological principles, representing the inductive sum of verifiable generalization permit to the topic. The proposition of a theory must be constituent and they must be such that existing generalization can be derived deductively. Further, a theory must be empirically tested to be useful.

Tradition: A social situation process in which elements of cultural heritage are transmitted from generation to generation by contact of continuity, or the non-material culture content so transmitted, having the prestige sanction of antiquity. By extension, in an institution where personnel turns over more frequently than once a generation, practices, ideas and lore transmit over a series of such "turnovers" are called traditions.

Training: A process by which individuals are helped to acquire certain specific skills related to given operations/contexts, it ensures mechanical and methodical replication of certain "roles" operations; without further training for changed contexts, a trainee will not be able to engage in further self-learning or acquiring higher order values required for changing needs/or for developing "Integrated Personality".

Tribe: A social group, usually comprising a number of sibs, bands villages or other sub-groups which is normally characterized by the possessions of a definite territory, a distinct dialect, a homogenous and distinctive culture and either a unified political organization or at least some sense of commonsolidarity as against outsiders.

Unit: A unit in science means a single act, or a single social situation, or a person or phenomenon, viewed as a whole. A unit of study can be small or large, a simple or complex, obvious or illusive.

Universe: Refers to the total population from which a sample has been chosen for statistical purposes.

Urbanization: The process by which a society became urban, Urban growth is usually identified as urbanization and it means. The movement of people from rural to industrial residence and movement of people from agricultural to non-agricultural work.

Utility: Is the satisfaction, pleasure or need-fulfillment derived from consuming some quantity of a good and is basically a psychological concept incapable of direct measurement in terms of absolute units.

Validity of Data: Extent to which data correspond with some criterion, which is an acceptable measures of the phenomena being studied.

Value: It is used in almost every field of study and activity. Inspite of some differences in usage, there is one element of meaning which is uniform. It is the idea of worth. The value of anything is, what it worths to someone.

Value System: These are the briefs on good and bad aspects of behaviour held by a culture or social class.

Variables: A variable is any quantity or characteristics, which may possess different numerical values or categories. Time, age, price, score, or intelligence tests, political party preference and sex are examples of variables are the conditions or characteristics that the experimenter manipulates, controls or observes.

Variable Continuous: Continuous variable take continuous values and can be measured on a continuum, e.g. height, weight and income.

Variable Discrete: Discrete variable is defined only for a series of specified values with no possibility of variate values between those points. Thus each value of discrete variable is distinct and separate.

Variance: It is the extent to which a set of scores is scattered or dispersed.

Variance Analysis: It is a technique used for analyzing the relationships between different parameters in a complex business or organizational situation.

Verbalism: Thought based on verbal association.

Verifiability: The conclusion drawn through a scientific method are subjected to verification at any time. Verifiability presupposes that the phenomena must be capable of being observed and measured.

Vernacular Language: Refers to the natural or common language of a society.

Village Adoption Scheme: It was to encourage adoption of village by industrial and mercantile houses for development. The objective of improvement of scheme is to increase the available resources for improving modern technology in agriculture and assisting the rural population in obtaining other facilities. It also stipulates, that small agro based industries should be set up by the adopting industrial/mercantile houses so that the benefits would flow to the agricultural labourer by way of creating supplementary occupation and employment.

Vocational Education: Refers to activities designed to contribute to occupational proficiency.

Vocational Training: Training designed to teach the skills and knowledge needed for particular kinds of work.

Wealth: Composite holding of seals, property, monetary assets, durables and personal possessions, as well as the characteristic pattern of such holdings.

Welfare: An interest directed to the well being of persons or groups.

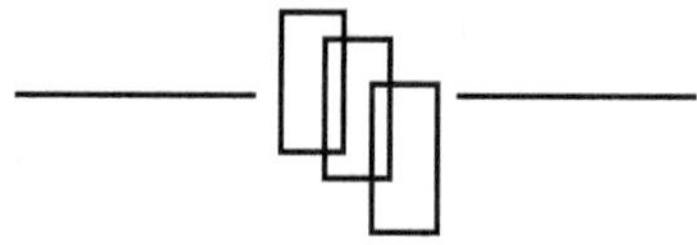

References

Abraham Maslow (1954). Motivation and personality. Harper and Row. New York.

Adeboye, T.O. (1977). International tranfer of technology: A comparitive study of differences in innovative behaviour. Un-published dissertation, Harvard university, Cambridge. Massachusetts.

Agrawal, A.N. and Singh, S.P. (1975). The economics of under develoment. Oxford university press. New Delhi.

Alagh, Y.K., (1991). Inaugural Address, National Workshop Performance of FFD Freshwater Aquaculture, Ahmedabad.

Alan D. Meyer (1984). Mingling decision making metaphors. Academy of management review. pp: 6-17.

Barrett, S.C.H. (1989). Water weed Invasion *Scientific American,* October 1989, p.92

Bernard A. Goldhirsch (1985). Leadership and performance beyond expectations. The free press. Newyork

Bhanot, K.K., et al. (1998). Fisheries Technologies for Women. Indian Farming, Oct. 1998: 46-48.

Bhaskar, G and Hendre, P. 1999. agriculture information system : A case study of KVK, Babhaleswar.

Bose, D K. 2000. Rural-India wired up. http;//www.google.com

Brooks, H. (1966). National science policy and technology transfer. Proceedings of the coference on technology transfer and innovation. National science foundation. Washington D.C.

Bryan, F (1967). Performance goals as determinants of level of performance and boredom. Journal of applied psychology. pp: 120-130.

Buckley, P.J. and Casson, M. (1976). The future of the multinational enterprise, London. Macmillan.

Cairncross, A.K. (1967). Factors in economic development, London.

Chambers Robert (1974). Managing rural development. Ideas and experience from east Africa. Scandinavian Institute of African studies.

Chester I Barnard (1964). The functions of the executive. Harvard University press.

Choudry, S. 1999. Wireless local loop, Computers Today. 15-31 july'1999, http://www.india-today.com/ctoday/19990715

CIFA (1998). National Aquaculture Development Plan, Central Institute of Freshwater Aquaculture, Bhubaneswar: 60p (mimeo).

Cohen Goel (2004). Technology transfer: strategic management in developing countries. Chapter – 3. Sage publications. New Delhi.

Daniel T. Carroll (1983). A disappointing search for excellence. Harvard business review. pp: 78-88.

Edwards, P and Demaine, H. (1997). Rural Aquaculture: Overviews and framework for country reviews. FAO/RAP publication 1997/36, Bangkok, pp:61.

Ensminger, Douglas (1968). An evolving strategy for India's agricultural development. The ford foundation. New Delhi.

Everett m. Rogers (1995). Diffusion of innovation. 4th edition. The free press. New York.

FAO 1999. Virtual Extension Research Communication, on http://www.fao.org/WAICENT/vercon/ as on July 8'1999. Rome: FAO.

FAO. 1993. The Potentials of Microcomputers in support of Agricultural extension, education and training. Rome: FAO.

Frame, J.D. (1983). International business and global technology. Lexington books.

Gaikwad, B. H. 2000. Participatory extension : A new paradigm.

Gaur, A and Yadav, S K. 2000. Harit Gyan (IT enabled agriculture), JIL

George, P.C and Sinha, V.R.P. (1975). Ten Year Aquaculture Development Plan for India 1975-84. FAO Workshop Aquaculture Planning in Asia, Bangkok, 1975.

Ghose, K.K and Sinha, V.R.P. (1991), Policy Issues in Development of Brackishwater Fish Culture, National Seminar: Productivity Constraints in Coastal Areas, Jadavpur University, Calcutta

GOI (1998). Economic survey, Ministry of finance. Government of India. India. Government of India.

Gupta, S. 1995. A research project on the commercial uses of the world wide

Harold Koontz (1961). The management theory jungle. Journal of the academy of management. pp: 174-188.

Harbinson, I and Meyer, I (1961). Organisation and Management. The free press. New york.

Henry Mintzberg (1973). The nature of managerial work. Harper & Row. New York. Chapter – 4.

ICAR. 1998. - Information Systems Development: Development strategies and implementation plans under NATP. ICAR 1998.

Jagannath Mohanty (2005). Modern trends in educational technology. Neelkamal publications Pvt. Ltd. Hyderabad.

Javed Jabbar. 1999. The observations in first plenary session of Asia Information Poor to Information Rich : Strategies for 21st century, Chennai on July 1, 1999 by Mr. Javed Jabbar, founder Chairman, South Asia Media Association and former Minister / Senator of Pakistan.

Jhingran, V.G (1975). Systems of polyculture of fishes in the inland waters of India, Proceedings 13th Pacific Science Congress Abstract: 38.

Joseph L. Massie (2000) Essentials of management. Prentice-Hall of India. New Delhi.

Kahen Goel (2001). Strategic assessment of high technology transfer: A managerial approach for energy technologies. Proceedings of the international symposium on environment engineering, University of technology. Tehran.

Kahen, G and Griffiths, C (1995). Human factors, technology transfer and information technology in the socio-economic development of the third world, University of Johannsberg. Johannsberg.

Katar Singh (1999). Rural development: principles, policies and management. 2nd edition, Sage publications. New Delhi.

Keith Davis and John Newstrom (1985). Human behaviour. 5th edition, Mcgraw-Hill Book company. New York. pp: 308.

Keith Davis and William C. Frederick (1984). Business and society. 5th edition, Mcgraw-Hill Book company. New York. pp: 564.

Kleiman, H.S. and Jamieson, W.M (1978). Two faces of international technology transfer. Battle today. No. 10. pp: 3-6.

Kolds, E (1981). Controlling international technology transfer, New York. Peragon.

Koontz, H and Weinrich, H (1985). Essentials of management. 5th edition, Mcgraw-Hill Book company. New York.

Korten, David (1981). The management of ocial transformation. Public administration review. pp: 610.

Kurien, V (1978). Food aid in the form of dairy products. 20th International dairy congress. Paris.

Lall, S (1982). Te chnological learning in the third world: some implications of technology exports. The economics of new technology in developing countries. London. pp: 159-179.

Leeuwis, C. 1993. Computers, myths and modeling: The social construction of diversity, knowledge, information and communication technologies in Dutch horticulture and agricultural extension. Wageningen Studies in Sociology, 36. Wageningen Agricultural University.

Lewis W. Arthur (1955). The theory of economic growth. Jhon Weilly. New York.

Malone, T; Yates j and Benjamin, R. 1988. Electronic markets and hierarchies.

Maunder, A H. 1973. Agricultural Extension: A Reference Manual (Abridged edition). Rome: FAO.

Mehta, S.R. (1984). Rural development policies and programmes. Sage publications. New Delhi.

Merrill, R (1972). The role of technology I cultural evolution, Social biology. Vol. 19. pp: 219-223.

Morris, David (1982). Measuring the conditions of India's poor: The physical quality of life index. Promilla company publishers. New Delhi.

NABARD (1997). National Bank for Agriculture and Rural Development. Annual report.

NDDB (1997). National Dairy Development Board. Annual report.

NIRD (1996). National Institute of Rural Development. Rural development statistics.

Peter F. Ducker (1954). The practice of management. Harper & Brothers. New York. pp: 63.

Price Monroe E. 1999. Research for Strategies for 21st Century, Dr. Price Monroe E, Director, Centre for Socio-legal studies, Wolfson College, University of Oxford, UK, paper presented at AMIC, 8th Conference July 1-3, 1999.

Rada, J.F. A third word perspective: strategies for better or worse. New York, Pergamon. pp: 203-231.

Radheyshyam (1997) Rural Aquaculture in Orissa: Sarkana Success story, CIFA Bhubanswar: 1-7.

Radheyshyam and Tripathy, N.K. (1992). Aquaculture as a nucleus for integrated rural development: an experience, Fishing Chimes, 12(9): 37-49.

Radheyshyam and Tripathy, N.K. (1988). Netweaving technology for gainful employment of rural women folk, All India workshop on gainful employment for women in fisheries field, CIFT Cochin, 104-111.

Rodrique, M.V. (1992). Perspective of communication and communicative competence. Concept publishing company. New Delhi.

Rogers E.M. & shoemakers, C (1971) : Diffusion of Innovations, The free press, New york.

Samanta R K. 1993. Extension Strategy for Agricultural Development in 21st Century. Mittal Publications Delhi.

Shagufta Jamal and Arya, H.P.S. (2004). Participatory rural appraisal in agriculture and animal husbandry: A training manual. Concept publishing company. New Delhi.

Sharif, N (1992). Technological dimensions of international cooperation and sustainable development. Technological forecasting and social change. Vol. 45 (2): 367-83.

Sharma, G.R.K. (2006). Cyber Livestock Communication: Extension Education. Concept Publishing company. New Delhi. Chapter – 5.

Simon, D (1991). International and the trans-border movement of technology. Technology transfer in international business. Oxford university pres. Oxford. pp: 5-29.

Sinha, V.R.P. (1972). Composite Fish Culture in India, Indian Farming, 22(6): 18-19.

Sinha, V.R.P. (1991) Role of women in Aquaculture. Women in Agriculture-Technological Perspective. Ed. C.Prasad and Shri Ram, IFWA, ICAR, New Delhi: 315-327

Sinha, V.R.P. (1996) Shrimp Culture, Trade and the Environment in India. INFOSYS, Kualalumpur (Mimeo), 31p

Steven Alter (1980). Decision support system: current practice and continuing challenges. Addison-wesley publishing company. pp: 30-40.

Thakur, N.K. et. al (1997), Whether shrimp farming is polluting the environment - A scientific appraisal. Fishing Chimes 17(4); 28-30

Thompson, J (1967). Organization in action. Mcgraw-Hill. New York.

Thompson, R (1993). Learning and technological change. Mcgraw-Hill. New York.

Turban, E; Mclean, E and Wetherbe, J. 2001. Information Technology for management : Making connections for strategic advantage. Jhon weilly & sons inc., Newyork.

Van den Ban A.W. Hawkins, H.S. (1997) Agricultural extension. Blackwell science, Netherlands.

Verma, S S. 2000. Detrimental social effects of Internet Technology, Science and culture, vol.66(7 & 8), pp: 247-249.

William H. Peace (1986). I thought I knew what good management was. Harvard business review. pp: 59-65.

William L. Glueck and Lawrence R. Jauch (1984). Business policy and strategy management. 5th edition, Mcgraw-Hill Book company. New York. Chapter – 2.

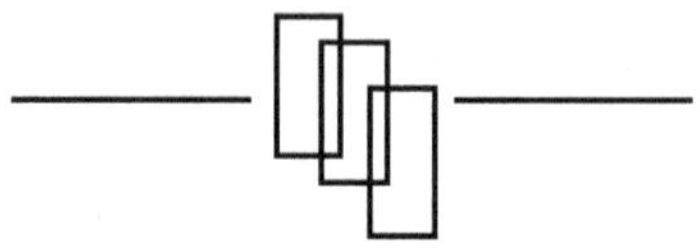